# Circular Economy and Environmental Resilience

Pardeep Singh · Shikha Daga · Kiran Yadav ·
Vandana Mishra
Editors

# Circular Economy and Environmental Resilience

## Solutions for a Sustainable Tomorrow, Volume 1

Springer

*Editors*
Pardeep Singh
Department of Environmental Studies
PGDAV College
University of Delhi
New Delhi, Delhi, India

Shikha Daga
Department of Commerce
PGDAV College
University of Delhi
New Delhi, Delhi, India

Kiran Yadav
Department of Commerce
PGDAV College
University of Delhi
New Delhi, Delhi, India

Vandana Mishra
Department of Environmental Studies
University of Delhi
New Delhi, Delhi, India

ISBN 978-3-031-93090-4 ISBN 978-3-031-93091-1 (eBook)
https://doi.org/10.1007/978-3-031-93091-1



This Springer imprint is published by the registered company Springer Nature Switzerland AG
The registered company address is: Gewerbestrasse 11, 6330 Cham, Switzerland

# Preface

*Circular Economy and Environmental Resilience* lay the foundational understanding of the circular economy about environmental resilience, offering a comprehensive exploration of how circular practices can contribute to sustainable development in the face of escalating global environmental pressure. As the world grapples with the complex challenge of climate change, resource depletion, biodiversity loss, and ecosystem degradation, there is an urgent need to rethink traditional linear economic models that rely on a "take-make-dispose" approach. This volume seeks to address that need by delving into the transformative potential of the circular economy—a model that emphasizes resource efficiency, waste minimization, and the regenerative capacity of natural systems. This book examines the theoretical constructs that underpin the circular economy, providing a detailed analysis of key concepts such as closed-loop systems, cradle-to-cradle design, industrial symbiosis, and regenerative economics. This volume, *Circular Economy and Environmental Resilience*, explores the transformative role of circular economy principles in addressing global sustainability challenges. Bringing together diverse perspectives from scholars worldwide, the book delves into conceptual foundations, digital innovations, policy frameworks, and macroeconomic implications of circular practices. By examining the intersection of circular economy, environmental resilience, and sustainable development, this volume provides valuable insights for researchers, policymakers, and industry leaders striving to create a more sustainable and regenerative future.

The book opens with a foundational discussion on the circular economy and its role in enhancing environmental resilience, setting the stage for deeper explorations into global sustainability challenges. It then examines the green transition, the integration of digital technologies in circular practices, and the evolving macroeconomic implications of climate change on economic and agricultural growth. Critical barriers to circular adoption, the role of institutions and leadership in driving sustainability, and regional case studies—ranging from Europe to Africa and Asia—offer a holistic understanding of circularity in diverse contexts.

Through these chapters, the volume comprehensively analyzes the pathways to achieving sustainable development through circular practices. By bridging theoretical insights with real-world challenges, this book serves as a valuable resource for academics, policymakers, and industry practitioners committed to shaping a more resilient and regenerative future.

This volume establishes the foundational concepts of the circular economy and its potential to drive environmental resilience, offering a comprehensive exploration of how circular practices can foster sustainability amidst global environmental challenges.

New Delhi, India

Pardeep Singh
Shikha Daga
Kiran Yadav
Vandana Mishra

# Contents

# Beyond the Take-Make-Dispose Model—Unlocking the Power of Circular Economy for an Environmentally Resilient Future

**Shikha Daga, Kiran Yadav, Pardeep Singh, and Vandana Mishra**

**Abstract** The circular economy (CE) has emerged as a transformative approach, aiming to optimize resource utilization, minimize waste, and foster long-term economic and environmental resilience. This paper explores the evolution, principles, and growing academic discourse surrounding CE, emphasizing its role in addressing global challenges such as resource depletion, climate change, and economic instability. It highlights the transition from the "take-make-dispose" model to a closed-loop system that integrates reuse, remanufacturing, and recycling, ensuring a more sustainable and resilient industrial framework. The study uses bibliometric analysis, with the help of the Biblioshiny software of R studio, to map the academic growth of circular economy research, revealing an exponential increase in scholarly interest since 2017. This research represents key concepts related to the *Circular Economy* by emphasizing interconnected themes such as *sustainable development, recycling, waste management, sustainability, and environmental impact.* This paper also provides a glimpse of all the chapters included in this book. This study highlights that integrating Circular Economy principles with Sustainable Development Goals can drive economies toward a low-carbon, resource-efficient, and resilient future, reinforcing the global sustainability agenda. It emphasizes that a circular economy is not only a viable but also a necessary approach to achieving global sustainability goals. Additionally, the study underscores the need for further interdisciplinary research to bridge existing gaps between knowledge and practice in this field.

---

S. Daga · K. Yadav (✉)
Department of Commerce, PGDAV College, University of Delhi, Delhi, India
e-mail: id-kiran.yadav@pgadav.du.ac.in

P. Singh
Department of Environmental Studies, PGDAV College, University of Delhi, Delhi, India

V. Mishra
Department of Environmental Studies, Bioresources and Environmental Biotechnology Laboratory, University of Delhi, Delhi, India

Centre for Interdisciplinary Studies On Mountain & Hill Environment (CISMHE), University of Delhi, Delhi, India

 
P. Singh et al. (eds.), *Circular Economy and Environmental Resilience*,
https://doi.org/10.1007/978-3-031-93091-1_1

**Keywords** Circular economy · Environmental resilience · Bibliometrics · Sustainable development goals

## 1 Introduction

Severe environmental changes have forced economies worldwide to identify alternatives to traditional modes of production and consumption. While there is the implementation of sustainable development goals at the macro level, the micro-level promotion of circular economy (CE) principles has taken center stage to attain the goal of green growth (Daga et al., 2024). CE refers to an economic model that aims to promote the continuous use of resources, thereby eliminating waste and generating sustainable growth. It focuses on three R's, namely reuse, remanufacture, and recycle, thereby making a paradigm shift from the linear—take, make, dispose approach to the closed loop system where the resources are retained in the system for as long as possible (Agrawal et al., 2025). The evolution of CE dates back to the 1960s and 1970s wherein environmentalists questioned the sustainability of rapid industrialization and resource depletion. It gave rise to various theories like the Kenneth Boulding's "Spaceship Earth" (1966), Walter Stahel's "Cradle to Cradle" approach, and the Ellen MacArthur Foundation's modern advocacy that have shaped the development of CE into a globally recognized framework.

The traditional production-consumption model being linear is considered unsustainable as it leads to the wastage of resources. It has led to the global issues of resource depletion, waste accumulation and climate degeneration (Aithal & Aithal, 2023). The linear production-consumption model has not only added to environmental deterioration but has also led to environmental deterioration. It has made the supply chains vulnerable as resources are being extracted, used, and depleted at an alarming rate. The throwaway culture has to the issue of exponential accumulation across various landfills with a huge rise in plastic accumulation and changes in the overall environment. The carbon footprint associated with the production and disposal of goods has further contributed to greenhouse gas emissions, exacerbating global warming. Not only does the traditional production-consumption model impact environment, it also brings great shortcomings for the overall economic stability. This is because overuse of resources brings huge fluctuations in prices thereby increasing macroeconomic uncertainty and disrupting the overall supply chains. Recognizing these challenges, transitioning from a linear to a circular economy is crucial for achieving long-term environmental and economic sustainability.

The circular economy aims to create industrial systems that are restorative and regenerative by design, ensuring that resources are used efficiently, waste is minimized, and ecosystems are protected (Kennedy & Linnenluecke, 2022).

Circular Economy integrates concepts of resource efficiency that help in reusing materials, recycling waste, and hence creating a closed loop system. The adoption of circular economy principles help economies to promote the goals of sustainable development and contribute towards a resilient and greener future. However, not all economies have the infrastructure, funds, and technology to foster circular economic principles. This is the reason why certain economies are ahead in sustainability promotion as compared to others. On the other hand, environmental resilience refers to the capacity of ecosystems and human societies to absorb shocks and disturbances while maintaining their essential functions. Industries extract continuous pressure by their activities on natural systems through resource extraction, pollution, and emissions, the need for a resilient approach to economic growth has become evident (Kennedy & Linnenluecke, 2022).

The implementation of the principles of circular economy, like narrowing resource loops, i.e., improving resource efficiency, suggested by research (Bimpizas-Pinis et al., 2021), slowing resource loops, i.e., extending product life cycles (Lei et al., 2021), and closing resource loops, i.e., recycling and repurposing materials, helps businesses and policymakers to contribute to the resilience of social-ecological systems (Akhimien et al., 2021). These practices not only reduce environmental degradation but also enhance economic stability by mitigating supply chain disruptions and resource scarcity (Velenturf & Purnell, 2021).

Despite its benefits, the intersection of circular economy practices and environmental resilience remains underexplored. This chapter gives a brief analysis of the academic growth in the area of circular economy (CE) over the years, and part II gives a short glimpse of all the chapters included in the book. Figure 1 highlights how the interest in the domain of CE has seen exponential growth only since 2017. Though the concept has been there for ages still its relevance in promoting sustainable development has garnered academic attention only recently, capturing the unexplored area. By integrating CE principles with sustainable business models, economies can achieve a low-carbon, resource-efficient, and resilient future while advancing the global sustainability agenda.

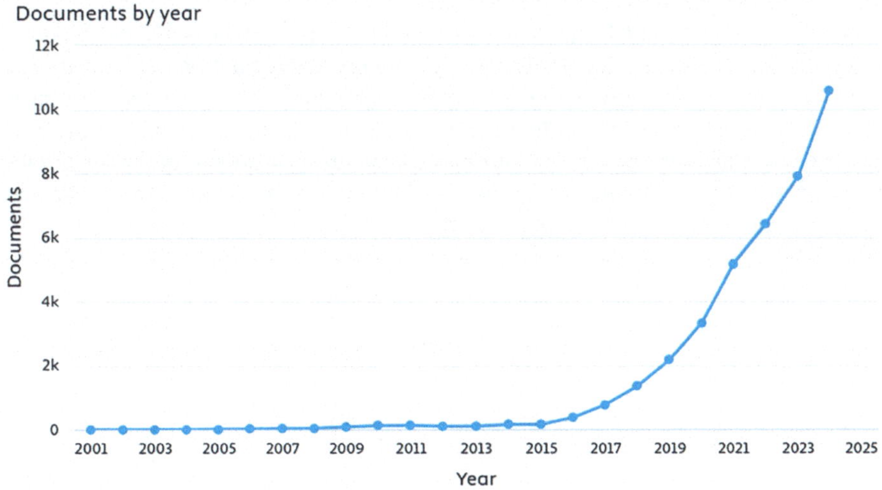


**Fig. 1** Documents per year. *Source* SCOPUS

The adoption of CE principles across economies is widely recognized as there are various firms like Apple, Tesla, Unilever, and others have integrated the same in their supply chains (Lei et al., 2021). Scholars have raised concerns that some CE strategies, such as excessive efficiency measures, may inadvertently reduce system resilience by eliminating redundancies and buffers needed to withstand external shocks (Bimpizas-Pinis et al., 2021). This highlights the importance of integrating resilience thinking into CE frameworks to ensure long-term sustainability. There is, however, still a huge gap in the academics as highlighted in the bibliometrics analysis done in the chapter and highlighted in the next section.

## 2 Existing Literature: A Bibliometric Overview

The importance of CE in promoting sustainability and the attainment of sustainable development goals (SDGs) is an area of interest in recent scenarios. The study by Castro-Lopez et al. (2025) documented the role of CE across the European Union in promoting SDGs along with highlighting the rise of CE-enabled startups while the study by (Long et al., 2025) asserts the role of technology in promoting CE towards the Industry 4.0 revolution.

The last few years have witnessed a radical shift towards sustainable development and adherence to its important antecedents, acceptance of CE as an important constituent has been realized across both developed and emerging economies as can be seen from rising interest across economies presented in Fig. 2. While Italy tops the chart in terms of academic work on CE, it is being followed by developed counterparts of the United Kingdom the United States, the most populated and the largest emitter China, with the world's most densely populated economy India followed by Spain, Germany, Australia, Brazil and others.

The rising interest in CE across Italy, the UK, the USA, Germany, and Spain is mainly because of the mandatory and strict environmental regulation implementation in these economies. In addition to regulatory requirements, other incentives for businesses in terms of enhanced public image and lesser scrutiny also lead to enhanced CE adoption. Italy has topped the chart because of strong academic-industry collaboration. Emerging economies of China and India not only top the world charts in terms of population and emissions but are also ahead in the adoption of CE principles. This is to ensure their alliance with sustainable development goals and to satisfy various stakeholders' goals. Even resource-rich economies like Australia and Brazil too have realized the importance of CE adoption and have integrated the same in their production-consumption process. The increased attention to CE principles across diverse economies reflects a paradigm shift towards sustainable development.

The second component of the bibliometric analysis is to identify the areas of interest that are mainly focus on principles of CE.

Bibliometric analysis is a scientific tool that helps in analyzing and describing the main areas of research. One important aspect of the same is the world cloud also known as the tag cloud. It helps in analyzing the literature by providing the

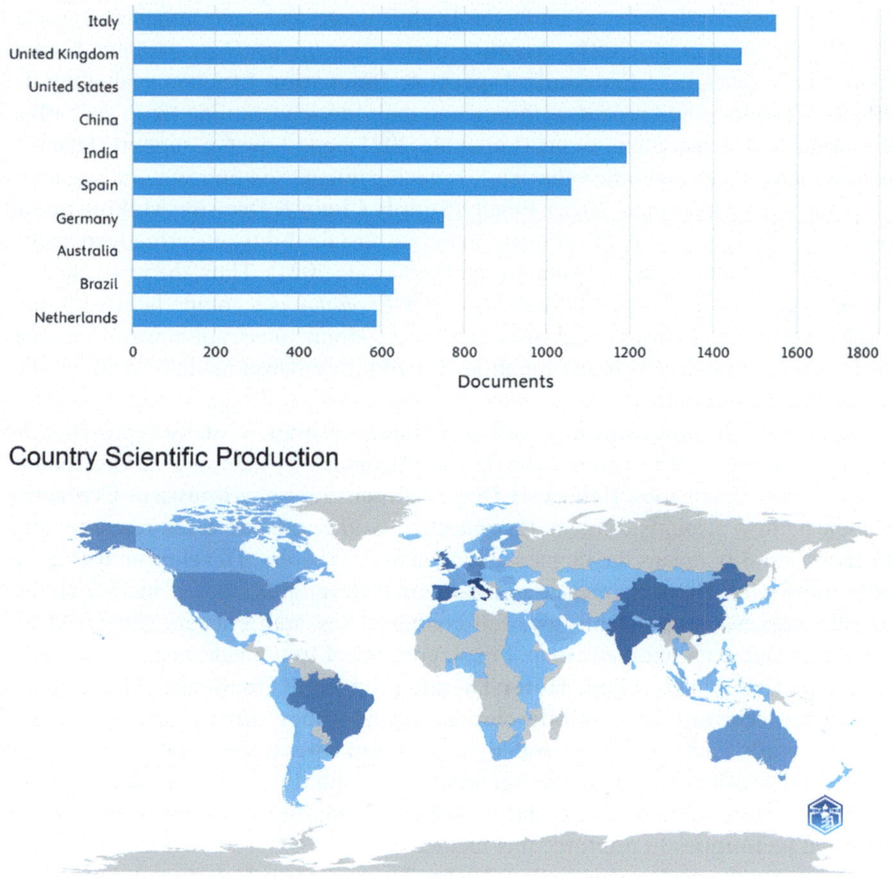


**Fig. 2** Documents by country. *Source* Biblioshiny

most frequently used words in the existing literature (Patil et al., 2023). Figure 3 provides the word cloud of CE concepts where there is found to be a close integration between CE, sustainable development, waste management, m sustainability, and environmental impact. To get the word cloud we have used author's keywords to form the clusters and here the word circular economy is the one that most repeatedly used in the literature with its other counter parts. This is because CE has a great role in reducing waste, reusing materials and hence making a radical shift towards cleaner and greener economy. Another prominent theme, **Sustainable Development & Sustainability** emphasizes the goal of CE in ensuring economic growth while minimizing environmental harm (Shehata et al., 2022). Sustainable development aligns with long-term strategies for resource efficiency, social well-being, and reuse. The next prominent theme is **Waste Management & Recycling** CE strategies heavily rely on *waste reduction* (D'Adamo et al., 2022)*, recycling*, and *proper resource utilization* to minimize landfill waste and environmental degradation (Salmenperä et al., 2021). **Another theme** like *life cycle assessment & environmental impact, and* **Economic Aspects & Innovation** focus on evaluating the effects of products and processes throughout their life span, aiming for lower carbon footprints and reduced emissions (Lei et al., 2021), and how CE supports business models that create economic value while reducing reliance on finite natural resources (Tapaninaho & Heikkinen, 2022). Finally, **Supply Chain & Decision Making theme** involves redesigning *supply chains* to integrate sustainability, ensuring responsible sourcing, production, and consumption (Hazen et al., 2021). Thus, the word cloud of circular economy highlights how the *Circular Economy* is a comprehensive framework connecting environmental sustainability, economic innovation, waste management, and policy-driven transformation to build an environmentally resilient and resource-efficient future.

The **strategic diagram, also called a thematic map,** visually represents the thematic evolution of research in the Circular Economy by mapping various **themes** based on two dimensions. **Relevance Degree dimensions, also known as Centrality it** indicate how strongly a theme is connected to other themes in the research field. Higher centrality means the theme is fundamental. Whereas, **Development Degree, also known as Density,** measures the internal development of a theme. Higher density suggests well-established, self-contained research clusters (Pessin et al., 2022). The strategic/thematic map, Fig. 4, consists of four quadrants:

The first quadrant is called**- Motor Themes (Top Right Quadrant)-**These themes are **highly relevant and well-developed**, meaning they drive research progress. The keywords **"article," "controlled study," and "waste disposal"** indicate that **structured studies on waste management** are a central force in Circular Economy research. Future studies can expand on **policy implications and innovative waste disposal techniques** to reinforce this theme**.**

**The second quadrant is called–Niche Themes (Top Left Quadrant)-**These topics are **highly developed but less central**, indicating specialized research areas. The terms **"nonhuman," "biomass," and "wastewater treatment"** suggest that

**Fig. 3** Word cloud of circular economy. *Source* Biblioshiny

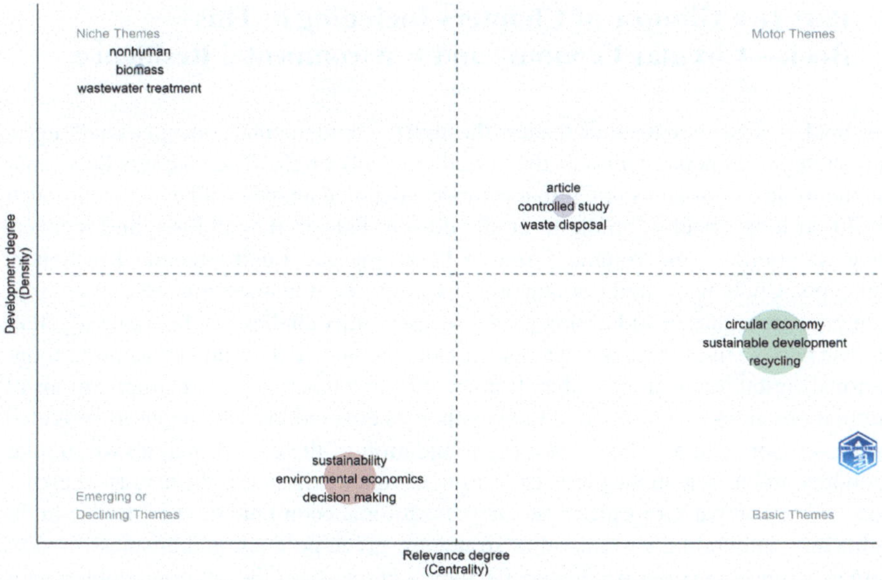


**Fig. 4** Strategic map/thematic map of circular economy. *Source* Biblioshiny

these themes are well-researched but not yet widely connected to broader Circular Economy discussions. This highlights the need for **integrating bio-based waste solutions** into mainstream Circular Economy policies.

The third quadrant is called **Basic Themes (Bottom Right Quadrant)**-These themes are **fundamental and widely relevant** but still evolving. The presence of **"circular economy," "sustainable development," and "recycling"** confirms their foundational importance in research. These topics require further exploration, particularly in terms of **innovative business models and digital transformation** in Circular Economy practices.

The fourth quadrant is called **Emerging or Declining Themes (Bottom Left Quadrant)**-These themes are either **gaining traction or losing relevance** over time. The terms **"sustainability," "environmental economics," and "decision-making"** indicate areas that are either **underexplored in Circular Economy research or becoming outdated**. Given the rising importance of ESG (Environmental, Social, and Governance) and regulatory frameworks, research should **revitalize these themes** by integrating them into policy-driven Circular Economy models.

## 3 Part II-a Glimpse of Chapters Including in This Book—Circular Economy and Environmental Resilience

The book is structured to guide readers through the fundamental concepts, challenges, and strategic implementation of the varied concepts of CE. The chapters delve into the theoretical constructs of CE that provide a detailed analysis of key concepts such as closed-loop systems, cradle-to-cradle design, industrial symbiosis, and regenerative economics. This volume, Circular Economy and Environmental Resilience: Concepts, Challenges, and Frameworks, explores the transformative role of circular economy principles in addressing global sustainability challenges. Bringing together diverse perspectives from scholars worldwide, the book delves into conceptual foundations, digital innovations, policy frameworks, and macroeconomic implications of circular practices. The outline of the key themes covered in each chapter is provided in this section: Chap. 2 lays down the foundation of the core principles of circular economy and its role in the green transition. Chapter 3 discusses the historical evolution of CE, comparing its principles with traditional economic models. It also highlights how integrating sustainability strategies, green policies, and business models help in promoting systemic change. Chapter 4 emphasizes the environmental degradation caused by industrialization, deforestation, and waste accumulation. It outlines how transitioning from a linear economy to a circular model can mitigate these issues by reducing pollution and promoting responsible consumption. Chapters 5 and 6 explore the relationship between CE and environmental resilience. By adopting CE principles such as decarbonization, waste reduction, and sustainable business models, societies can enhance their adaptive capacities and mitigate climate risks. Chapter 7 address the concept of environmental degradation through circular practices along with highlighting the consequences of resource depletion, pollution, and climate change, positioning circular economy as a viable solution. Chapter 8 investigates diverse challenges that hinder widescale adoption of CE, including economic, political, and technological constraints. The influence of greenwashing, cost considerations, and geopolitical disruptions is also examined. Chapter 9 highlights the role of institutions and leadership in promoting sustainable development. Policy interventions, regulatory frameworks, and the impact of lobbying efforts on CE adoption are analyzed in this chapter. Chapter 10 discusses the interplay between environmental economics and circular economy. In this chapter, economic models that incorporate environmental costs, incentives for sustainable practices, and market-based mechanisms to promote circularity are explored. The role of technology in promoting CE cannot be overlooked and hence Chap. 11 discusses the role of digital Transformation in promoting Circular Economy. The adoption of digital technologies—such as AI, IoT, and blockchain—in CE strategies is discussed in this chapter. Chapter 12 provides regional and sectoral applications of circular economy. It's an in-depth case study of CE implementation in Bulgaria, highlighting policy gaps, sectoral challenges, and lessons learned. Additionally, Chap. 13 examines the relationship between climate change, economic growth, and agricultural resilience in Africa, emphasizing the macroeconomic implications of CE. Lastly, Chap. 14 contributes to the book's broader narrative by illustrating how CSR strategies in the fashion industry can support circular economy principles, enhance sustainability reporting, and promote environmental resilience.

The book thus offers diverse perspectives of CE principles across different domains thereby providing an exhaustive readership material for all those who are interested in understanding and adopting the principles of CE.

## 4 Conclusion

The transition to a circular economy is not merely an option but a pathway to enhancing environmental resilience and achieving sustainable development. As identified in this study, circular economy strategies, including waste reduction, resource efficiency, and closed-loop systems, contribute to mitigating environmental degradation while fostering economic growth (D'Adamo et al., 2022; Tapaninaho & Heikkinen, 2022). However, despite the growing global adoption of circular economy principles, significant challenges remain, particularly in emerging economies where regulatory frameworks and technological capabilities may be inadequate. Through a glimps of the chapter included in this book, we are sure that the book opens with a foundational discussion on the circular economy and its role in enhancing environmental resilience, setting the stage for deeper explorations into global sustainability challenges and providing a holistic understanding of CE's principles, challenges, and applications, equipping policymakers, business leaders, and researchers with the knowledge to drive meaningful change. By fostering interdisciplinary collaboration, innovative business models, and supportive policy frameworks, societies can build resilient and sustainable economies that balance economic growth with environmental stewardship.

The following chapters delve deeper into the mechanisms and strategies that make CE a viable and necessary approach to achieving global sustainability goals. Through an exploration of case studies, theoretical frameworks, and empirical insights, this book aims to serve as a valuable resource for navigating the complexities of circularity and environmental resilience.

Future research could explore the intersection of circular economy strategies and digital transformation, assessing how emerging technologies such as artificial intelligence and blockchain can facilitate circular business models. Additionally, more empirical studies are needed to evaluate the long-term economic and environmental impacts of circular economy initiatives in diverse geographic contexts. Though the literature has discussed about various infrastructural, cost and economy related barriers that delay the adoption of CE principles across different economies. Its detailed analysis considering diverse economic groups can be undertaken in future studies. For instance, countries that have financial constraints and/or are technologically not very advanced can face delays in CE adoption. Similarly geopolitical tensions and economic uncertainty can also shift focus away from the adoption of CE and hence it can be undertaken in future studies. There are studies that highlight how mandatory regulations can lead to forceful adoption of CE across economies. this too can be an interesting area of study wherein voluntary and mandatory adoption of CE can be compared. The results of the same can be very useful for policy-making. In nutshell., CE adoption requires efforts at the micro, macro and global level. By strengthening global cooperation and adopting context-specific circular economy frameworks, nations can work towards a resilient, resource-efficient, and sustainable future.

## References

Agrawal, S., Jain, V. K., & Verma, H. (2025). Circular economy! Sustainable growth and achieving zero net emissions – a way ahead for building sustainable tomorrow by optimising consumption. *International Journal of Sustainable Agricultural Management and Informatics*, 11(1), 72–96. Inderscience Publishers. https://doi.org/10.1504/IJSAMI.2025.143112

Aithal, S., & Aithal, P. S. (2023). Importance of circular economy for resource optimization in various industry sectors—a review-based opportunity analysis. *International Journal of Applied Engineering and Management Letters, 1*, 191–215. https://doi.org/10.47992/ijaeml.2581.7000.0182

Akhimien, N. G., Latif, E., & Hou, S. S. (2021). Application of circular economy principles in buildings: A systematic review. *Journal of Building Engineering, 38*, Article 102041.

Bimpizas-Pinis, M., Bozhinovska, E., Genovese, A., Lowe, B., Pansera, M., Alberich, J. P., & Ramezankhani, M. J. (2021). Is efficiency enough for circular economy? *Resources Conservation and Recycling, 167*, Article 105399.

Castro-Lopez, A., Entrialgo, M., Liao, C. T., & Santos-Vijande, M. L. (2025). Towards circular economy through innovation: The role of entrepreneurial orientation and human resource management. *International Entrepreneurship and Management Journal, 21*(1). Springer. https://doi.org/10.1007/s11365-024-01032-x

D'Adamo, I., Mazzanti, M., Morone, P., & Rosa, P. (2022). Assessing the relation between waste management policies and circular economy goals. *Waste Management, 154*, 27–35.

Daga, S., Yadav, K., Lakshmi, V., & Singh, P. (2024). Growth vs. green: Unpacking the economic–environmental dilemma in major carbon emitters with panel ARDL analysis. *Discover Environment, 2*(1), 143. https://doi.org/10.1007/s44274-024-00173-8

Hazen, B. T., Russo, I., Confente, I., & Pellathy, D. (2021). Supply chain management for circular economy: Conceptual framework and research agenda. *The International Journal of Logistics Management, 32*(2), 510–537.

Kennedy, S., & Linnenluecke, M. K. (2022). Circular economy and resilience: A research agenda. *Business Strategy and the Environment, 31*(6), 2754–2765. https://doi.org/10.1002/bse.3004

Lei, H., Li, L., Yang, W., Bian, Y., & Li, C.-Q. (2021). An analytical review on application of life cycle assessment in circular economy for built environment. *Journal of Building Engineering, 44*, Article 103374.

Long, W., Ng, S. T., Lu, W., Mora, L., & Bao, Z. (2025). Deciphering how digital functions enable circular economy practices in construction: A critical review of recent progress and future outlook. *Environmental Impact Assessment Review, 113*. Elsevier Inc. https://doi.org/10.1016/j.eiar.2025.107889

Patil, R. R., Kumar, S., Rani, R., Agrawal, P., & Pippal, S. K. (2023). A bibliometric and word cloud analysis on the role of the internet of things in agricultural plant disease detection. *Applied System Innovation, 6*(1), 27.

Pessin, V. Z., Yamane, L. H., & Siman, R. R. (2022). Smart bibliometrics: An integrated method of science mapping and bibliometric analysis. *Scientometrics, 127*(6), 3695–3718.

Salmenperä, H., Pitkänen, K., Kautto, P., & Saikku, L. (2021). Critical factors for enhancing the circular economy in waste management. *Journal of Cleaner Production, 280*, Article 124339.

Shehata, N., Obaideen, K., Sayed, E. T., Abdelkareem, M. A., Mahmoud, M. S., El-Salamony, A.-H.R., Mahmoud, H. M., & Olabi, A. G. (2022). Role of refuse-derived fuel in circular economy and sustainable development goals. *Process Safety and Environmental Protection, 163*, 558–573.

Tapaninaho, R., & Heikkinen, A. (2022). Value creation in circular economy business for sustainability: A stakeholder relationship perspective. *Business Strategy and the Environment, 31*(6), 2728–2740. https://doi.org/10.1002/bse.3002

Velenturf, A. P., & Purnell, P. (2021). Principles for a sustainable circular economy. *Sustainable Production and Consumption, 27*, 1437–1457.

# Navigating Circular Horizons: Introduction to the Concept of Circular Economy and Environmental Resilience

**Chukwuemeka Kingsley John, Fidelis Odedishemi Ajibade, Temitope Fausat Ajibade, Pankaj Kumar, Bashir Adelodun, and James Rotimi Adewumi**

**Abstract** The circular economy marks a transformative shift from the traditional "take, make, dispose" model to a regenerative approach focused on sustainability and resource efficiency. This strategy aims to decrease waste, increase product lifespans, and establish closed-loop systems—where materials are continuously recycled, repurposed, or remanufactured—instead of throwing away materials. "Navigating Circular Horizons" examines the fundamental ideas of the circular economy and how it might be used to solve urgent global issues including environmental degradation, resource depletion, and climate change. The circular economy promotes a more robust and flexible socioeconomic structure by lowering reliance on limited resources and minimising ecological damage. This chapter explores essential strategies for integrating circular principles throughout industries, including encouraging waste reduction, creative design, and sustainable consumer behaviour. It also looks at how government legislation, corporate social responsibility, and public awareness have influenced this change. Finally, the conversation emphasises how crucial multidisciplinary cooperation and innovation are to promoting the ideas of the circular economy and creating a sustainable future.

---

C. K. John
Department of Construction and Civil Engineering, University Centre Jersey Highlands College, St Saviour Jersey, UK

F. O. Ajibade (✉) · T. F. Ajibade · J. R. Adewumi
Department of Civil and Environmental Engineering, Federal University of Technology, Akure, Nigeria
e-mail: foajibade@futa.edu.ng

P. Kumar
School of Environmental Studies, Maa Shakumbhari University, Saharanpur, India

Research and Development Division, Society for AgroEnvironmental Sustainability, Dehradun, India

B. Adelodun
Arusha Climate and Environmental Research Centre, Aga Khan University, Arusha, Tanzania

Department of Agricultural and Biosystems Engineering, University of Ilorin, Ilorin, Nigeria

 
P. Singh et al. (eds.), *Circular Economy and Environmental Resilience*,
https://doi.org/10.1007/978-3-031-93091-1_2

**Keywords** Circular economy · Environmental resilience · Sustainability · Resource efficiency · Waste minimization

## 1 Introduction

The circular economy characterizes a paradigm shift from the traditional linear economic model, which is characterized by a 'take, make, dispose' approach that leads to significant resource depletion and environmental degradation (Croitoru et al., 2024; Skawińska & Zalewski, 2018). By design, the circular economy seeks to increase the longevity of resources, materials, and products while preserving a regenerative methodology. Three fundamental tenets form its foundation: reducing pollution and waste, preserving goods and resources, and re-establishing natural ecosystems. Essentially, the circular economy encourages closed-loop systems that maximise available resources and reduce waste. Innovative approaches are needed to do this, such as creating durable, repairable, and recyclable items and implementing resource-efficient business models like sharing and leasing platforms. The ultimate objective is to reduce environmental effect and develop an economic system that functions within the ecological bounds of the planet (Efe, 2024; Kirchherr et al., 2023; Patti, 2023). The circular economy also offers significant economic and social benefits, making it increasingly relevant in the context of global sustainability challenges. The circular economy reduces reliance on finite materials and improves resource utilisation, hence minimising the risks of resource shortages, supply chain interruptions, and volatile commodity prices. Along with generating new market opportunities and employment in sectors like recycling, remanufacturing, and sustainable services, it also stimulates innovation in business strategies, manufacturing techniques, and product design. The circular economy improves resilience by reducing pollution and waste, two key causes of environmental deterioration and climate change. Businesses and governments that embrace circular ideas not only promote social well-being and economic stability, but they also help create a better environment. This change sets the stage for a more equitable and sustainable future (Awan & Sroufe, 2022; Nygaard, 2023; Yang et al., 2023).

The ability of natural systems to tolerate, adjust to, and bounce back from disturbances—whether acute shocks like natural catastrophes or long-term stresses like habitat loss and climate change—is known as environmental resilience (Fiksel & Bakshi, 2023). The concept is important because ecosystems supply basic services that humans depend on for existence and well-being, like food, clean water, and air. Ecosystems that are resilient can continue to function in the face of stress, supporting biodiversity and the ecological processes that sustain life on Earth. Environmental resilience is influenced by several elements, such as ecological complexity, biodiversity, interconnection, and adaptability. Because they are better able to adapt to changes, ecosystems with a variety of species are typically more robust (Correia & Lopes, 2023; Grambow et al., 2022; Ward et al., 2022, 2024). Given the increasing pressures that pollution, urbanisation, and deforestation are putting on ecosystems, it is more crucial than ever to comprehend and improve environmental resilience. The ability of many ecosystems to withstand shocks has decreased, increasing their susceptibility to climate change. For example, many locations are now more vulnerable to extreme weather events due to the loss of wetlands, which act as natural flood barriers (Abbass et al., 2022). In a similar vein, coral reefs have been harmed by ocean acidification and warming sea temperatures, endangering marine ecosystems and the livelihoods of millions of people who rely on them. Restoring and safeguarding ecosystems and lessening human influence are both necessary to improve environmental resilience. This include encouraging environmentally friendly farming practices, preserving natural areas, reducing greenhouse gas emissions, and enhancing local people' capacity to adjust to shifting environmental conditions. Making resilience a top priority will enable ecosystems to sustain human existence and function even in the face of major setbacks (Althea, 2023; Cooke et al., 2023; Obaisi et al., 2022).

Sustainable resource management and ecosystem stability meet at the nexus of environmental resilience and the circular economy to form systems that are resilient to environmental shocks and efficient at the same time. The circular economy strengthens environmental resilience by reducing the demand for natural resources and minimizing ecological damage through strategies like waste reduction, material reuse, and resource efficiency. It reduces the need to extract virgin resources, which is one of the primary causes of pollution, habitat damage, and biodiversity loss, by keeping goods and materials in use for as long as feasible. For instance, recycling and reusing materials lessens the need to extract raw resources, which lessens the environmental effect of mining and deforestation, two practices that upset ecosystems. Ecosystems are more likely to remain resilient, bounce back from shocks, and continue to deliver vital functions when natural systems are not under as much stress (Cook et al., 2022; Fiksel & Bakshi, 2023; Yang et al., 2023). Environmental resilience and the circular economy have a common emphasis on long-term sustainability over immediate profits. Because circular practices make materials and products survive longer, they help create more resilient systems. One example of this is the construction of lasting, adaptive products. Easy-to-repair or update modular items, for example, have a longer lifespan and use fewer resources. Because it enables systems to react to and bounce back from unforeseen environmental, social, or economic changes, adaptability is essential to resilience. Additionally, circular economy initiatives that focus on restoring natural systems such as regenerative agriculture or ecosystem restoration, directly contribute to environmental resilience by improving soil health, increasing biodiversity, and re-establishing natural cycles. Through the integration of circular economy concepts with initiatives to improve ecosystem resilience, we can build a more flexible global system that can endure environmental problems in the future (Awan & Sroufe, 2022; Ballinas et al., 2022; Zhang et al., 2022; Zupancic, 2023).

The importance and relevance of the circular economy and environmental resilience in today's world cannot be overstated, particularly as we face escalating environmental challenges such as climate change, resource depletion, and biodiversity loss (Ahmad et al., 2022). The conventional linear economic model is proven to be unsustainable since it depends on the ongoing extraction of limited resources and produces a large amount of waste. This model increases the frequency and intensity of climate-related disasters and exacerbates environmental deterioration. To mitigate these effects, we must radically rethink how we produce, consume, and manage resources by implementing a circular economy. Through waste minimization, material conservation, and natural system regeneration, the circular economy provides a means of disentangling economic expansion from environmental degradation and preventing human activity from surpassing the planet's ecological carrying capacity. This strategy improves the ability of businesses and communities to adjust to and recover from environmental shocks in addition to supporting the preservation of ecosystems and the reduction of greenhouse gas emissions (Datzberger et al., 2023; Sarkhoshkalat et al., 2024; Upreti, 2023; Ward et al., 2024). The circular economy and environmental resilience are becoming more important for maintaining long-term economic stability and social well-being in addition to resolving environmental challenges. The lack of raw materials is becoming a serious problem as the world's population rises and resource demand rises. By establishing closed-loop processes where materials are continuously reused, recycled, or regenerated, a circular economy lessens dependency on virgin resources. This reduces the demand for natural resources while simultaneously encouraging innovation and opening new business opportunities in sectors like sustainable product design, recycling, and remanufacturing. Furthermore, by using circular practices to improve environmental resilience, communities, and organizations will be better equipped to endure and adjust to the effects of climate change, including resource shortages, supply chain disruptions, and extreme weather events. Through the integration of circular economy principles with environmental resilience-building activities, more resilient, egalitarian, and sustainable systems that can withstand future challenges can be developed. Achieving sustainable development goals and ensuring a viable future for future generations depends on this dual strategy (França et al., 2022; Kandpal et al., 2024; Moroni et al., 2022; Yang, 2022).

This chapter seeks to give a thorough examination of how environmental resilience strategies, and the concepts of the circular economy might be successfully combined to address urgent global issues including waste management, resource depletion, and climate change. The chapter aims to demonstrate how these integrated concepts can be applied across different industries to create resilient and sustainable systems. The first goal is to clarify the fundamental ideas that underlie both the circular economy and environmental resilience. The second goal is to analyse the synergies between these two frameworks and how they can complement each other in practice. This chapter will also discuss the opportunities and difficulties of putting these ideas into practice, as well as provide predictions for future developments. Particular attention was paid to how important innovation and policy are to hastening the shift to a more robust and sustainable global economy.

# 2 Foundations of Circular Economy

## *2.1 Historical Development and Evolution of Circular Economy*

The notion of the circular economy has developed over several decades, stemming from an increasing recognition of the constraints of the conventional linear economy, which relies on a 'take, make, dispose' framework (Rashid & Malik, 2023). Several important concepts and initiatives that arose in reaction to growing environmental concerns in the middle of the twentieth century are responsible for the historical development of the circular economy. An early source of inspiration was Kenneth Boulding's 1966 essay "The Economics of the Coming Spaceship Earth," in which he contended that Earth needs to be viewed like a spaceship, a closed system with limited resources and that the economy ought to change from being open and exploitative to being closed and cyclical (Piorkowsky, 2023). The basis for considering economics in terms of waste reduction and resource efficiency was established by this concept. The environmental movement gained momentum in the 1970s due to studies like the Club of Rome's "Limits to Growth" study, which emphasized the unfeasibility of continued economic growth in an environment with limited resources. The conversation around sustainability and the necessity for an economic structure that could function within planetary bounds was greatly influenced by these early contributions (Anaruma et al., 2022; Piorkowsky, 2023; Priewe, 2022). Another CE concept was seen in Germany during World War II, when circularity emerged through retail recycling programs like Max Bahr's invention of reusable shopping baskets. Similar measures were taken by American supermarkets in the 1950s and 1960s, who offered shopping trolleys to cut down on packaging waste. In the 1970s, Walter Stahel coined the phrase "cradle-to-cradle," which highlights sustainable product lifecycles and contrasts with the conventional "cradle-to-grave" strategy (Nagamalini & Wesley, 2024; Shell, 2014). Canadian ecologist C.S. Holling coined the term "environmental resilience" in 1973 to describe an ecosystem's ability to withstand shocks and adapt to change. Holling's work influenced later ecology and environmental management studies by laying the groundwork for understanding how natural systems endure in the face of environmental variability (Holling, 1973). The convergence of CE and environmental resilience concepts has led to integrated approaches in sustainable development. The circular economy notion gained more momentum in the 1980s and 1990s as researchers and industry professionals looked for methods to make it practical. Around this time, the discipline of industrial ecology came into being, promoting the creation of industrial systems that resemble closed-loop processes in nature, in which waste from one process is used as an input for another (Montag, 2023). William McDonough and Michael Braungart also developed the "Cradle to Cradle" design concept around this time, emphasizing the need to create items that could be constantly recycled into new products or safely returned to the environment (Adenle et al., 2024). With the formalization of the circular economy as a comprehensive framework for sustainability in the early 2000s, these ideas were

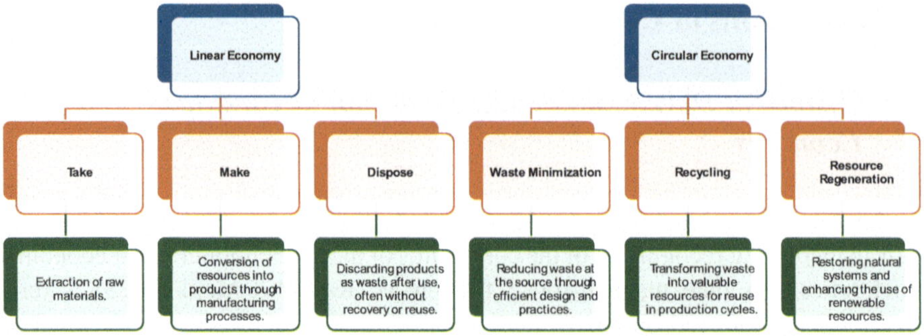


**Fig. 1** Shift from a traditional linear economic model to a circular economy

further developed. Established in 2010, the Ellen MacArthur Foundation has been instrumental in popularising and furthering the circular economy through its development of frameworks, research publications, and partnerships with governments and businesses to put circular principles into practice. As a result of its transformation from a theoretical idea to a widely accepted practical method, the circular economy is now acknowledged as a crucial tactic for solving environmental issues and attaining sustainable development (Fig. 1) (Awan & Sroufe, 2022; Guzzo et al., 2022; Kirchherr et al., 2023).

## *2.2 Key Principles of Circular Economy*

The fundamental tenet of the circular economy is to establish a regenerative and sustainable economic structure. Prioritizing waste and pollution reduction in design is the primary principle. To eliminate waste and reduce the environmental effect from the start, this principle highlights the need to reconsider how things are made, utilized, and designed. Circular economy approaches do not view waste as an inherent result; rather, they view it as a resource that can be removed by strategic design. This entails designing for durability, reparability, and upgradability; choosing materials that are either entirely recyclable or biodegradable; and making sure that goods are simple to disassemble and reprocess. Companies can greatly lessen their environmental impact and produce goods that support a closed-loop system in which resources are continuously cycled through the economy by tackling waste and pollution from the design stage (Awan & Sroufe, 2022; Shahhoseini et al., 2023; Viles et al., 2022). Another fundamental principle of the circular economy is keeping products and materials in use for as long as possible. Using techniques including reuse, repair, refurbishing, remanufacturing, and recycling, this philosophy aims to extend the lifecycle of objects. The circular economy minimizes the demand for new resource extraction, which lowers energy consumption and environmental degradation by making the most use of products and resources. This concept also encourages the development of novel business models, such as product-as-a-service, in which customers purchase a product's services rather than the actual item. This aligns corporate incentives with sustainability goals by reorienting the focus from selling more things to preserving them for longer. Furthermore, the concept promotes the growth of secondary markets for second hand goods, where items can be refurbished or resold to obtain a second life. By ensuring that goods and materials are kept in circulation, these tactics foster a more robust and sustainable economic system that is advantageous to society and the environment (Kandpal et al., 2024; Kaplan, 2024; Voulvoulis, 2022).

## 2.3 *Models and Frameworks for Circular Economy*

The circular economy's models and frameworks offer organized methods for applying its ideas at various dimensions and sectors, assisting businesses and decision-makers in making the shift from linear to circular systems. The Butterfly Diagram from the Ellen MacArthur Foundation is one well-known model that separates the biological and technical processes. Products created from organic materials that can be safely reintroduced into the environment through procedures like anaerobic digestion or composting are included in the biological cycle (Kirchherr et al., 2023). The regrowth of natural systems and nutrient recycling are highlighted in this cycle. The technological cycle, on the other hand, minimizes waste by keeping valuable resources in circulation and focuses on materials and products made for longevity, reuse, and recycling. This model is frequently used to show how various products and materials should be handled throughout their life cycles to encourage resource efficiency and sustainability (Morsink-Georgali et al., 2022). The ReSOLVE framework is another important framework that provides a thorough approach to circular economy practices. Regenerate, Share, Optimise, Loop, Virtualise, and Exchange are referred to as ReSOLVE. Every element of the framework focuses on tactics to promote circularity. Regenerate is centered on using techniques like regenerative agriculture to rebuild ecosystems and natural systems (Smol et al., 2023; Tedesco et al., 2022). Share places a strong emphasis on business models like tool libraries and automobile sharing that encourage sharing and collaboration above ownership. Optimizing entails using cutting-edge technology and procedures to cut waste and increase resource efficiency. Designing systems and products to facilitate closed-loop material flows in which obsolete goods are recycled or repurposed is known as "looping." Utilizing digital technology, such as virtual goods or services, to minimize the use of physical resources is known as virtualization. Exchange, the last point, emphasizes how crucial it is to exchange goods and resources both within and between industries to optimize resource utilization. By implementing these tactics, businesses can move towards circular processes in a methodical manner and reap major financial and environmental rewards (Awan et al., 2022; Mishra et al., 2023).

A prominent paradigm that concentrates on creating completely regenerative systems and products is the Cradle to Cradle (C2C) framework. This framework, which was created by William McDonough and Michael Braungart, promotes a design philosophy in which items are imagined with their end of life in mind. The Cradle-to-Cradle methodology places a strong emphasis on designing items that can be recycled indefinitely into new ones without sacrificing quality, or that are biodegradable and safely return to the environment. The framework offers a certification process that assesses the environmental performance of items based on social justice, renewable energy consumption, material health, and material reutilization. This strategy fosters innovation and ongoing improvement in the development of closed-loop systems in addition to sustainable product design. Businesses may increase material efficiency and contribute to a circular economy where resources are continuously cycled through the economy in a way that promotes sustainability and environmental health by implementing the Cradle-to-Cradle principles (Adenle et al., 2024; Sharma, 2024).

## *2.4 Case Studies: Successful Circular Economy Practices*

The successful implementation of circular economy practices is exemplified by various case studies across different industries and regions, showcasing how these principles can lead to significant environmental and economic benefits.

The government of the Netherlands has set an ambitious aim to transition to a fully circular economy by 2050, and its national policy is one notable example of this. The Dutch approach is centered on cutting waste, increasing innovation in industries like manufacturing, agriculture, and construction, and lowering resource usage. Within this paradigm, the city of Rotterdam stands out as an example of success since it has embraced circular approaches for waste management and urban development. The city has started programs such as the "BlueCity" project, which transforms an abandoned swimming facility into a bustling hub for circular enterprises. Utilizing one another's waste streams as resources, BlueCity companies work together to create a closed-loop system that minimizes waste and maximizes resource efficiency. Through the support of startups and businesses that prioritize sustainability, this project not only revitalises urban landscapes but also promotes economic growth (Chioatto & Sospiro, 2023; Möslinger et al., 2023). IKEA, a multinational firm, is a remarkable case study as well. Its business strategy incorporates circular economy principles to address the environmental impact of its products. By 2030, IKEA aims to become a fully circular firm by emphasizing the design of items that are recyclable, reusable, and repaired. The "Furniture Take-Back" program is one of its effective circular initiatives; buyers can return unwanted IKEA furniture for recycling if refurbishment is not possible, or for resale. IKEA has also created a variety of items with recycled materials, like the "Kungsbacka" kitchen fronts composed of PET plastic bottles and repurposed wood. These initiatives are part of a larger plan to cut carbon emissions, encourage sustainable consumption, and minimize waste. IKEA can lessen its environmental impact and meet consumer demand for sustainable items by incorporating circular economy techniques into its operations.

This would improve its brand reputation and competitiveness in the market (Singh, 2024).

Through its "Refactory" project in Flins, France, the Renault Group has led the way in the automotive sector in implementing circular economy techniques. This facility's mission is to extend the lifespan of automobiles and their parts by recycling, remanufacturing, and reconditioning them. By 2030, the Refactory project hopes to process more than 45,000 automobiles annually, greatly lowering waste and resource usage. The facility houses remanufacturing equipment that uses less energy and material than making new components to restore used automotive parts to their original state. Additionally, Renault has created a program for vehicle refurbishing that increases the lifespan of used automobiles by preparing them for the market. This program not only lessens the impact on the environment but also opens new job opportunities and revenue streams in sustainable businesses. A good example of how circular economy principles may be effectively incorporated into intricate industrial processes to benefit the environment, and the economy is the Refactory project (Granata & Petrides, 2022; Mihai & Ulman, 2023;). The apparel brand Patagonia is another example of a successful firm that has incorporated circular economy ideas into its operations by providing repair services, reselling used goods, and creating sustainable clothing designs. Through its "Worn Wear" initiative, Patagonia encourages users to fix rather than throw away their clothing, prolonging its useful life and cutting down on waste. With this strategy, the company's operations have had less of an adverse effect on the environment and have improved its market position and brand loyalty. Philips is one of the electronics companies that has embraced circular economy principles by providing lighting as a service, where users pay for the use of lights instead of purchasing the items. With the use of this model, Philips is encouraged to create durable, energy-efficient lighting fixtures that can be updated or repurposed, cutting down on waste and resource usage. These case studies demonstrate how circular economy principles can be successfully applied in a variety of settings, resulting in sustainable environmental practices as well as financial rewards (Olwoch et al., 2023; Stathis, 2023). These case studies show how circular economy principles can be implemented successfully in a variety of industries, resulting in significant decreases in waste production, resource usage, and environmental effects. These projects, which adopt the principles of the circular economy, not only foster sustainability but also stimulate innovation, provide economic prospects, and bolster resilience against worldwide obstacles.

# 3 Environmental Resilience: Concepts and Applications

## 3.1 Defining Environmental Resilience

The ability of natural systems, including ecosystems, landscapes, and biodiversity, to withstand shocks, adjust to changes, and bounce back from unfavorable incidents without compromising their core operations is known as environmental resilience. It includes the degree to which these systems are resilient to shocks from the environment, including pollution, natural catastrophes, climate change, and human activity. The ability of resilient ecosystems to maintain their ecological processes and structural integrity in the face of stress (Gupta et al., 2023). This implies that these systems can carry on delivering essential services like clean air, clean water, fertile soil, and wildlife habitat even in the face of disruptions. To ensure that ecosystems can support biodiversity and human well-being in the face of current and future environmental stressors, it is crucial to understand environmental resilience when creating policies to protect and restore them (Adla et al., 2022; Kattel, 2022). The dynamic interaction of ecological, social, and economic elements is another aspect of environmental resilience. Ecosystems with a variety of species and intricate interactions are more likely to be resilient because these characteristics help the ecosystems adjust to changing conditions. But resilience is a constant state of adaptation and change rather than a static one. For instance, wetlands that have suffered from pollution or development may find it difficult to carry out their natural duties, which include controlling flooding and filtering water. Wetlands that have been preserved or restored with ecological connectedness and biodiversity, on the other hand, are more resilient to disturbances and can adjust to changing environmental conditions. Consequently, a complete perspective that incorporates ecological health, adaptive capacity, and the interplay between natural and human systems is necessary to define environmental resilience. This approach offers a thorough knowledge of how ecosystems can maintain their functionality in the face of adversity (Mina et al., 2022; Wang et al., 2022a).

## 3.2 Factors Influencing Environmental Resilience

Biodiversity is one of the primary factors impacting environmental resilience, as it offers ecosystems the capacity to absorb disturbances and continue functioning. Species redundancy, or the presence of a diversity of species in ecosystems carrying out comparable ecological roles, is ensured by high biodiversity (Kattel, 2022). Due to this redundancy, ecosystems can remain stable even in the face of environmental changes that affect some species, such as pollution, habitat destruction, or changes in the climate. For example, several species of algae and herbivorous fish work together to regulate the growth of algae in a coral reef environment, thus no single species is entirely in charge of this process (Tebbett et al., 2024). The resilience and general health of the reef can be preserved if one species disappears because other species can fill the void. Additionally, by improving processes that are essential to the stability and productivity of ecosystems, like pollination, soil formation, and nutrient cycling, biodiversity boosts ecosystem resilience (Pascual et al., 2022; Tebbett et al., 2024). An ecosystem's resilience is also significantly influenced by its complexity and structure. Complex ecosystems, such as forests with several canopy layers, a wide variety of plant species, and a rich understorey, provide a range of niches and habitats and are thus generally more resilient. A greater variety of species, which can interact in many ways to maintain ecological functions, are supported by this structural complexity. Because it is less likely that a single pathogen will harm every species present, a forest with a variety of plant species and strata can tolerate pests and diseases better. Ecosystems with complex food webs, in which species have several feeding interactions, are also more resistant to the extinction of a single species. Due to the interconnectedness and redundancy included in these intricate networks, the ecosystem can sustain itself in the face of major disruptions (Artime et al., 2024; Uhl et al., 2024; Marini et al., 2022).

The degree of connectedness between various ecosystems in a landscape, or ecosystem connectivity, is essential for preserving environmental resilience. Connectivity makes it possible for genes, species, and ecological processes to travel across different landscapes, which is essential for adaptation and recovery after shocks. As conditions change, animals may migrate to more favourable habitats in a connected landscape. For instance, animals may move to higher altitudes in response to rising temperatures. Through this migration, genetic diversity is preserved, and ecosystems can adjust to changing environmental conditions. Furthermore, the resilience of more extensive biological processes like seed dispersal in forests or water flow in river systems—which are critical to maintaining life across geographical boundaries, is bolstered by interconnected ecosystems. On the other hand, when species and ecological processes are constrained, fragmented landscapes where ecosystems are isolated by human development or agriculture often show less resilience, which results in a loss of biodiversity and a decline in ecosystem services (Anderson et al., 2023; Chen et al., 2022; López et al., 2024; Parvizi et al., 2022). An ecosystem's adaptive capacity, which is greatly impacted by human management techniques, is its capability to respond to and adjust to changes in the environment. Higher adaptive capacities are typically found in ecosystems that are managed using techniques that mimic natural processes, such as holistic grazing, conservation agriculture, or sustainable forestry. By protecting biodiversity, preserving soil health, and preserving water supplies, these approaches enhance ecosystem resilience. For example, managed burns in habitats that are suited to fire, such as savannas or specific kinds of forests, support ecological balance and inhibit the accumulation of combustible material, lowering the likelihood of devastating wildfires. Moreover, by facilitating ecosystems' more efficient recovery from shocks, adaptive management techniques that are adaptable and sensitive to shifting environmental conditions can increase resilience. For instance, during droughts, water management systems that adapt to seasonal fluctuations in rainfall can more effectively manage resources, thereby meeting the demands of both humans and the environment (Parwada et al., 2022; Shang & Xie, 2024; Twidwell et al., 2022).

Environmental resilience is significantly impacted by climate variability, which includes variations in temperature, precipitation patterns, and harsh weather conditions that have the power to strain ecosystems to the point where they can no longer recuperate (Abbass et al., 2022). Rapid changes may be difficult for ecosystems that have developed under certain climatic conditions, which can result in changes to the species makeup of the ecosystem, the spread of invasive species, and altered ecological functions. Drought stress, for example, can result in decreased water quality, extinction of aquatic species, and a reduction in ecosystem services like fish production and water purification when droughts become more frequent and intense (Han et al., 2022). The timing of biological activities like flowering and migration can also be affected by rising temperatures, which can have an impact on species interactions and the stability of ecosystems. Ecosystems possessing high resilience are generally better equipped to handle these climate shifts because of their adaptive management strategies, diversity, and connectivity. Nonetheless, persistent climate change poses a serious threat, necessitating the integration of resilience-building techniques into environmental management and conservation initiatives (Gaines et al., 2022). Environmental resilience is greatly impacted by human activity, especially when it comes to land use change, which modifies natural habitats and the processes that support them. Several factors, including industrial growth, deforestation, agriculture, and urbanization, can damage ecosystems, lower biodiversity, and interfere with natural processes including nutrient and water cycles. An ecosystem's capacity to bounce back from shocks and adjust to novel environments may be weakened by these modifications. Deforestation, for instance, diminishes the resilience of tropical ecosystems by eliminating important habitats, lowering species diversity, and changing the hydrological and climate cycles that sustain them. On the other hand, due to urbanization, impermeable surfaces are created, which prevents natural water infiltration, increases runoff, and decreases groundwater recharge, increasing the risk of flooding and the consequences of heat islands in metropolitan areas. Enhancing environmental resilience requires addressing the effects of human activity through conservation, habitat restoration, and sustainable land use planning (Anderson et al., 2023; Ferreira et al., 2022; Kumar et al., 2022). Environmental resilience is also significantly influenced by socioeconomic factors, such as cultural norms, governance systems, and community economic standing. Communities that depend significantly on natural resources like fishing or agriculture for their economic well-being are especially susceptible to changes in the environment and may find it more difficult to adjust to disruptions. However, by promoting sustainable resource management techniques, robust governance, community involvement, and traditional ecological knowledge can improve resilience. For example, in areas like the Amazon and portions of Africa, community-led conservation programs that combine traditional knowledge with contemporary conservation techniques have been successful in managing resources sustainably and preserving ecosystem resilience. Furthermore, by coordinating socioeconomic development with environmental sustainability, policies that uphold fair access to resources, encourage sustainable development, and foster community involvement in environmental decision-making can enhance the resilience of human communities as well as ecosystems (Das et al., 2022; Gupta et al., 2022; Li et al., 2022; Raimi et al., 2022).

### *3.3 Measuring and Assessing Resilience in Environmental Systems*

The creation of particular measures and indicators that can adequately convey the intricate and ever-changing character of ecosystems is necessary for the measurement and evaluation of resilience in environmental systems. In this sense, resilience describes an ecosystem's ability to take in shocks, adjust to shifting circumstances, and bounce back from them without losing its vital services and functions. It is essential to select important indicators that represent the stability, variety, and health of an ecosystem to measure this capability effectively. These metrics may include species richness, ecosystem productivity, biodiversity, and the degree of functional redundancy, the existence of several species that can carry out comparable ecological roles among them. Furthermore, understanding how effectively an ecosystem can tolerate and recover from disturbances like pollution, climate change, or habitat fragmentation depends on measures of ecosystem connectivity and the integrity of natural processes like water cycling and nutrient fluxes. Researchers and practitioners can provide a thorough evaluation of the resilience of an ecosystem by combining these indicators (Correia & Lopes, 2023; Mayar et al., 2022; Wu et al., 2024). Many tools and approaches, from sophisticated mathematical modelling to qualitative assessments, have been created to effectively measure resilience in environmental systems. Ecological modelling is a popular method that mimics how ecosystems react to diverse stresses over time, enabling researchers to forecast possible outcomes under various scenarios. The resilience of ecosystems under various management approaches can be assessed by these models by incorporating data on species interactions, climate, and human influences. An additional crucial instrument is the resilience assessment framework, which offers an organized method for assessing ecosystems' resilience by considering elements like system boundaries, important change agents, and possible thresholds that ecosystems may cross before degrading to an undesirable state. As they incorporate local knowledge and guarantee that the assessment accurately reflects the social and ecological reality of the place, participatory approaches which involve stakeholders in the assessment process are also beneficial. Through the integration of these instruments and approaches, resilience may be assessed with greater precision and specificity, enabling informed management and policy choices (Cheng et al., 2022; Bene et al., 2023; Power et al., 2024).

Using resilience assessments in environmental management is essential to developing plans that improve ecosystems' capacity to endure shocks and bounce back. Following the establishment and evaluation of resilience measurements and indicators, adaptive management strategies that give priority to the preservation and improvement of ecosystem resilience can be informed by the findings. Resilience evaluations, for instance, can be used to identify important sites under protected area management that, because of their sensitivity to stressors such as climate change, require restoration or more strict protection. Resilience assessments in agricultural landscapes can direct the application of sustainable techniques, such as rotational grazing or agroforestry, that support soil health and biodiversity and make the system more resilient to pest outbreaks or droughts. Moreover, resilience evaluations can help guide policy choices by pointing out crucial boundaries that must be maintained to prevent ecosystem damage that cannot be reversed. Decision-makers can create more effective policies that safeguard ecosystems while also improving their capacity to adapt to and flourish in the face of environmental changes by including resilience evaluations in environmental management (Gupta et al., 2022; Bene et al., 2023; Wu et al., 2024).

## *3.4 Case Studies: Building Environmental Resilience*

Several case studies illustrate successful efforts to build environmental resilience through various strategies and interventions.

A. The Netherlands: Integrated Water Management and Flood Resilience

The Netherlands offers a fascinating example of how coordinated water management may strengthen environmental resilience. The Netherlands has long been at the forefront of cutting-edge flood management techniques due to its low-lying topography and history of severe flooding. This strategy is best shown by the "Room for the River" program, which was started in response to the increased risk of floods brought on by climate change. The scheme incorporates natural processes by permitting rivers to overflow into designated regions, creating floodplains that absorb excess water, as an alternative to merely depending on conventional engineering solutions like dikes and dams. This strategy not only lessens the likelihood of floods in crowded regions but also enhances biodiversity and restores natural habitats. The Netherlands has created a more resilient environment that is better able to withstand the growing effects of climate change by adopting a comprehensive, nature-based strategy for flood management (Lambrechts et al., 2023; van Doorn-Hoekveld et al., 2022).

B. Japan: Forest Management and Disaster Resilience

The management of forests in Japan provides another illuminating case study on enhancing environmental resilience. Japan's forest management techniques are intended to lessen the effects of calamities like typhoons, tsunamis, and earthquakes, which are common in the country. The nation's massive reforestation initiatives, especially in the wake of World War II's devastation, have been essential in halting the usual occurrence of landslides and soil erosion in hilly areas. When it rains heavily, forests serve as organic barriers that stabilize the soil and absorb excess water, lessening the severity of landslides and floods. Furthermore, Japan's satoyama landscapes—traditional rural settings with coexisting agricultural and natural ecosystems—showcase how sustainable land use may boost biodiversity while supplying ecosystem services that strengthen resilience to disasters. These approaches have produced multifunctional landscapes that promote ecological and human well-being in addition to restoring degraded regions (Edgington, 2022; Hayashi, 2023; Xu and Sakai, 2024).

C. Rwanda: Community-Based Conservation and Climate Resilience

Particularly considering climate change, Rwanda's accomplishments in community-based conservation and its influence on enhancing environmental resilience offer a compelling example. The goal of the nation's massive reforestation and landscape restoration initiatives, like the Gishwati-Mukura National Park initiative, is to restore vital ecosystems that have been weakened by agricultural growth and deforestation. Strong legislative support and community involvement—local residents participate in conservation efforts and gain firsthand advantages from eco-tourism and sustainable agricultural methods have propelled these initiatives. Rwanda has improved agricultural output, secured water resources, and increased the resilience of its forests and wetlands to climate-related pressures like floods and droughts. Local communities are included in these conservation activities to guarantee their sustainability and benefit sharing, which promotes social and environmental resilience. To create resilience to environmental difficulties, Rwanda's strategy emphasizes the value of integrating policy support, community engagement, and ecosystem restoration (Kayitesi et al., 2024; Nsengimana et al., 2024; Raimi et al., 2022).

D. The Everglades National Park in Florida, USA

The restoration of Florida's Everglades National Park is one well-known example. Development, pollution, and methods of water management have had a major effect on the Everglades, a unique wetland habitat. Enhancing water flow, lowering nutrient contamination, and re-establishing natural habitat connections have been the main goals of restoration projects. Through reviving its ecological functions such as flood control and wildlife habitat and promoting the recovery of native species, these initiatives seek to increase the Everglades' resilience. The restoration effort serves as an example of how focused interventions can reconstruct ecological processes and enhance an ecosystem's capacity for future adaptation (Casanova et al., 2023; Mitsch et al., 2023).

E. The reforestation of the Loess Plateau in China

China's reforestation of the Loess Plateau is another good example. After being badly damaged by deforestation and soil erosion, this area has undergone significant land management and reforestation initiatives with the goals of lowering erosion, enhancing soil health, and restoring vegetation cover. Land productivity, water retention, and biodiversity have all significantly improved due to the reforestation operations. The Loess Plateau is now more resistant to environmental stresses like erosion and drought because of the adoption of sustainable land management techniques and the restoration of native vegetation. This example demonstrates how integrated methods of land management and restoration can improve environmental resilience and benefit nearby populations and ecosystems in several ways (Kumar et al., 2022; Xu et al., 2022).

F. The restoration of mangrove forests in Southeast Asia

The regeneration of mangrove forests in Southeast Asia is an additional example of enhancing environmental resilience in coastal areas. Mangroves are essential for protecting coastlines because they act as natural barriers against erosion and storm surges while also providing habitat for a variety of marine animals. The goal of projects in nations like the Philippines and Indonesia has been to rebuild mangrove forests that have been harmed by coastal development, agriculture, and logging. Through the improvement of natural buffers against extreme weather events and the promotion of fish population recovery, these restoration activities have increased coastal resilience. The effective restoration of mangroves serves as an example of how ecosystem-based strategies can improve resilience and support community well-being as well as environmental conservation (Ahmadzai et al., 2022; Nguyen et al., 2022).

# 4 Synergies Between Circular Economy and Environmental Resilience

## *4.1 Circular Economy as a Tool for Enhancing Resilience*

The circular economy (CE) is a revolutionary strategy that improves resilience by moving away from the "take-make-dispose" paradigm of the linear economy and towards a closed-loop system. This shift minimizes waste production and lessens reliance on limited resources, which is essential for constructing resilience. The circular economy helps protect against resource scarcity and supply chain disruptions by encouraging material reuse, refurbishment, and recycling, which increases economies' resilience to shocks to the environment and the economy. For instance, materials are used if feasible in a robust circular system, lowering the vulnerability brought on by erratic raw material markets. Through the continuous flow of materials and resources, this system not only protects natural resources but also helps communities and companies recover from shocks like natural disasters and economic downturns (Kandpal et al., 2024; Truong et al., 2024).

Rooted in their common objectives of sustainability, resource efficiency, and adaptive capacity, the circular economy and environmental resilience have synergy. Reducing environmental degradation and protecting ecosystems are two ways that a circular economy naturally promotes environmental resilience. The circular economy eliminates waste and pollution, which relieves strain on natural systems and permits them to regenerate and continue operating over time. Additionally, circular methods that increase environmental resilience include cutting carbon footprints, increasing biodiversity through sustainable agriculture, and employing renewable energy. For example, restoring soil health using biodegradable materials and encouraging natural cycles in agriculture might increase agricultural systems' resilience to climate change-related events like droughts and floods. This integrated strategy makes sure that economic activity not only supports itself but also helps natural habitats adapt and flourish in the face of changing circumstances (Kennedy & Linnenluecke, 2022; Roy et al., 2024a). A holistic approach to sustainable development involves combining environmental resilience tactics with the ideas of the circular economy (Fig. 2). A systematic strategy is necessary for this integration, addressing the social, environmental, and economic facets all at once. Communities, corporations, and policymakers must work together to develop frameworks that support circular activities and increase the resilience of human and environmental systems. For example, creating circular supply chains with an emphasis on local production and consumption can lower carbon emissions and make communities more resilient to interruptions in the global supply chain. Additionally, urban areas can be made more resilient to the effects of climate change by investing in circular infrastructure, such as renewable energy grids and water recycling systems. This integrated strategy creates a more resilient and sustainable world by encouraging a culture of innovation and adaptability, which not only promotes the sustainable use of resources but also makes sure that human and environmental systems are prepared to handle future problems (de Oliveira & Oliveira, 2023; Sylva & Iyer-Raniga, 2023).

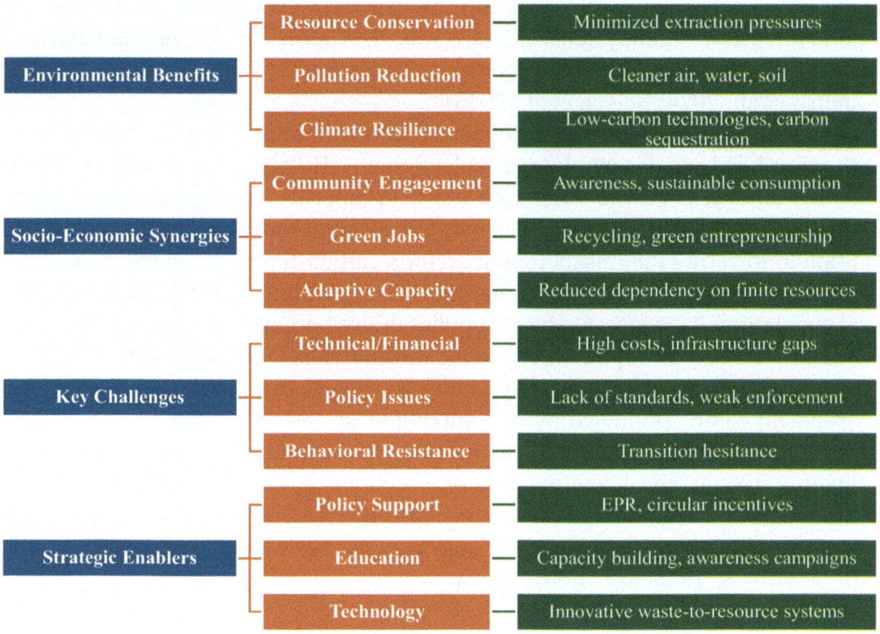


**Fig. 2** Synergies between circular economy and environmental resilience

### 4.2 Designing Resilient Circular Systems

Achieving synergies between environmental resilience and the circular economy requires designing resilient circular systems. These systems are designed to be flexible so that resources are used effectively, and they can continue to function even in the face of disturbances. The emphasis on modularity and flexibility in product development is a crucial component of resilient circular design since it makes maintenance, repairs, and upgrades simpler and increases product life cycle length while lowering waste. Products may, for instance, be made with replaceable parts, which makes it simpler to fix broken pieces rather than having to throw the entire thing away. This strategy improves overall resilience by ensuring that systems can swiftly adjust to shifting climatic conditions or supply chain interruptions while also conserving resources (De Angelis, 2022; Kennedy & Linnenluecke, 2022). The relationship between circular economy principles and environmental resilience is further strengthened by including ecosystem services in the architecture of resilient circular systems. The health and stability of the environment depend on ecosystem services including pollination, carbon sequestration, and water filtration. We can develop circular economies that reduce waste and increase the resilience of ecosystems by building systems that complement and align with these natural processes. In agriculture, for example, combining circular economy techniques with permaculture principles can improve biodiversity and soil health, resulting in more resilient food production systems. Because these systems are more resilient to environmental stressors like extreme weather, long-term sustainability is ensured. Businesses and communities can build systems that are robust in terms of the environment and economically by valuing and incorporating ecosystem services into circular design (Kandpal et al., 2024; Selvan et al., 2023).

Another essential element in building resilience within a circular economy framework is localized and decentralized circular systems. By managing resources such as materials, energy, and waste locally, these systems improve local resilience and lessen reliance on global supply chains. They also give priority to local resource cycles. In addition to lowering the transportation sector's carbon footprint, community-based energy production, urban farming, and localized recycling facilities guarantee that local populations have access to basic supplies in an emergency. Response to interruptions like natural catastrophes or economic downturns is facilitated by decentralized systems' inherent flexibility and adaptability. Resilience in terms of the economy and environment can be promoted by customizing these systems to the unique requirements and circumstances of the community by concentrating on local solutions. By taking this strategy, local environments and economies are better equipped to adapt to changing circumstances and bounce back from setbacks. It also helps the circular economy's objective of cutting waste and conserving resources (Awan & Sroufe, 2022; Watabe & Yamabe-Ledoux, 2023).

### 4.3 Overcoming Challenges in Integrating Circular Economy and Resilience

The move from linear to circular systems has several technical and infrastructure demands, which make integrating the concepts of the circular economy with environmental resilience a significant problem. Waste is viewed as an inevitable result in many modern sectors and infrastructures that are firmly rooted in linear models with single-use products. A significant investment in new technologies is necessary to transition to a circular economy. Examples of these technologies include sophisticated recycling techniques, advances in materials science, and digital platforms that track and manage resources across their entire existence. To facilitate circular processes, infrastructure must also be redesigned. Examples of this include creating industrial symbiosis networks or building neighborhood recycling centers, where trash from one industry is used as a resource for another. Governments, corporations, and academic institutions must work together to develop and expand these technologies and infrastructures to make them viable and accessible for wider adoption (Cimprich et al., 2023; Kennedy & Linnenluecke, 2022). Integrating resilience techniques and the circular economy is also hampered by behavioral and economic hurdles. Businesses and consumers used to the cost structures of the linear economy may find it difficult to make the upfront investments necessary to make the transition to circular systems. Products with longer lifespans or modular designs, for instance, could cost more up front, which deters people from buying them if they don't fully comprehend the long-term financial and environmental advantages. Furthermore, it is imperative that consumer behavior change because the implementation of circular models relies on a broad embrace of recycling, reusing, and sharing. In addition to financial incentives like grants, tax breaks, and subsidies to promote circular practices, overcoming these obstacles calls for intensive public awareness and education efforts. To change people's attitudes and behaviors and encourage more sustainable consumption habits, these ads must emphasize the long-term advantages of circularity, both economically and environmentally (George & Baskar, 2024; Kandpal et al., 2024).

The integration of circular economy principles with environmental resilience can be aided or hindered by policy and regulatory frameworks. Businesses and communities trying to implement circular practices face challenges because many of the policies now in place are designed with linear economic models in mind. For example, product standards may not take modularity or recyclability into account, and waste management laws may not sufficiently encourage or promote material reuse. Policymakers must update and modify regulatory frameworks to better fit with circular and resilient goals to overcome these obstacles. This entails formulating regulations that support extended producer responsibility (EPR), encouraging the use of sustainable materials, and establishing challenging goals for resource efficiency and waste reduction. Policies could also encourage innovation by lowering red tape and establishing precise standards for circular economy procedures. International collaboration is also essential since cross-border legislation and standards are needed to address global supply chains and environmental issues. Governments may facilitate the integration of resilience and the circular economy by tackling these legislative and regulatory obstacles, accelerating the shift towards a more robust and sustainable future (Awan & Sroufe, 2022; Kirchherr et al., 2023; Wang et al., 2022b).

## *4.4 Policy and Governance: Supporting Circular and Resilient Economies*

Policy is essential to facilitating the shift to resilient and circular economies because it creates the legal and regulatory frameworks required for systemic transformation. Policies like extended producer responsibility (EPR), which holds producers responsible for the full lifecycle of their products, including end-of-life disposal, are examples of how governments might encourage the adoption of circular practices. These regulations encourage manufacturers to create items that are simpler to maintain, recycle, and repair, which lowers waste and conserves resources. Furthermore, laws that encourage resource efficiency such as mandates for recycling, waste reduction goals, and prohibitions on single-use plastics directly support environmental resilience by limiting environmental deterioration and lessening the burden on natural ecosystems. Governments may promote the broad adoption of circular economy concepts and build a more resilient economy that is more resilient to shocks from the environment and the economy by establishing and enforcing these policies (De Angelis, 2022; Guzzo et al., 2022). Integrating circular economy ideas with environmental resilience requires strong governance frameworks. These frameworks must make it easier for diverse sectors of government, businesses, and communities to collaborate with one another and guarantee that programs and policies are coordinated. For example, creating intergovernmental organizations or task forces devoted to resilience and the circular economy can aid in coordinating efforts across regions, guaranteeing that national, regional, and local policies are consistent and reinforce one another. Furthermore, public–private partnerships in which corporations, governments, and civil society collaborate to create and execute circular economy strategies should be promoted by governance frameworks. These collaborations can promote creativity, distribute risks and benefits, and guarantee the scalability and sustainability of circular activities. In addition to fostering social cohesion and long-term sustainability, transparent and inclusive governance systems that involve all stakeholders can also foster trust and guarantee that the advantages of resilient and circular economies are spread fairly (De Angelis, 2022; Kandpal et al., 2024).

International collaboration and policy alignment are essential for fostering circular and resilient economies, given the global nature of environmental concerns and the interconnectivity of economies. Global frameworks and accords, such as the United Nations Sustainable Development Goals (SDGs) and the Paris Agreement on climate change, give nations a starting point for working together on circular economy projects that also improve environmental resilience. International standards can help harmonize practices across borders, for instance in the areas of waste management, recycling, and product design. This can facilitate worldwide trade in recycled materials and lessen the environmental impact of production and consumption. International collaboration can also be used to mobilize technological know-how and financial resources, especially for developing nations that might not have the infrastructure or capacity to adopt resilient and circular practices. Countries can push the shift to circular economies which are more robust to the challenges of the twenty-first century and more sustainable, together by coordinating international cooperation and coordinating national policies with global objectives (Awan & Sroufe, 2022; Rossi & Morone, 2023; Roy et al., 2023).

# 5 Sectoral Applications and Case Studies

## *5.1 Circular Economy in Agriculture and Food Systems*

The circular economy in agriculture focuses on closing nutrient loops, which is essential for reducing waste and enhancing the sustainability of food production systems. The widespread use of synthetic fertilizers and pesticides in traditional linear agriculture practices can result in nutrient runoff, degraded soil, and water body pollution. A circular approach to agriculture, on the other hand, aims to recycle nutrients inside the system to minimize the demand for outside inputs and lessen its negative effects on the environment. Composting organic waste, planting cover crops to fix nitrogen in the soil, and incorporating livestock into crop production to use manure as a natural fertilizer are among the methods that can help achieve this. These techniques not only lessen the need for chemical inputs but also enhance biodiversity and soil health, both of which are vital for agricultural ecosystems to remain resilient. Circular agriculture can counteract the negative consequences of conventional farming practices while increasing productivity by sealing nutrient loops (Kara et al., 2022; Scavo et al., 2022). Cutting food waste at every point of the supply chain, from production to consumption, is essential to creating a circular economy in agriculture and food systems. Roughly one-third of all food produced is lost or wasted in linear food systems, making food waste a serious problem. In addition to wasting precious resources, this trash releases greenhouse gases during its decomposition in landfills when it breaks down organic material. Repurposing excess food for human use, enhancing transportation and storage to lower post-harvest losses, and other measures are only a few of the ways that circular economy concepts try to minimize food waste, and turning food waste that isn't edible into useful byproducts like compost, biofuel, or animal feed. By guaranteeing that a larger portion of the food produced reaches consumers, these strategies not only improve food security but also lessen the environmental impact of food systems. A circular food system creates a more resilient and sustainable food supply by reorienting the emphasis from maximizing production to maximizing the value that is taken from each unit of food (Ajayi, 2023; Batista et al., 2021; Dobermann et al., 2022).

Through the restoration and maintenance of ecosystem services, regenerative agriculture—a technique closely associated with the concepts of the circular economy plays a critical role in bolstering the resilience of food systems. In contrast to traditional agricultural practices that frequently exhaust natural resources, regenerative agriculture prioritizes restoring soil health, boosting biodiversity, and enhancing water cycles, all of which support agricultural landscapes' long-term viability. Regenerative agriculture relies heavily on techniques like polyculture, agroforestry, and no-till farming since they replicate natural processes and encourage ecosystem regeneration. These methods not only improve agricultural systems' resilience to natural shocks like droughts, floods, and pest outbreaks, but they also absorb carbon and lower greenhouse gas emissions. Food systems can be made into engines of environmental restoration by incorporating regenerative agriculture into the circular economy framework. This will provide essential ecosystem services while producing food in a resilient and sustainable way (Mrabet, 2023; Wang et al., 2022b).

### 5.2 *Circular Economy in Manufacturing and Industry*

The incorporation of circular economy principles into manufacturing and industry commences with waste minimization and material efficiency. Conventional manufacturing procedures frequently adopt a linear paradigm in which raw materials are harvested, converted into products, and then discarded as waste. This method causes substantial environmental contamination in addition to the depletion of natural resources. But in the manufacturing sector, a circular economy centres on creating goods and procedures that optimize resource utilization while reducing waste. This can be accomplished by utilizing advanced manufacturing processes like 3D printing, which minimizes material waste by making components layer by layer, as well as strategies like optimizing material utilization during production and designing goods for durability and repairability. Further reducing waste and resource consumption is industrial symbiosis, which uses trash from one production process as an input for another. Through increasing material efficiency, circular manufacturing decreases the environmental impact of production and contributes to the resilience of industrial ecosystems (Bjørnbet et al., 2021; Orji & Ojadi, 2022; Roy et al., 2024b). The extension of product life through remanufacturing, refurbishing, and repair is another essential component of the circular economy in manufacturing. Circular production aims to preserve a product's value for as long as feasible, as opposed to the conventional approach that discards things after usage. In remanufacturing, used products are disassembled, parts are replaced and restored, and then they are reassembled in an as-new state. In addition, reusing resources and energy when compared to the production of new goods, this procedure lessens the manufacturing process's environmental impact. Analogously, renovating and fixing items prolongs their lifetime, postponing the requirement for fresh materials and cutting down on wastage. Businesses that implement these strategies can develop new business models, such as leasing things rather than selling them outright or offering products as services. This change in consumer preferences pushes producers to create products that require less upkeep and have longer lifespans, balancing financial gain with resilience and sustainability of the environment (Awan & Sroufe, 2022; Farhana et al., 2022; Zhang et al., 2021).

In the manufacturing sector, circular economy concepts also stress the creation of circular supply chains and the resource recovery from products that have reached the end of their useful lives. Rather than letting materials go to waste, circular supply chains are made to loop them back into the production cycle. To do this, closed-loop systems that recover, recycle, and repurpose materials to create new goods must be established. Companies can set up take-back programs, for instance, in the electronics sector, which gathers old electronics, extracts important metals and components, and reintegrates them into the production process. By doing this, the raw material consumption is decreased and the environmental damage caused by the extraction and processing of virgin resources is minimized. By lowering carbon emissions and dependency on fossil fuels, the integration of renewable energy sources into supply chains also improves circularity. Circular supply networks, which close the energy and material loops, help create a more sustainable industrial sector where environmental stewardship and resource recovery are essential to daily operations (Hung & Tsai, 2024; Sarkhoshkalat et al., 2024).

## *5.3 Circular Economy in Urban Planning and Infrastructure*

Urban planners and infrastructure designers must prioritize resource efficiency, longevity, and adaptability when creating systems and structures that align with the concepts of the circular economy. Conventional urban infrastructure frequently follows a linear strategy, using materials only once before discarding them, which results in substantial waste and resource depletion. On the other hand, circular urban planning prioritizes building infrastructure that is easily adaptable and can be used again in the future. For instance, buildings may be expanded, modified, or demolished with little waste because of modular construction processes. In a similar vein, using strong, long-lasting materials that can tolerate wear, and tear lowers the need for replacements and repairs. Circular urban infrastructure increases the resilience of urban environments while conserving resources by placing an emphasis on flexibility and long-term durability, thereby allowing them to adapt to changing needs and conditions (Liu & Ramakrishna, 2021; Rahla et al., 2021). Urban waste management and resource recovery can be revolutionized by the application of circular economy ideas, which aim to minimize waste and maximize material reuse. Collected, transported, and disposed of waste in landfills are common practices in traditional waste management systems, which can lead to resource loss and contamination of the environment. Recycling, composting, and resource recovery are all part of a circular approach to urban waste management, which aims to close the loop. To compost organic waste and use it as a resource for urban agriculture or landscaping, cities can, for example, set up mechanisms for segregating and collecting organic trash. Recycle programs and cutting-edge sorting systems can also recover valuable items like metals and plastics from trash streams, which the manufacturing process can recombine. Urban systems can lessen their influence on the environment, save resources, and help create a more resilient and sustainable urban environment by using these strategies (Liu & Ramakrishna, 2021; Rahla et al., 2021; Wang et al., 2021).

To improve resilience and sustainability, green infrastructure—which integrates natural systems into urban environments—is a crucial way that circular economy ideas are applied in urban development. Green infrastructure uses elements like parks, green roofs, urban forests, and permeable pavements to manage rainwater, enhance air quality, and create recreational places. This contrasts with typical grey infrastructure, which frequently consists of concrete and steel. Urban parks may increase biodiversity and provide communal areas, while green roofs can improve building insulation, absorb rainfall, and lessen the impacts of heat islands. Through their ability to lessen the effects of extreme weather events like heat waves and torrential rain, these natural systems not only provide vital ecosystem services but also help make metropolitan areas more resilient. Urban planners may strengthen the concepts of a circular economy and environmental resilience by integrating green infrastructure into their plans to build places that are not only more resilient to climate change but also more sustainable (Bjørnbe et al., 2021; de Oliveira et al., 2022).

## *5.4 Circular Economy in Waste Management and Resource Recovery*

Advanced recycling technologies that improve resource recovery and lessen environmental effects are essential, according to circular economy principles in waste management. Conventional waste management methods frequently concentrate only on basic sorting and disposal, paying little attention to maximizing the value of recovered materials. Alternatively, a circular approach incorporates advanced recycling techniques that can manage a wide variety of materials and guarantee the efficient recovery and repurposing of important resources within the production cycle. The effective separation and processing of complex materials, such as electronic trash, mixed polymers, and composite materials, is made possible by technologies like automated sorting systems, chemical recycling, and closed-loop recycling systems. These innovative approaches decrease contamination and resource loss while simultaneously increasing the quantity and quality of recycled materials. Waste management systems can more effectively promote a circular economy by putting money into and utilizing such technologies, which will also minimize waste, conserve resources, and lessen the need for virgin materials (Mhatre et al., 2021; Rashid & Shahzad, 2021). Garbage-to-energy (WtE) technologies, which convert garbage into useful energy and materials, and resource recovery facilities are essential elements of circular economy waste management systems. Anaerobic digestion, gasification, and incineration are some of the methods used by WtE plants to transform non-recyclable waste into energy, such as heat or electricity. This lessens the amount of waste that is dumped in landfills and offers a sustainable energy source to replace burning fossil fuels. Resource recovery facilities can harvest and repurpose valuable materials from waste streams in addition to recovering energy. For example, materials that can be recovered, cleaned, and processed for reuse in new products include glass, metals, and some types of plastic. Cities may achieve greater levels of resource efficiency and promote the ideas of a circular economy, which minimizes waste and preserves the value of materials, by incorporating WtE and resource recovery into waste management systems (Chen et al., 2023; Sharma et al., 2024; Yadav et al., 2022).

The goals of the circular economy are greatly advanced by Extended Producer Responsibility (EPR) programs, which transfer the management of end-of-life items from consumers and communities to producers and manufacturers. Manufacturers are required by EPR programs to consider all aspects of a product's lifecycle, including recycling and disposal. This encourages the creation of items that are simpler to reuse, recycle, or repair, which lowers waste production and increases the rates at which resources are recovered. Take-back programs and deposit-return systems, for instance, have been established because of EPR initiatives for electronics and packaging, making it easier to collect and recycle unwanted goods. EPR initiatives improve the effectiveness of waste management systems and foster innovation in product design and materials management by holding manufacturers responsible for the end-of-life management of their products. By encouraging trash reduction, resource conservation, and the development of a more robust and sustainable waste management infrastructure, this strategy is consistent with the concepts of the circular economy (Ezeudu, 2024; Mihai et al., 2024).

### *5.5 Environmental Resilience in Coastal and Marine Ecosystems*

Restoration of coastal ecosystems is essential for improving environmental resilience because it restores natural processes and habitats that act as a buffer against environmental pressures. Important functions offered by coastal ecosystems include storm surge protection, erosion control, and habitat for a variety of marine species. Examples of these ecosystems include mangroves, salt marshes, and seagrass beds. For example, the restoration of mangrove forests along Bangladeshi and Thai coastlines has shown to be highly beneficial in terms of lowering coastal erosion and shielding residents from storm surges. Extreme weather events are lessened by these ecosystems because they serve as natural barriers by absorbing and diffusing wave energy. Furthermore, by filtering pollutants out of the water and providing vital homes for fish and bird species, restored coastal wetlands promote biodiversity. Coastal areas can become more robust to the effects of climate change, such as increasing sea levels and an increase in the frequency of severe weather events, by giving priority to the restoration and preservation of these ecosystems (Sahavacharin et al., 2022; Sunkur et al., 2023). By preserving important habitats and encouraging sustainable fisheries management, Marine Protected Areas (MPAs) play a crucial role in enhancing the resilience of marine ecosystems. By offering safe havens where marine animals can flourish and recuperate from fishing pressures and other human activity, MPAs aid in the maintenance of ecological balance. An illustrative example of how a well-managed MPA has resulted in the recovery of coral reef ecosystems and related marine life is the Great Barrier Reef Marine Park in Australia. MPAs can aid in preventing overfishing, protecting biodiversity, and enhancing the resilience of marine ecosystems by enforcing limits on fishing, tourism, and other activities. Rebuilding fish stocks is facilitated by efficient fisheries management in MPAs, and this is essential for preserving food security and the means of subsistence for communities who depend on marine resources. Key tactics for strengthening the resilience of maritime habitats to the stresses of climate change and human impact include the creation and enforcement of MPAs, scientific research, and community involvement (Grorud-Colvert et al., 2021; Villasante et al., 2023).

Resilience is increased when circular economy ideas are included in marine and coastal management, with an emphasis on waste reduction and sustainable resource usage. For example, by creating mechanisms for gathering, recycling, and repurposing plastic trash, the idea of circularity can be used to manage plastic contamination in coastal settings. Creative solutions can lessen the buildup of marine waste and its detrimental impact on marine life. Examples include using recycled ocean plastics to make new products or establishing circular supply chains for fishing gear. Furthermore, preserving marine resources and minimizing waste, encouraging sustainable seafood practices, and lowering bycatch through better-designed fishing gear and techniques are consistent with the ideas of the circular economy. To execute these circular strategies—which not only address the immediate problems of marine pollution but also support the long-term resilience of coastal and marine ecosystems—governments, corporations, and local communities must work together. Coastal and marine management can create more resilient and sustainable outcomes by implementing a circular economy approach, protecting the health and vitality of these crucial areas (Grorud-Colvert et al., 2021; Sharma et al., 2021; Sunkur et al., 2023).

## *5.6 Environmental Resilience in Water Resource Management*

Integrated Water Resource Management (IWRM) is a comprehensive approach that improves environmental resilience by managing water resources in an organized manner across different sectors and scales. Water resource management, or IWRM, seeks to strike a balance between competing needs and sustainable water use while preserving ecosystem health. To manage water resources in the face of fluctuating demands and climate unpredictability, for instance, governments, farmers, and environmental organizations have worked closely together in the Murray-Darling Basin of Australia using the IWRM strategy. The IWRM method has contributed to increased agricultural productivity and the preservation of important environmental flows by putting into practice strategies including water trading, increasing irrigation efficiency, and restoring wetlands. By considering both natural requirements and human uses, this integrated strategy not only increases the water systems' resistance to floods and droughts but also promotes the long-term sustainability of water resources (Cacal et al., 2023; Mersha et al., 2021). The use of nature-based solutions (NbS) to address water-related issues and increase resilience in water resource management is becoming more widely acknowledged. Natural processes or systems are used in NbS to provide water-related services and reduce risks including erosion, flooding, and water scarcity. For example, replenishing riparian buffers and reforesting watersheds can boost groundwater recharge, lessen runoff, and prevent soil erosion. An important example of how natural floodplains and wetlands are used in river management to accept floodwaters and lower flood hazards is the Netherlands' "Room for the River" initiative. By re-establishing the natural cycles of water and improving ecosystems' ability to control water flows, the overall health and resilience of water supplies are enhanced by nature-based solutions, which help shield communities from severe weather occurrences (Beceiro et al., 2022; van Doorn-Hoekveld et al., 2022).

Building systems that optimize the reuse and recycling of water and related resources is a key component of applying circular economy concepts to the management of water resources. By cleaning and recycling wastewater, circular water management seeks to minimize waste and close the loop, lowering pollution and freshwater extraction. For instance, complex water recycling systems have been put in place in cities like Singapore and Tokyo, which clean and reuse wastewater for uses other than potable water, like industrial processes and irrigation. Freshwater supplies are under much less pressure thanks to Singapore's NEWater initiative, which recycles treated wastewater into high-quality water and enhances water security. To further minimize overall water usage and waste, circular methods advocate for the adoption of water-efficient technology and practices in industry and agriculture. Water resource management can become more resilient by adopting these circular solutions, which lessen reliance on limited water supplies, lessen the effects of pollution, and guarantee a more sustainable and dependable supply of water for a variety of applications (Liu & Ramakrishna, 2021; Rahla et al., 2021).

# 6 Innovation and Technology in Circular Economy and Environmental Resilience

## 6.1 Role of Technology in Driving Circular Economy

Cutting-edge recycling technologies maximize resource efficiency and reduce waste, which is essential to the circular economy. The kinds of materials that can be processed efficiently and the quality of the products that can be recycled are two areas where traditional recycling methods frequently have limitations. This environment is changing due to innovations like chemical recycling and digital sorting technology. Chemical recycling enables the recycling of materials that were previously thought to be non-recyclable by disassembling complex polymers and composites into their basic constituents. Artificial intelligence and machine learning are used by digital sorting systems to increase the precision and effectiveness of segregating various elements in waste streams. For instance, the exact identification and separation of different types of plastics and metals is made possible using AI-powered optical sorters in recycling facilities, which raises the purity and value of recovered materials. These developments guarantee that resources are continually recycled back into production, which not only increases the efficacy of recycling programs but also moves the economy closer to a more circular one (Reddy et al., 2023; Shanker et al., 2023). The circular economy is also propelled by technological advancements in product design and manufacturing, which prioritize the creation of long-lasting, recyclable, and repairable goods. One such innovation is design for disassembly, in which items are made with the end user's life cycle in mind, making recycling and disassembly simpler. This strategy is further supported by technologies like 3D printing and modular manufacturing, which make it possible to produce easily replaceable or improved parts rather than having to throw away entire goods. For example, businesses such as IKEA are investigating furniture designs that are modular in nature, enabling users to upgrade or replace individual parts instead of buying whole sets. Furthermore, manufacturers are assisted in tracking the environmental impact of their products and optimizing design decisions for sustainability by digital technologies like product lifecycle management (PLM) software. Through the incorporation of these technical advancements into the design and production processes, businesses may produce products that are in line with the principles of the circular economy, thereby decreasing waste and prolonging the lifespan of their products (Chan & Lim, 2023; Dan et al., 2023).

As resource efficiency and improved material and product management become more feasible, smart systems and data-driven strategies play a critical role in promoting the circular economy. Smart products, supply chain dynamics, and resource flows can all be better understood by utilizing blockchain technology, big data analytics, and Internet of Things (IoT) sensors. Real-time product and infrastructure status monitoring is made possible by IoT sensors, which may provide data that boosts recycling process efficiency, optimizes resource usage, and forecasts maintenance requirements. By using sensors to monitor garbage bin fill levels, for instance, smart waste management systems can lower operating costs and enable more effective collection routes. Blockchain ensures that items are recycled properly and that materials are procured responsibly by improving supply chain transparency and traceability. Businesses and governments can create data-driven plans that promote environmental resilience, resource management, and system circularity by utilizing these technologies (Awan et al., 2021; Li et al., 2023).

By creating sustainable materials with less of an impact on the environment and longer product lifespans, material science innovations play a critical role in improving environmental resilience. By using recyclable or renewable materials, innovative materials and technologies aim to minimize waste and lessen reliance on finite resources. Bio-based plastics and composites, for instance, made from algae or agricultural waste, provide substitutes for conventional petroleum-based plastics, which worsen pollution and deplete resources. Technological advancements like self-healing materials, can automatically fix little damage, increase product longevity, and lessen the need for frequent replacements. Furthermore, developments in nanotechnology make it possible to create materials with improved qualities, including greater energy efficiency or durability. By encouraging the use of sustainable materials, these innovations not only support the tenets of a circular economy but also increase environmental resilience by lowering the ecological footprint of production and consumption (Mtibe et al., 2021; Ur Rahim et al., 2021). Urban design and development are changing due to smart infrastructure technologies, which also greatly increase environmental resilience. For instance, smart grids optimize energy distribution, incorporate renewable energy sources, and cut down on energy waste by utilizing real-time data and sophisticated analytics. This technique reduces risk and increases efficiency, strengthening the resilience of urban infrastructure. Similarly, smart water management systems make better use of water resources by monitoring water usage, identifying leaks, and utilizing sensors and data analytics. Climate-responsive design and green infrastructure are examples of adaptable urban planning techniques that are being incorporated into development projects more and more. To control stormwater, lessen the effects of heat islands, and improve biodiversity, for example, cities are implementing green roofs, permeable pavements, and urban wetlands. Cities may increase their resilience to environmental pressures and climate change by utilizing smart infrastructure and adaptable urban planning, which also fosters sustainability and circularity (Krishnan et al., 2022; Yang et al., 2021).

Circular business models and increased resource efficiency made possible by digital platforms are pushing innovations that improve environmental resilience. Online marketplaces for sharing economies, product take-back programs, and repair services make it easier to extend product lifecycles and cut waste. To prolong the life of their products and cut down on the need for fresh resources, businesses such as Patagonia and ThredUp, for instance, use internet platforms to provide repair services and resale possibilities for apparel. Digital solutions that assist firms in assessing the environmental impact of their products and streamlining their supply chains include environmental management systems and lifecycle assessment (LCA) software. Additionally, supply chains are made more transparent and traceable by blockchain technology, which guarantees that materials are supplied ethically and that goods are managed sustainably. Digital platforms facilitate the shift towards a circular economy and improve the resilience of environmental systems by reducing waste and improving resource management. They also enable circular business models and offer data-driven insights (Awan & Sroufe, 2022; Ferrari et al., 2021).

## 6.2 Digital Tools and Platforms for Circular Economy

Lifecycle Assessment (LCA) software is an essential digital tool for improving the circular economy by assessing the environmental impact of products throughout their entire lifecycle. Businesses can evaluate how resources are used, how much energy is consumed, how much waste is generated, and how much waste is extracted from products before they are disposed of by using LCA tools. For example, programs like SimaPro or GaBi offer comprehensive information about how environmentally friendly a product is, enabling businesses to pinpoint resource consumption hotspots and potential improvement areas. Through a thorough analysis of these variables, companies may make well-informed decisions on design, materials, and production methods that reduce environmental impact and improve sustainability. Furthermore, life cycle assessment (LCA) tools make it easier to compare various product designs or processes, which aids in the creation of more circular products that adhere to waste reduction and resource efficiency standards. Businesses can increase their environmental resilience and better connect with the ideals of the circular economy by incorporating LCA into their operations (Farooq et al., 2024; Ferrari et al., 2021; Khanna et al., 2022). Through the promotion of product sharing and leasing models that decrease waste and prolong the lifecycle of goods, digital platforms are redefining the circular economy. Instead of buying things like apparel, electronics, and equipment altogether, customers can rent or share them through platforms like ShareGrid and Rent the Runway. This methodology minimizes resource usage and trash generation in addition to lowering the demand for new items. By using technology to track the condition of shared objects, manage inventories, and make transactions easier, these platforms improve the sharing economy's efficiency. For instance, these solutions may anticipate maintenance requirements, guarantee on-time repairs, and maximize item availability by utilizing data analytics and IoT sensors. Digital platforms promote circular economy principles and enhance environmental resilience by lowering total resource consumption and related environmental effects by facilitating product access without ownership (McCoy & Chi, 2022; Liu and Ramakrishna, 2021; Liu et al., 2021). In the circular economy, blockchain technology is beginning to emerge as a game-changer, providing improved supply chain traceability and transparency. Blockchain helps businesses and customers track the source of materials, confirm the legitimacy of products, and guarantee sustainability requirements are being followed by offering a decentralized, irreversible log of transactions. Blockchain is being used, for instance, by IBM and Everledger to track the origin and lifecycle of items like electronics and diamonds, guaranteeing that they are sourced ethically and disposed of correctly. Blockchain makes it possible to manage returns, recycling procedures, and material reuse effectively, which helps to promote circular supply chains. Businesses may enhance their capacity to close the material loop, cut waste, and guarantee that parts and products are recycled and reused in accordance with circular economy principles by incorporating blockchain technology. Supply chains are encouraged to adopt more sustainable and responsible practices due to increased accountability and openness (Fernandes et al., 2023; Kshetri, 2021).

### *6.3 Future Trends: Emerging Technologies and Resilient Circular Economies*

Artificial intelligence (AI) has the potential to revolutionize circular economy optimization by improving decision-making and efficiency in a variety of industries. Large-scale data analysis can be performed using AI technologies like machine learning and predictive analytics to find patterns and insights that lead to resource optimization and waste reduction. For example, AI algorithms can predict demand more precisely, which improves inventory control and lowers manufacturing overproduction. Furthermore, by increasing the precision of material sorting and separation through computer vision and robotic systems, AI-powered systems can improve recycling procedures. AI is being used by businesses like AMP Robotics to automate and enhance recyclable material sorting, resulting in increased recovery rates and decreased contamination. With AI's further development, circular economy techniques will become more accurate and effective, enhancing resource efficiency and environmental resilience (Li et al., 2023; Oluleye et al., 2023). Through the integration of sensors and networked devices to monitor and control resource flows in real time, the Internet of Things (IoT) is revolutionizing resource management. IoT technology makes it possible to track objects and materials at every stage of their lifespan, giving important information about their location, usage, and condition. Smart sensors, for instance, may keep an eye on how well industrial machinery is operating and identify inefficiencies or maintenance needs before they result in breakdowns. IoT sensors in trash cans can monitor fill levels and plan the best possible collection routes for garbage management, which lowers operating costs and increases recycling rates. IoT provides real-time data on product performance and usage, enabling improved maintenance and end-of-life management. This helps to enable the development of circular business models, including product-as-a-service. Businesses may support the goals of the circular economy, improve resource management procedures, and increase resilience to resource-related problems by using the power of IoT (Fernandes et al., 2023; Li et al., 2023).

Since biotechnologies provide environmentally friendly methods of resource recovery and waste management, they are becoming a major force behind advances in the circular economy. Complex materials that were previously challenging to treat are now able to be broken down and recycled because of biotechnology advancements like microbial and enzymatic processes. For example, scientists have created engineered bacteria that can break down plastic trash and produce new materials or other useful byproducts. Similarly, textiles may be effectively broken-down using enzyme-based technologies, which makes it possible to recover and recycle the fibres from old garments. Enzyme-based recycling of PET plastics into reusable raw materials is being led by companies such as Carbios. The circular economy may improve environmental resilience through creative and sustainable solutions, lessen dependency on non-renewable resources, and manage trash more efficiently by utilizing biotechnologies (Adil et al., 2022; Güell Barcons et al., 2023). The circular economy is being reshaped by decentralized manufacturing and 3D printing, which facilitate local production and cut waste through on-demand manufacture. Compared to traditional subtractive manufacturing processes, 3D printing technology greatly reduces material waste by enabling the fabrication of products and components layer by layer from digital designs. Moreover, 3D printing can make it easier to use bio-based or recycled materials, which is further consistent with the circular economy principles. Decentralized production models, in which goods are produced near their point of use, minimize the need for lengthy supply chains and the emissions that go along with them. This technology helps these models as well. Businesses are investigating the use of 3D printing, for instance, to produce spare parts as needed, lower the need for inventory, and prolong the life of current items. More flexible, environmentally friendly, and circular manufacturing processes will be made possible by 3D printing technology as it develops, fostering innovation and environmental resilience in the circular economy (Bjørnbet et al., 2021; Ferreira et al., 2023).

## 7 Challenges and Opportunities

### *7.1 Barriers to Implementing Circular Economy and Environmental Resilience*

The financial and economic limitations that corporations, governments, and communities confront are a major obstacle to putting a circular economy into practice and improving environmental resilience. The shift from a conventional linear economy, in which goods are produced, utilized, and then discarded, to a circular economy necessitates large initial investments in new infrastructure, technology, and procedures. These expenses can be exorbitant for many firms, especially small and medium-sized organizations (SMEs), which discourages innovation and the adoption of circular practices. Furthermore, because of their established market structures and cheaper beginning expenses, linear business models are frequently preferred in the present economic systems. Due to perceived risks and uncertainties, such as the unclear ROI or the protracted payback times linked with circular initiatives, financial institutions may also be hesitant to invest in circular economy projects. Financial incentives such as tax breaks, subsidies, or low-interest loans must be established to overcome these obstacles and promote investment in circular economy practices. To further encourage investment and broad adoption, new business models that show the financial viability and long-term sustainability of circular processes must be created (Awan & Sroufe, 2022; Morales et al., 2022). Another major obstacle to the adoption of environmental resilience and circular economy methods is the regulatory and policy frameworks. Since many current laws are based on linear economic models, circular practices are not sufficiently supported or encouraged. For instance, waste management laws in some areas might restrict material recycling or reuse, which would make it more difficult for companies to finish the resource loop. Furthermore, for businesses that operate in several countries, the absence of uniform standards and laws can be confusing and unsettling. The dispersion of regulations may impede the advancement of worldwide supply networks for the circular economy and deter companies from allocating resources toward circular practices. Governments must rewrite current laws and establish regulatory frameworks that support circular economy projects to address these issues. This entails creating precise norms and standards for circular activities, such as designing durable and recyclable items and creating extended producer responsibility (EPR) frameworks that make producers responsible for their products' whole lifecycles. Furthermore, promoting global collaboration and harmonizing rules can aid in resolving cross-border issues and speeding up the implementation of circular economy principles worldwide (Cimprich et al., 2023; Hartley et al., 2023).

Significant obstacles to the implementation of environmental resilience and circular economy techniques can come from cultural and behavioral reasons. Both firms and consumers are frequently locked into linear consumption patterns that put short-term cost savings, convenience, and disposability ahead of sustainability and long-term value. In many civilizations, there is a strong "throwaway culture" where objects are made with planned obsolescence or single-use in mind, and the idea of recycling, reusing, or mending things is not entirely accepted. It can be difficult to bring about the profound changes in consumer attitudes and corporate policies needed to alter these deeply ingrained behaviors. Campaigns to raise public awareness and educate the public are crucial to highlighting the advantages of a circular economy and supporting more environmentally friendly consumption habits. Furthermore, companies can have a significant impact by creating products that are recyclable, robust, and repairable as well as by providing services like product take-back programs, repair services, and leasing models that encourage circular behaviour. Supporting programs that encourage sustainable lifestyles, such as community repair workshops, second-hand markets, and circular economy hubs, is another way that governments and non-governmental organizations (NGOs) may help (Krajnc et al., 2022; Mhatre et al., 2023). Enhancing environmental resilience and implementing circular economy strategies face major challenges due to technological and infrastructure constraints. Circular models may not be easily adapted to many industrial processes and technologies now in use, as they are designed for linear production. For instance, low recovery rates and low-quality recovered materials may result from present recycling methods' inability to handle the complexity of modern materials. Closing the resource loop can also be hampered by a lack of infrastructure for the collection, sorting, and processing of waste materials. In certain areas, especially developing nations, recyclable materials are frequently lost to the environment or landfills due to a lack of strong waste management systems. To overcome these obstacles, research and development expenditures must be made to create new, more easily recyclable materials, enhance material recovery procedures, and progress recycling technology. It also entails developing the infrastructure such as material recovery facilities, repair centers, and reverse logistics systems necessary to support circular economy operations. By combining resources, knowledge, and creativity, public–private partnerships can be extremely helpful in overcoming these obstacles and creating the infrastructure and technology required for a circular economy (Bianchi et al., 2023; Kara et al., 2022).

Significant obstacles to the adoption of a circular economy and the improvement of environmental resilience include supply chain and market constraints. In many cases, the supply chains for circular products are not as well developed or efficient as those for linear products, leading to higher costs, limited availability, and logistical challenges. For instance, because of problems like uneven quality, unpredictability in the supply, and a lack of developed markets, procuring recycled materials or secondary raw materials may be more costly and complicated than finding virgin resources. Moreover, the worldwide scope of numerous supply chains poses challenges to the adoption of circular methods, since goods and materials may traverse several nations with disparate laws, norms, and marketplaces. Collaboration across industries and countries is necessary to establish stable markets for secondary materials, standardize quality requirements, and build effective logistics systems that can manage material reverse flows to build more robust and circular supply chains. Additionally, it necessitates that businesses reconsider their supply chain strategy, shifting their emphasis from efficiency and cost reduction to sustainability, resource security, and long-term resilience (Kara et al., 2022; Mhatre et al., 2023; Nguyen & Le, 2022). A further obstacle to putting circular economy principles into practice and strengthening environmental resilience is the lack of expertise among stakeholders. Many companies—SMEs may lack the technical know-how and comprehension required to switch to circular models. This includes being aware of sustainable materials, closed-loop manufacturing techniques, and circular design ideas. It also includes having the necessary abilities to put these practices into practice. Furthermore, legislators and regulators might not be aware of or comprehend the advantages of a circular economy, or the laws required to promote its growth. Targeted education and training initiatives that give communities, businesses, and policymakers the information and resources they need to implement circular practices are necessary to close these knowledge and skill gaps. This includes measures to enhance capacity and industry-specific training programs in addition to formal education programs offered by universities and vocational institutions. Disseminating case studies, best practices, and innovations in resilience-building and the circular economy can be greatly aided by knowledge-sharing networks and platforms (Awan & Sroufe, 2022; Carissimi et al., 2023).

A major obstacle to the adoption of circular economy principles is resistance to change, which is frequently caused by institutional inertia and vested interests. The adoption of circular economy methods by large organizations may be hindered by established protocols, legacy systems, and an emphasis on immediate financial results. A fundamental shift in organizational culture and business strategy is frequently necessary to facilitate the transition to a circular economy, in addition to operational and technological advancements. Stakeholders who are used to linear business models may oppose this because they think the change is dangerous or disruptive. Institutional impediments that include limited political will, incoherent policies, and strict regulatory frameworks can also make it difficult to establish and carry out circular economy initiatives. It takes strong leadership and a distinct vision that combines the aims of the circular economy with long-term corporate goals to overcome resistance to change. It also necessitates assembling a large coalition of interested parties that are dedicated to advancing the shift to an environmentally resilient circular economy, including corporations, governments, NGOs, and communities (Awan & Sroufe, 2022; Castro-Lopez et al., 2023).

### 7.2 *Opportunities for Innovation and Growth*

The creation and application of new technologies is one of the most important avenues for innovation and expansion in the framework of a circular economy. The management, tracking, and reuse of resources throughout supply chains can be completely transformed by cutting-edge technologies like blockchain, artificial intelligence (AI), and the Internet of Things (IoT). AI can, for example, predict the stages of a product's lifecycle and enable more intelligent design and waste-free manufacturing procedures to maximize resource efficiency. IoT solutions may help businesses monitor material flows in real-time, which enables them to spot inefficiencies, follow products through their whole lifecycle, and put just-in-time recycling or reuse plans in place. On the other side, resources are purchased, used, and recycled in accordance with sustainability norms thanks to blockchain's transparency and traceability features. These technological advancements boost environmental resilience and financial success by boosting not just the efficacy and efficiency of circular practices but also by generating new business models based on data-driven resource management (Fernandes et al., 2023; Liu et al., 2021). The shift towards a circular economy presents a plethora of prospects for the creation of novel business models that give precedence to sustainability, product durability, and resource efficiency. For instance, the transition from traditional ownership to service-based models, like leasing or product-as-a-service, permits businesses to keep ownership of their materials and goods, which incentivizes them to create long-lasting and recyclable items. This business approach generates ongoing revenue streams and long-term client relationships in addition to cutting waste. Remanufacturing, refurbishing, and extending the life of a product are examples of circular business models that allow enterprises to reach new markets and client segments by offering superior, environmentally friendly substitutes for newly developed items. Offering eco-friendly solutions helps businesses stand out in the marketplace and grow their market share and brand loyalty in response to the growing customer demand for sustainable products. Moreover, the circular economy promotes cooperation between industries, resulting in the development of new partnerships and value chains that spur economic growth and innovation (Agrawal et al., 2023; Awan & Sroufe, 2022).

Additionally, there is a rare chance to improve environmental sustainability and resilience globally through innovation in circular economic techniques. Businesses may drastically lessen their environmental footprint, preserve natural resources, and lessen the effects of climate change by creating systems and products that are regenerative by nature. For example, advances in materials science may result in the creation of entirely recyclable or biodegradable materials, which lessen the load on landfills and cut greenhouse gas emissions. In a similar vein, developments in sustainable agriculture, such as regenerative farming techniques, can improve the resilience of food systems to climatic fluctuation, boost biodiversity, and restore soil health. Additionally, the circular economy encourages the effective use of energy and water, which lessens the burden on these vital resources and strengthens the resilience of industrial and urban ecosystems. The circular economy provides a route to a more resilient and sustainable future where economic development does not come at the expense of the health of the earth by balancing environmental sustainability with economic growth (Cacal et al., 2023; Ezeudu, 2024). The circular economy offers tremendous potential for economic expansion and the creation of jobs, especially in the fields of recycling, remanufacturing, and sustainable product design. There is an increasing need for qualified people in fields like waste management, product design, and green technologies as more companies implement circular business practices. For instance, the increase in material recovery and recycling facilities generates jobs in sorting, processing, and logistics, while the remanufacturing sectors demand specialized labour in quality control and engineering. Furthermore, the emergence of novel cyclical business models, like product leasing and repair services, opens doors for small business development and entrepreneurship. By lowering reliance on limited resources and minimizing the risks connected with resource scarcity and price volatility, the circular economy also fosters economic resilience. The circular economy promotes long-term economic stability and growth by encouraging a more diverse and sustainable economy, which strengthens and stabilizes the labour market (Farhana et al., 2022; Farooq et al., 2024; Fernandes et al., 2023).

As the potential of the circular economy becomes more widely acknowledged, national and international legislative and regulatory support is developing, opening new avenues for innovation and expansion. Incentives for circular activities are being introduced by governments in the form of extended producer responsibility (EPR), which mandates that producers bear liability for the full product lifetime, including disposal at the end of the product's useful life. Innovation in product creation and manufacturing processes is also being fuelled by legislation that supports eco-design, waste reduction, and the use of recycled materials. Public procurement regulations that give priority to circular goods and services open new markets for companies that use sustainable business methods. International agreements and partnerships, such as the Circular Economy Action Plan and the European Green Deal, are also promoting a global circularity movement by facilitating cross-border cooperation and the exchange of best practices. Businesses may take advantage of the opportunities provided by the circular economy by coordinating their business plans with these policy frameworks. This will help them stay competitive and drive innovation and growth in a market that is changing quickly (Husgafvel et al., 2022; Leclerc & Badami, 2024). Research and education developments are directly associated with growth and innovation opportunities in the circular economy. New circular economy theories, technology, and practices are being developed in large part by academic institutions, research groups, and think tanks. Businesses and governments can foster innovation and overcome technological obstacles to realize the full potential of the circular economy by allocating resources towards research and development (R&D). Research into novel materials, such as improved composites or bio-based plastics, for instance, can result in the creation of goods that are simpler to recycle or naturally decompose. Programs for education and training are equally crucial for developing the abilities and know-how needed to apply circular practices across industries. To give the upcoming generation of workers and leaders the skills they need to succeed in a circular economy, universities, and vocational institutions are offering more and more courses and certifications in environmental management, sustainable design, and circular economy principles. The circular economy can develop into a dynamic and ever-evolving field that promotes resilience in the face of global challenges and sustainable growth by cultivating a culture of ongoing learning and innovation (Awan & Sroufe, 2022; Ezeudu, 2024; Hartley et al., 2023).

## *7.3 Global Perspectives: Circular Economy and Resilience in Different Regions*

**European Union: Leading the Global Circular Economy Movement**

The European Union (EU) has taken the lead in advancing the circular economy movement worldwide and has established itself as a pioneer in fusing environmental resilience with circular principles. Part of the European Green Deal, the EU's Circular Economy Action Plan lays out a thorough plan to move the area towards a more resource-efficient and sustainable economy. This plan places a strong emphasis on the development of closed-loop supply chains, waste minimization, and sustainable product design. Aiming to become carbon neutral by 2050, the EU's approach to the circular economy is intimately tied to its climate ambitions. The EU aims to lessen its reliance on finite resources and cut greenhouse gas emissions by encouraging circular behaviors including recycling, reusing, and remanufacturing. The region's strong regulatory framework, which includes laws like the Eco-design Directive and the Extended Producer Responsibility (EPR), encourages innovation and establishes international norms for sustainable production and consumption. This demonstrates the region's dedication to circularity (Cristina-Maria & Ioana-Raluca, 2023; Pouikli et al., 2022).

**Asia: Rapid Growth and Diverse Approaches**

With notable differences across various nations and regions, Asia offers a varied environment when it comes to the implementation of circular economy methods. Some rapidly industrializing countries, such as China, deliberately incorporate circularity ideas into their national development policies after realizing the benefits circularity offers to the economy and environment. China is committed to converting its industrial base into one that is more resource-efficient and environmentally friendly, as evidenced by the 2008 Circular Economy Promotion Law and the more recent "Made in China 2025" project. The main goals of these programs are to improve recycling, cut down on resource usage, and stimulate the development of sustainable technology. Conversely, nations such as South Korea and Japan, possessing developed economies and sophisticated technological capacities, have long been adopters of circular practices, especially regarding resource recovery and waste management. One excellent example is the "Sound Material-Cycle Society" program in Japan, which uses stringent regulations and innovative technology to minimize environmental impact, encourage recycling, and reduce waste. However, due to problems including poor infrastructure, a lack of funding, and varying degrees of policy implementation, other regions of Asia—particularly those in South and Southeast Asia—face serious difficulties in putting circular economy principles into practice. These areas are frequently distinguished by unofficial recycling industries and differing degrees of governmental backing for circular economy initiatives (van de Gronden, 2022; Türkeli et al., 2022).

**Africa: Circular Economy as a Tool for Sustainable Development**

The circular economy is becoming more widely acknowledged in Africa as an essential instrument for promoting sustainable development and reducing poverty. The continent's distinct socioeconomic environment, which is marked by high levels of resource dependency, urbanization, and rapid population expansion, offers opportunities as well as problems for the adoption of circular practices. African countries are starting to investigate circular economy models, which can provide economic opportunity while addressing important challenges including resource shortages, waste management, and environmental degradation. For instance, nations like Kenya and South Africa have made progress in creating circular economy plans that emphasize energy recovery, trash recycling, and sustainable agriculture. Through recycling programs, waste-to-energy projects, and the encouragement of sustainable purchasing habits, South Africa's National Waste Management Strategy promotes the growth of a circular economy. Conversely, Kenya has emerged as a frontrunner in the region regarding the prohibition of single-use plastics and the advancement of sustainable substitutes, demonstrating an increasing consciousness about the ecological consequences of trash. However, there are some obstacles to the adoption of circular economy principles in Africa, such as inadequate technological availability, inadequate infrastructure, and budgetary limitations. Building capability, collaborating internationally, and creating creative business models that are tailored to the local environment are all necessary to meet these problems (Morais et al., 2022; Hagos et al., 2022; Koech et al., 2023).

**North America: Market-Driven Circular Economy Initiatives**

The circular economy is mostly being pushed by corporate initiatives and market forces in North America, with different degrees of government backing. Growing interest in circular economy methods has been observed in the US and Canada, especially in manufacturing, waste management, and sustainable design. With businesses like Apple, Nike, and General Motors using circular business models that prioritize resource efficiency, material recovery, and product life extension, the private sector has been instrumental in furthering circularity in the United States. The need for sustainable products from consumers, corporate sustainability objectives, and the possibility of cost savings through resource optimization are the main forces behind these initiatives. The United States has not yet established a comprehensive federal policy on the circular economy, so progress in this area is uneven. Instead, states and municipalities are spearheading the implementation of circular practices through local laws, such as those pertaining to extended producer responsibility or the outlawing of single-use plastics. The Canadian government has demonstrated a more robust endorsement of the circular economy, especially regarding trash reduction and recycling initiatives. For instance, Canada's "Zero Plastic Waste" effort seeks to eradicate plastic waste by combining industry collaborations, legislative measures, and public awareness campaigns. Notwithstanding these initiatives, there are still obstacles in the way of North America's complete shift to a circular economy, such as fragmented regulations, low consumer awareness, and the requirement for more funding for circular infrastructure (Fichte, 2023; Osobajo et al., 2022; Tleuken et al., 2022).

**Latin America: Emerging Circular Economy Movements**

Although the circular economy is still a relatively new idea in Latin America, its potential to alleviate the region's social and environmental problems is becoming more widely acknowledged. Several nations, including Brazil, Chile, and Colombia, are investigating circular economy approaches to cut waste, preserve resources, and advance sustainable development. The National Solid Waste Policy in Brazil, which was implemented in 2010, encourages recycling, waste reduction, and the incorporation of informal waste pickers into official waste management systems. This creates the foundation for a circular economy. With its "Roadmap to a Circular Economy," which offers tactics for cutting waste, encouraging eco-design, and bolstering circular business models across a range of industries, Chile has also achieved noteworthy progress. Colombia has also started campaigns to support sustainable mining methods, water management, and circular agriculture because it understands how important it is to protect natural resources from the effects of climate change and economic expansion. However, socioeconomic inequality, a lack of infrastructure, and unstable economies frequently make it difficult to apply circular economy principles throughout Latin America. Stronger regulatory frameworks, more funding for circular technologies, and the creation of inclusive business models that benefit all societal segments especially the most vulnerable are required to remove these obstacles (Abu-Bakar & Charnley, 2024; Jain, 2023; Lino et al., 2023; Negrete-Cardoso et al., 2022; Ulloa-Murillo et al., 2022).

**Middle East and North Africa (MENA): Circular Economy in Resource-Scarce Environments**

Arid climate, resource scarcity, and economic dependence on natural resources like oil and gas mean that the Middle East and North Africa (MENA) area presents special potential and problems in the shift to a circular economy. Circular economy strategies have gained popularity due to a rising understanding of the necessity of diversifying economies and minimizing environmental effects in recent years. As part of their larger objectives for economic diversification, the United Arab Emirates (UAE) and Saudi Arabia, for example, have started large-scale projects to advance sustainability and circularity. The development of renewable energy sources, waste management, and resource efficiency are all emphasized in Saudi Arabia's "Vision 2030" and the United Arab Emirates' "Vision 2021." These nations are investigating circular economy models that centre on waste-to-energy technology, environmentally friendly building practices, and water conservation to lessen their environmental impact and generate new economic opportunities. However, there are other obstacles to putting circular methods into practice in the MENA region, such as political unpredictability, water scarcity, and low public awareness. Strong government leadership, financial support for environmentally friendly technology, and collaborations between the public and private sectors are necessary to address these issues and foster innovation and the development of resilient, sustainable economies in the area (Al-Addous et al., 2024; Alotaibi et al., 2024; Almulhim & Al-Saidi, 2023).

## 7.4 *Strategies for Overcoming Challenges and Maximizing Opportunities*

The creation and incorporation of thorough policy frameworks and regulations is one of the best ways to overcome obstacles in the implementation of circular economy and environmental resilience programs. Governments are essential in establishing the framework for circular economies because they enact rules that promote environmentally friendly behaviour and penalize wasteful spending. This involves putting laws into place like extended producer responsibility (EPR), which mandates that producers collect their goods at the end of their useful lives so they can be recycled or disposed of properly. Governments can also impose eco-design requirements on products, such as making them recyclable, long-lasting, and repairable, to guarantee that circular principles are incorporated from the beginning of the design process. Levelling the playing field by coordinating policies across various industries and geographical areas might also encourage companies to embrace circular processes without worrying about losing out on market share. National and regional policies can promote the broad adoption of circular economy ideas, so boosting economic growth and environmental resilience, by coordinating with global sustainability targets, such as the Sustainable Development targets (SDGs) of the United Nations (Abu-Bakar & Charnley, 2024; Cimprich et al., 2023; Grorud-Colvert et al., 2021). Collaborating across the public and private sectors is a crucial approach to fully using the potential that a circular economy presents. To support circularity and environmental resilience, public–private partnerships (PPPs) can stimulate innovation, share risks, and combine resources. To create cutting-edge recycling infrastructure or waste-to-energy facilities that minimize landfill usage and produce renewable energy, for instance, governments can work with private businesses. Governments can fund and promote developments in sustainable materials, circular business models, and waste management technologies through research and development (R&D) collaborations. When businesses from many industries collaborate to maximize resource utilization and reduce waste over a product's whole lifecycle, it is also essential to create circular value chains. PPPs can open new avenues for circular economy efforts by dismantling silos and promoting cross-sector collaboration, resulting in synergies that build more resilient and sustainable systems (Liu & Ramakrishna, 2021; Rahla et al., 2021).

Circular economy principles are greatly aided by technological advancement, and taking advantage of these developments is crucial to maximizing potential and overcoming obstacles. Digital technologies like blockchain, artificial intelligence (AI), and the Internet of Things (IoT) have the potential to completely change how resources are monitored, managed, and reused in a variety of businesses. IoT devices, for instance, may track the location and state of goods in real-time, enabling predictive maintenance and prolonging the life of products. Blockchain technology may guarantee that resources are sourced, processed, and recycled in accordance with the principles of the circular economy by bringing transparency and traceability to supply chains. Artificial intelligence (AI) can improve manufacturing operations by cutting down on material waste, increasing energy efficiency, and enabling more accurate recycling methods. Furthermore, developments in materials science, including the creation of biodegradable plastics and closed-loop recycling systems, can help remove some of the main obstacles to circularity, such as the challenge of recycling complicated materials. Businesses and governments can drive both economic and environmental resilience by embracing technology innovation and investing in it to overcome technical obstacles and open new potential for circular economy projects (Fernandes et al., 2023; Li et al., 2023). It is essential to raise awareness and educate people about the circular economy and environmental resilience to promote broad adoption and overcome obstacles. The advantages of circular practices and their part in building a sustainable future must be understood by businesses, consumers, and legislators alike. The implementation of circular economy principles in school curricula and the provision of specialized training programs for professionals in manufacturing, construction, and waste management are just a few examples of the diverse ways that educational initiatives can be implemented. Campaigns for public awareness can also be quite effective in influencing consumer behaviour by enticing consumers to purchase recyclable, long-lasting, and repairable goods. Furthermore, highlighting the financial and ecological advantages of circular processes can assist companies in changing their perspective from one of immediate profitability to one of long-term sustainability. Stakeholders may establish the cultural and societal underpinnings required for the widespread adoption of circular economy ideas by raising awareness and educating people, making the world more resilient and sustainable (Awan & Sroufe, 2022; Cimprich et al., 2023; Hartley et al., 2023; Morales et al., 2022).

For businesses, particularly small and medium-sized organizations (SMEs), funding access is frequently a major obstacle to their efforts to shift to circular business models. To solve this, financial support systems must be established to remove obstacles and take full advantage of the circular economy's potential. Businesses that implement circular processes might receive incentives from governments and financial institutions in the form of grants, low-interest loans, and tax advantages. Furthermore, creating green investment funds and promoting private investment in circular economy initiatives can supply the funding required to expand prosperous programs. Additional players in the circular economy movement can be drawn in via financial mechanisms that mitigate risks associated with investments in cutting-edge circular technologies and business models. Public money, for instance, can be utilized to lower the risk associated with early-stage investments in circular start-ups, thereby increasing the appeal of these businesses to venture investors. Stakeholders can ensure that the economic advantages of the shift to a circular economy are widely shared while removing the financial obstacles that impede development by constructing strong financial support systems (Chowdhury et al., 2022; Cimprich et al., 2023; Ezeudu, 2024). For the circular economy to succeed and sustain over time, it is imperative that the shift is egalitarian and inclusive. It is possible to solve issues with inequality and resource access by implementing strategies that prioritize social fairness and inclusivity. This entails creating circular economy programs and policies that take marginalized and disadvantaged communities' needs into account to guarantee that they profit from the business opportunities created by circular practices. For instance, integrating unofficial waste collectors into regular waste management systems can boost recycling initiatives' efficacy and efficiency while also improving the collectors' working conditions and financial stability. Furthermore, projects centered around the circular economy can be developed to provide employment in marginalized areas, assisting in the fight against poverty and advancing economic growth. Making sure the shift to a circular economy is equitable and inclusive will improve social resilience and increase support for circular practices, which will increase their acceptance and long-term success (Chowdhury et al., 2022; Kaplan, 2024; Kirchherr et al., 2023; Sunkur et al., 2023).

## 8 Conclusion

The challenges of resource scarcity, waste management, and climate change necessitate the integration of circular economy concepts and environmental resilience. The main conclusions raised emphasise how urgently systems of production and consumption must be rethought to cut waste and boost resource efficiency. By encouraging sustainable practices across industries, stimulating innovation, and completing product lifecycles, circular economy models provide a route to more resilient and sustainable communities. Gaining a better understanding of these ideas is essential to hastening the shift to a circular economy, which separates economic expansion from environmental deterioration and guarantees a healthier planet for coming generations. The ramifications are obvious: governments, businesses, and individuals must work together to embrace the circular economy's tenets and create efficient frameworks if progress is to be realised. While we navigate the promise of the circular economy, it is critical to prioritise collaboration, innovation, and legislation that promote sustainable behaviour at all societal levels. The full potential of circular economy methods can be unlocked by leveraging technological advancements, cultivating public–private collaborations, and allocating resources towards education and awareness. This strategy will support international efforts to tackle climate change, advance social justice, and improve environmental sustainability while also generating economic opportunities. Future developments in environmental resilience and the circular economy have enormous potential to build a more sustainable and equitable world. Circular processes will probably be adopted by more industries and geographical areas as knowledge rises and their advantages become more apparent, propelling the global transition to sustainability. For this goal to be realised, all stakeholders must continue to be committed, creative, and cooperative. Environmental resilience can be made a reality by consistently pushing the limits of what the circular economy can do, opening the door to a future in which humans and the environment coexist together.

## References

Abbass, K., Qasim, M. Z., Song, H., Murshed, M., Mahmood, H., & Younis, I. (2022). A review of the global climate change impacts, adaptation, and sustainable mitigation measures. *Environmental Science and Pollution Research, 29*(28), 42539–42559.

Abu-Bakar, H., & Charnley, F. (2024). Developing a strategic methodology for circular economy roadmapping: A theoretical framework. *Sustainability, 16*(15), 6682.

Adenle, Y. A., Haideri, S., & Sandouka, I. (2024). Understanding the best practices of cradle to cradle in furnishings, carpet, and textile industries–A case studies analysis and conceptual model. *Cleaner and Circular Bioeconomy, 8*, Article 100088.

Adil, M., Nasir, A., Sikandar, A., & Khan, N. M. (2022). Recent advances in circular bioeconomy. *Waste-to-Energy,* 59–84.

Adla, K., Dejan, K., Neira, D., & Dragana, Š. (2022). Degradation of ecosystems and loss of ecosystem services. In *One health* (pp. 281–327). Academic Press.

Agrawal, D., Dwivedi, A., Patil, A., & Paul, S. K. (2023). Impediments of product recovery in circular supply chains: Implications for sustainable development. *Sustainable Development, 31*(3), 1618–1637.

Ahmad, F., Saeed, Q., Shah, S. M. U., Gondal, M. A., & Mumtaz, S. (2022). Environmental sustainability: challenges and approaches. In *Natural resources conservation and advances for sustainability* (pp. 243–270).

Ahmadzai, M. R., Zaki, P. H., Ismail, M. H., Bawon, P., & Karam, D. S. (2022). The societal and economic impact of reforestation strategies and policies in Southeast Asia—A review. *Forests, 14*(1), 1.

Ajayi, A. P. (2023). Lean and circular economy relevance in post-harvest loss prevention: Nigerian retailers' perceptions. *International Journal of Integrated Supply Management, 16*(1), 82–105.

Al-Addous, M., Bdour, M., Rabaiah, S., Boubakri, A., Schweimanns, N., Barbana, N., & Wellmann, J. (2024). Innovations in solar-powered desalination: A comprehensive review of sustainable solutions for water scarcity in the Middle East and North Africa (MENA) region. *Water, 16*(13), 1877.

Almulhim, A. I., & Al-Saidi, M. (2023). Circular economy and the resource nexus: Realignment and progress towards sustainable development in Saudi Arabia. *Environmental Development, 46*, Article 100851.

Alotaibi, S., Martinez-Vazquez, P., & Baniotopoulos, C. (2024). Advancing circular economy in construction mega-projects: Awareness, key enablers, and benefits—case study of the kingdom of Saudi Arabia. *Buildings, 14*(7), 2215.

Althea, A. (2023). Impact of ocean acidification on coral reefs and the marine ecosystems in Phillipines. *International Journal of Natural Sciences, 3*(2), 48–60.

Anaruma, J. F. P., Oliveira, J. H. C. D., Anaruma Filho, F., Freitas, W. R. D. S., & Teixeira, A. A. (2022). The first two decades of Circular Economy in the 21st century: a bibliographic review. *Benchmarking: An International Journal, 29*(9), 2691–2709.

Anderson, M. G., Clark, M., Olivero, A. P., Barnett, A. R., Hall, K. R., Cornett, M. W., & Cameron, D. R. (2023). A resilient and connected network of sites to sustain biodiversity under a changing climate. *Proceedings of the National Academy of Sciences, 120*(7), Article e2204434119.

Artime, O., Grassia, M., De Domenico, M., Gleeson, J. P., Makse, H. A., Mangioni, G., & Radicchi, F. (2024). Robustness and resilience of complex networks. *Nature Reviews Physics, 6*(2), 114–131.

Awan, U., & Sroufe, R. (2022). Sustainability in the circular economy: Insights and dynamics of designing circular business models. *Applied Sciences, 12*(3), 1521.

Awan, U., Shamim, S., Khan, Z., Zia, N. U., Shariq, S. M., & Khan, M. N. (2021). Big data analytics capability and decision-making: The role of data-driven insight on circular economy performance. *Technological Forecasting and Social Change, 168*, Article 120766.

Awan, U., Sroufe, R., & Bozan, K. (2022). Designing value chains for industry 4.0 and a circular economy: A review of the literature. *Sustainability, 14*(12), 7084.

Ballinas, M. B. P., Arumugam, P., Verstand, D., Siegmund-Schultze, M. G. R., Keesstra, S. D., Chavez, L. G., & Voskamp, I. M. (2022). Discussion paper: Nature based solutions in food systems: Review of nature-based solutions towards more sustainable agriculture and food production.

Batista, L., Dora, M., Garza-Reyes, J. A., & Kumar, V. (2021). Improving the sustainability of food supply chains through circular economy practices–a qualitative mapping approach. *Management of Environmental Quality: An International Journal, 32*(4), 752–767.

Beceiro, P., Brito, R. S., & Galvão, A. (2022). Nature-based solutions for water management: Insights to assess the contribution to urban resilience. *Blue-Green Systems, 4*(2), 108–134.

Béné, C., Frankenberger, T. R., Nelson, S., Constas, M. A., Collins, G., Langworthy, M., & Fox, K. (2023). Food system resilience measurement: Principles, framework and caveats. *Food Security, 15*(6), 1437–1458.

Bianchi, S., Bartoli, F., Bruni, C., Fernandez-Avila, C., Rodriguez-Turienzo, L., Mellado-Carretero, J., & Coltelli, M. B. (2023). Opportunities and limitations in recycling fossil polymers from textiles. *Macromol, 3*(2), 120–148.

Bjørnbet, M. M., Skaar, C., Fet, A. M., & Schulte, K. Ø. (2021). Circular economy in manufacturing companies: A review of case study literature. *Journal of Cleaner Production, 294*, Article 126268.

Cacal, J. C., Taboada, E. B., & Mehboob, M. S. (2023). Strategic implementation of integrated water resource management in selected areas of Palawan: SWOT-AHP method. *Sustainability, 15*(4), 2922.

Carissimi, M. C., Creazza, A., Pisa, M. F., & Urbinati, A. (2023). Circular economy practices enabling circular supply chains: An empirical analysis of 100 SMEs in Italy. *Resources, Conservation and Recycling, 198*, Article 107126.

Casanova, M. T., Bachmann, M., Bagella, S., Bester, A., Lloyd, L., & Farnsworth, L. (2023). Just add water? Wetland and riparian restoration. *Ecological restoration: Moving forward using lessons learned* (pp. 195–239). Springer International Publishing.

Castro-Lopez, A., Iglesias, V., & Santos-Vijande, M. L. (2023). Organizational capabilities and institutional pressures in the adoption of circular economy. *Journal of Business Research, 161*, Article 113823.

Chan, T. H., & Lim, S. Y. (2023). The emergence of novel product uses: An investigation of exaptations in IKEA hacks. *Management Science, 69*(5), 2870–2892.

Chen, Z., Grossfurthner, L., Loxterman, J. L., Masingale, J., Richardson, B. A., Seaborn, T., & Narum, S. R. (2022). Applying genomics in assisted migration under climate change: Framework, empirical applications, and case studies. *Evolutionary Applications, 15*(1), 3–21.

Chen, T., Arshad, I., & Iqbal, W. (2023). Assessing the supply chain management of waste-to-energy on green circular economy in China: An empirical study. *Environmental Science and Pollution Research, 30*(44), 100149–100164.

Cheng, Y., Elsayed, E. A., & Huang, Z. (2022). Systems resilience assessments: A review, framework and metrics. *International Journal of Production Research, 60*(2), 595–622.

Chioatto, E., & Sospiro, P. (2023). Transition from waste management to circular economy: The European Union roadmap. *Environment, Development and Sustainability, 25*(1), 249–276.

Chowdhury, S., Dey, P. K., Rodríguez-Espíndola, O., Parkes, G., Tuyet, N. T. A., Long, D. D., & Ha, T. P. (2022). Impact of organisational factors on the circular economy practices and sustainable performance of small and medium-sized enterprises in Vietnam. *Journal of Business Research, 147*, 362–378.

Cimprich, A., Young, S. B., Schrijvers, D., Ku, A. Y., Hagelüken, C., Christmann, P., & Hool, A. (2023). The role of industrial actors in the circular economy for critical raw materials: A framework with case studies across a range of industries. *Mineral Economics, 36*(2), 301–319.

Cook, E., Velis, C. A., & Cottom, J. W. (2022). Scaling up resource recovery of plastics in the emergent circular economy to prevent plastic pollution: Assessment of risks to health and safety in the Global South. *Waste Management & Research, 40*(12), 1680–1707.

Cooke, S. J., Heger, T., Murphy, S. D., Shackelford, N., Febria, C. M., Rochefort, L., & Higgs, E. S. (2023). Ecological restoration in support of sustainability transitions: Repairing the planet in the Anthropocene. *Introduction to designing environments: Paradigms & approaches* (pp. 93–112). Springer International Publishing.

Correia, A. M., & Lopes, L. F. (2023). Revisiting biodiversity and ecosystem functioning through the lens of complex adaptive systems. *Diversity, 15*(8), 895.

Cristina-Maria, B. ĂT. U. ŞA. R. U., & Ioana-Raluca, S. B. Â. R. C. E. A. (2023). Harmonizing circularity: Romania'S progressive efforts towards circular economy integration in Europe. *Management of Sustainable Development, 15*(1), 62–74.

Croitoru, I. M., Grigoras, M. A., Popescu, A., & Grigoraş, B. A. (2024). Embracing the circular economy: a paradigm shift for sustainable prosperity. *Scientific Papers Series Management, Economic Engineering in Agriculture & Rural Development, 24*(2).

Dan, M. C., Ciortea, A., & Mayer, S. (2023). The refashion circular design strategy—Changing the way we design and manufacture clothes. *Design Studies, 88*, Article 101205.

Das, M., Das, A., Seikh, S., & Pandey, R. (2022). Nexus between indigenous ecological knowledge and ecosystem services: A socio-ecological analysis for sustainable ecosystem management. *Environmental Science and Pollution Research, 29*(41), 61561–61578.

Datzberger, S., Howard-Merrill, L., Iorfa, S. K., & Parkes, J. (2023). How do climate change and environmental degradation contribute to violence against children?

De Angelis, R. (2022). Circular economy business models as resilient complex adaptive systems. *Business Strategy and the Environment, 31*(5), 2245–2255.
de Oliveira, C. T., & Oliveira, G. G. A. (2023). What Circular economy indicators really measure? An overview of circular economy principles and sustainable development goals. *Resources, Conservation and Recycling, 190*, Article 106850.
de Oliveira, J. A. P., Bellezoni, R. A., Shih, W. Y., & Bayulken, B. (2022). Innovations in urban green and blue infrastructure: Tackling local and global challenges in cities. *Journal of Cleaner Production, 362*, Article 132355.
Dobermann, A., Bruulsema, T., Cakmak, I., Gerard, B., Majumdar, K., McLaughlin, M., & Zhang, X. (2022). Responsible plant nutrition: A new paradigm to support food system transformation. *Global Food Security, 33*, Article 100636.
Edgington, D. W. (2022). Planning for earthquakes and tsunamis: Lessons from Japan for British Columbia, Canada. *Progress in Planning, 163*, Article 100626.
Efe, A. (2024). Intersections of platform economy, sustainability, symbiosis, resource efficiency, renewable energy and the circular economy. *Toplum Ekonomi Ve Yönetim Dergisi, 5*(1), 144–173.
Ezeudu, O. B. (2024). Harnessing the drivers and barriers to implementation of extended producer responsibility for circular economy in Nigeria. *Circular Economy and Sustainability,* 1–26.
Farhana, K., Kadirgama, K., Mahamude, A. S. F., & Mica, M. T. (2022). Energy consumption, environmental impact, and implementation of renewable energy resources in global textile industries: An overview towards circularity and sustainability. *Materials Circular Economy, 4*(1), 15.
Farooq, A., Chairat, S., & Gheewala, S. H. (2024). Application of life cycle assessment for evaluating circular product systems. *Circular Economy and Sustainability,* 1–23.
Fernandes, L., Rosado da Cruz, A. M., Cruz, E. F., & Lopes, S. I. (2023). A review on adopting blockchain and IoT technologies for fostering the circular economy in the electrical and electronic equipment value chain. *Sustainability, 15*(5), 4574.
Ferrari, A. M., Volpi, L., Settembre-Blundo, D., & García-Muiña, F. E. (2021). Dynamic life cycle assessment (LCA) integrating life cycle inventory (LCI) and Enterprise resource planning (ERP) in an industry 4.0 environment. *Journal of Cleaner Production, 286*, 125314.
Ferreira, I. A., Godina, R., Pinto, A., Pinto, P., & Carvalho, H. (2023). Boosting additive circular economy ecosystems using blockchain: An exploratory case study. *Computers & Industrial Engineering, 175*, Article 108916.
Ferreira, C. S., Duarte, A. C., Kasanin-Grubin, M., Kapovic-Solomun, M., & Kalantari, Z. (2022). Hydrological challenges in urban areas. In *Advances in chemical pollution, environmental management and protection* (vol. 8(1), pp. 47–67). Elsevier.
Fichte, J. G. (2023). The butterfly effect. In *Next industrial revolution, the: A new age for innovation in industry*, p. 289.
Fiksel, J., & Bakshi, B. R. (2023). Designing for resilience and sustainability: An integrated systems approach. *Engineering and ecosystems: Seeking synergies toward a nature-positive world* (pp. 469–500). Springer International Publishing.
França, R., Nylén, E. J., Jokinen, A., & Jokinen, P. (2022). Filling the social gap in the circular economy: How can the solidarity economy contribute to urban circularity? In *Social and cultural aspects of the circular economy* (pp. 27–44). Routledge.
Gaines, W. L., Hessburg, P. F., Aplet, G. H., Henson, P., Prichard, S. J., Churchill, D. J., & Vynne, C. (2022). Climate change and forest management on federal lands in the Pacific Northwest, USA: Managing for dynamic landscapes. *Forest Ecology and Management, 504*, Article 119794.
George, A. S., & Baskar, T. (2024). Repairing the future: The global fight for accessible fixes and sustainable tech. *Partners Universal Innovative Research Publication, 2*(2), 71–88.
Grambow, M., Mauser, W., Flemming, H. C., Arzet, K., Disse, M., Völkel, J., & Wilderer, J. (2022). The key resources water, soil and intact ecosystems: In which world do we want to live in the future? *Strategies for Sustainability of the Earth System*, 119–141.
Granata, G., & Petrides, D. (2022). *Hydrometallurgical process design and economics with SuperPro designer*.

Grorud-Colvert, K., Sullivan-Stack, J., Roberts, C., Constant, V., Horta e Costa, B., Pike, E. P., & Lubchenco, J. (2021). The MPA Guide: A framework to achieve global goals for the ocean. *Science, 373*(6560), eabf0861.

Güell Barcons, P., Hammer, T. L., & Schmid, F. (2023). *Plastic entropy: an enzyme solution to plastic waste.*

Gupta, H., Nishi, M., & Gasparatos, A. (2022). Community-based responses for tackling environmental and socio-economic change and impacts in mountain social–ecological systems. *Ambio, 51*(5), 1123–1142.

Gupta, A. K., Goyal, M. K., & Singh, S. P. (Eds.). (2023). *Ecosystem restoration: Towards sustainability and resilient development*. Springer Nature.

Guzzo, D., Pigosso, D. C. A., Videira, N., & Mascarenhas, J. (2022). A system dynamics-based framework for examining circular economy transitions. *Journal of Cleaner Production, 333*, Article 129933.

Hagos, F., Gebrezgabher, S., Taron, A., & Odero, J. (2022). Investment climate assessment for circular bioeconomy-review of national policies and strategies in Kenya.

Han, H. Q., Yang, J. Q., Liu, Y., Zhang, Y. J., & Wang, J. W. (2022). Effect of the grain for green project on freshwater ecosystem services under drought stress. *Journal of Mountain Science, 19*(4), 974–986.

Hartley, K., Schülzchen, S., Bakker, C. A., & Kirchherr, J. (2023). A policy framework for the circular economy: Lessons from the EU. *Journal of Cleaner Production, 412*, Article 137176.

Hayashi, J. H. (2023). *Hydraulic Taiwan: Colonial conservation under Japanese Imperial and Chinese Nationalist Rule, 1895–1964* [Doctoral dissertation, Harvard University].

Holling, C. S. (1973, November). *Resilience and stability of ecological systems.*

Hung, S. H., & Tsai, K. W. (2024). Mapping E-waste distribution to form a circular supply chain–An investigation on e-waste towards circular economy in Taiwan. *Applied Environmental Research, 46*(2).

Husgafvel, R., Linkosalmi, L., Sakaguchi, D., & Hughes, M. (2022). How to advance sustainable and circular economy-oriented public procurement—A review of the operational environment and a case study from the Kymenlaakso region in Finland. In *Circular economy and sustainability* (pp. 227–277). Elsevier.

Jain, A. K. (2023). *Climate resilient, green and low carbon built environment*. Springer Nature.

Kandpal, V., Jaswal, A., Santibanez Gonzalez, E. D., & Agarwal, N. (2024). Circular economy principles: shifting towards sustainable prosperity. In *Sustainable energy transition: Circular economy and sustainable financing for environmental, social and governance (ESG) practices* (pp. 125–165). Springer Nature Switzerland.

Kaplan, R. (2024). "Rather than follow change, business must lead this transformation": Global business's institutional project to privatize global environmental governance, 1990–2010. *Organization Studies, 45*(1), 161–188.

Kara, S., Hauschild, M., Sutherland, J., & McAloone, T. (2022). Closed-loop systems to circular economy: A pathway to environmental sustainability? *CIRP Annals, 71*(2), 505–528.

Kattel, G. R. (2022). Climate warming in the Himalayas threatens biodiversity, ecosystem functioning and ecosystem services in the 21st century: Is there a better solution? *Biodiversity and Conservation, 31*(8), 2017–2044.

Kayitesi, N. M., Guzha, A. C., Tonini, M., & Mariethoz, G. (2024). Land use land cover change in the African Great Lakes Region: A spatial–temporal analysis and future predictions. *Environmental Monitoring and Assessment, 196*(9), 852.

Kennedy, S., & Linnenluecke, M. K. (2022). Circular economy and resilience: A research agenda. *Business Strategy and the Environment, 31*(6), 2754–2765.

Khanna, N., Wadhwa, J., Pitroda, A., Shah, P., Schoop, J., & Sarıkaya, M. (2022). Life cycle assessment of environmentally friendly initiatives for sustainable machining: A short review of current knowledge and a case study. *Sustainable Materials and Technologies, 32*, Article e00413.

Kirchherr, J., Yang, N. H. N., Schulze-Spüntrup, F., Heerink, M. J., & Hartley, K. (2023). Conceptualizing the circular economy (revisited): An analysis of 221 definitions. *Resources, Conservation and Recycling, 194*, Article 107001.

Koech, M. K., Munene, K. J., Kinoti, M. K., & Situma, D. K. (2023). Circular economy policies and innovations in Africa: Pillars for achieving sustainable development. In: *Circular economy adoption: Catalysing decarbonisation through policy instruments* (pp. 99–130). Springer Nature Singapore.

Krajnc, D., Kovačič, D., Žunec, E., Brglez, K., & Kovačič Lukman, R. (2022). Youth awareness and attitudes towards a circular economy to achieve the green deal goals. *Sustainability, 14*(19), 12050.

Krishnan, S. R., Nallakaruppan, M. K., Chengoden, R., Koppu, S., Iyapparaja, M., Sadhasivam, J., & Sethuraman, S. (2022). Smart water resource management using Artificial Intelligence—A review. *Sustainability, 14*(20), 13384.

Kshetri, N. (2021). *Blockchain and supply chain management*. Elsevier.

Kumar, R., Kumar, A., & Saikia, P. (2022). Deforestation and forests degradation impacts on the environment. *Environmental degradation: Challenges and strategies for mitigation* (pp. 19–46). Springer International Publishing.

Lambrechts, H. A., Paparrizos, S., Brongersma, R., Kroeze, C., Ludwig, F., & Stoof, C. R. (2023). Governing wildfire in a global change context: Lessons from water management in the Netherlands. *Fire Ecology, 19*(1), 6.

Leclerc, S. H., & Badami, M. G. (2024). Extended producer responsibility: An empirical investigation into municipalities' contributions to and perspectives on e-waste management. *Environmental Policy and Governance, 34*(2), 111–124.

Li, M., Cao, X., Liu, D., Fu, Q., Li, T., & Shang, R. (2022). Sustainable management of agricultural water and land resources under changing climate and socio-economic conditions: A multi-dimensional optimization approach. *Agricultural Water Management, 259*, Article 107235.

Li, J., Herdem, M. S., Nathwani, J., & Wen, J. Z. (2023). Methods and applications for Artificial Intelligence, big data, internet of things, and blockchain in smart energy management. *Energy and AI, 11*, Article 100208.

Lino, F. A., Ismail, K. A., & Castañeda-Ayarza, J. A. (2023). Municipal solid waste treatment in Brazil: A comprehensive review. *Energy Nexus*, 100232.

Liu, Z., Liu, J., & Osmani, M. (2021). Integration of digital economy and circular economy: Current status and future directions. *Sustainability, 13*(13), 7217.

Liu, L., & Ramakrishna, S. (Eds.). (2021). *An introduction to circular economy*. Springer Singapore.

López, D. R., Cavallero, L., Carranza, C., Easdale, M. H., & Peri, P. L. (2024). Resilience management at the landscape level: Fostering mitigation and adaptations to global change based on forest socio-ecosystems. *Integrating landscapes: Agroforestry for biodiversity conservation and food sovereignty* (pp. 161–197). Springer International Publishing.

Marini, L., Ayres, M. P., & Jactel, H. (2022). Impact of stand and landscape management on forest pest damage. *Annual Review of Entomology, 67*(1), 181–199.

Mayar, K., Carmichael, D. G., & Shen, X. (2022). Resilience and systems—a review. *Sustainability, 14*(14), 8327.

McCoy, L., & Chi, T. (2022). Collaborative consumption: A study of sustainability presentation in fashion rental platforms. *Sustainability, 14*(14), 8537.

Mersha, A. N., de Fraiture, C., Masih, I., & Alamirew, T. (2021). Dilemmas of integrated water resources management implementation in the Awash River Basin, Ethiopia: Irrigation development versus environmental flows. *Water and Environment Journal, 35*(1), 402–416.

Mhatre, P., Panchal, R., Singh, A., & Bibyan, S. (2021). A systematic literature review on the circular economy initiatives in the European Union. *Sustainable Production and Consumption, 26*, 187–202.

Mhatre, P., Gedam, V. V., Unnikrishnan, S., & Raut, R. D. (2023). Circular economy adoption barriers in built environment-a case of emerging economy. *Journal of Cleaner Production, 392*, Article 136201.

Mihai, F. C., & Ulman, S. R. (2023). Circular economy in green deal strategies. In *Sustainable and circular management of resources and waste towards a Green Deal* (pp. 49–63). Elsevier.
Mihai, F.C., Meidiana, C., Elagroudy, S., Ulman, S. R., Gutberlet, J., & Carvalho, C. (2024). Plastic waste management for zero waste to landfills: potential, challenges, and opportunities. *Zero Waste Management Technologies*, 97–136.
Mina, M., Messier, C., Duveneck, M. J., Fortin, M. J., & Aquilué, N. (2022). Managing for the unexpected: Building resilient forest landscapes to cope with global change. *Global Change Biology, 28*(14), 4323–4341.
Mishra, A., Dutta, P., Jayasankar, S., Jain, P., & Mathiyazhagan, K. (2023). A review of reverse logistics and closed-loop supply chains in the perspective of circular economy. *Benchmarking: An International Journal, 30*(3), 975–1020.
Mitsch, W. J., Zhang, L., Griffiths, L. N., & Bays, J. (2023). Contrasting two urban wetland parks created for improving habitat and downstream water quality. *Ecological Engineering, 192*, Article 106976.
Montag, L. (2023). Circular economy and supply chains: Definitions, conceptualizations, and research agenda of the circular supply chain framework. *Circular Economy and Sustainability, 3*(1), 35–75.
Morais, J., Corder, G., Golev, A., Lawson, L., & Ali, S. (2022). Global review of human waste-picking and its contribution to poverty alleviation and a circular economy. *Environmental Research Letters, 17*(6), Article 063002.
Morales, P., Flikkema, M., Castaldi, C., & de Man, A. P. (2022). The effectiveness of appropriation mechanisms for sustainable innovations from small and medium-sized enterprises. *Journal of Cleaner Production, 374*, Article 133921.
Moroni, I. T., Seles, B. M. R. P., Lizarelli, F. L., Guzzo, D., & da Costa, J. M. H. (2022). Remanufacturing and its impact on dynamic capabilities, stakeholder engagement, eco-innovation and business performance. *Journal of Cleaner Production, 371*, Article 133274.
Morsink-Georgali, P. Z., Kylili, A., Fokaides, P. A., & Papadopoulos, A. M. (2022). Compost versus biogas treatment of sewage sludge dilemma assessment using life cycle analysis. *Journal of Cleaner Production, 350*, Article 131490.
Möslinger, M., Ulpiani, G., & Vetters, N. (2023). Circular economy and waste management to empower a climate-neutral urban future. *Journal of Cleaner Production, 421*, Article 138454.
Mrabet, R. (2023). Sustainable agriculture for food and nutritional security. In *Sustainable agriculture and the environment* (pp. 25–90). Academic Press.
Mtibe, A., Motloung, M. P., Bandyopadhyay, J., & Ray, S. S. (2021). Synthetic biopolymers and their composites: Advantages and limitations—An overview. *Macromolecular Rapid Communications, 42*(15), 2100130.
Nagamalini, T., & Wesley, J. R. (2024). Pathway for a circular economy: facilitating circularity using the ADKAR model for waste management. *Management of Environmental Quality: An International Journal*
Negrete-Cardoso, M., Rosano-Ortega, G., Álvarez-Aros, E. L., Tavera-Cortés, M. E., Vega-Lebrún, C. A., & Sánchez-Ruíz, F. J. (2022). Circular economy strategy and waste management: A bibliometric analysis in its contribution to sustainable development, toward a post-COVID-19 era. *Environmental Science and Pollution Research, 29*(41), 61729–61746.
Nguyen, D. T., & Le, P. L. (2022). Twenty-year application of logistics and supply chain management in the construction industry. *Construction Management and Economics, 40*(10), 796–834.
Nguyen, N. T., Friess, D. A., Todd, P. A., Mazor, T., Lovelock, C. E., Lowe, R., & Huang, D. (2022). Maximizing resilience to sea-level rise in urban coastal ecosystems through systematic conservation planning. *Landscape and Urban Planning, 221*, Article 104374.
Nsengimana, V., Nzibaza, V., Nzarora, A., & Mugume, P. J. (2024). Conserving biodiversity and managing water resources in the Akagera River catchment, Eastern Rwanda: local engagement and knowledge cocreation through the adaptive planning process. *Socio-Ecological Practice Research*, 1–17.

Nygaard, A. (2023). The geopolitical risk and strategic uncertainty of green growth after the Ukraine invasion: How the circular economy can decrease the market power of and resource dependency on critical minerals. *Circular Economy and Sustainability, 3*(2), 1099–1126.

Obaisi, A. I., Adegbeye, M. J., Elghandour, M. M., Barbabosa-Pliego, A., & Salem, A. Z. M. (2022). Natural resource management and sustainable agriculture. *Handbook of climate change mitigation and adaptation* (pp. 2577–2613). Springer International Publishing.

Oluleye, B. I., Chan, D. W., & Antwi-Afari, P. (2023). Adopting Artificial Intelligence for enhancing the implementation of systemic circularity in the construction industry: A critical review. *Sustainable Production and Consumption, 35*, 509–524.

Olwoch, M. A., Sonnenberg, N. C., Reis, T. L., & Taljaard-Swart, H. (2023). Consumers' sustainable clothing consumption practices: Adopting the principles of voluntary simplicity. In *Consumer awareness and textile sustainability* (pp. 15–31). Springer Nature Switzerland.

Orji, I. J., & Ojadi, F. (2022). *The circular supply chain: Basic principles and techniques.* CRC Press.

Osobajo, O. A., Oke, A., Omotayo, T., & Obi, L. I. (2022). A systematic review of circular economy research in the construction industry. *Smart and Sustainable Built Environment, 11*(1), 39–64.

Parvizi, E., Fraser, C. I., & Waters, J. M. (2022). Genetic impacts of physical disturbance processes in coastal marine ecosystems. *Journal of Biogeography, 49*(10), 1877–1890.

Parwada, C., Chipomho, J., & Mandumbu, R. (2022). Building resilience to climate change by adopting conservation agriculture in the smallholder farming systems. In *Food security for African smallholder farmers* (pp. 281–294). Springer Nature Singapore.

Pascual, L. S., Segarra-Medina, C., Gómez-Cadenas, A., López-Climent, M. F., Vives-Peris, V., & Zandalinas, S. I. (2022). Climate change-associated multifactorial stress combination: A present challenge for our ecosystems. *Journal of Plant Physiology, 276*, Article 153764.

Patti, S. (2023). Circular economy and policy. Springer Books.

Piorkowsky, M. B. (2023). Kenneth Ewart Boulding: Economics and ecology. *Headstrong and unorthodox pioneers for a critical theory of consumption: contributions from economics and social sciences* (pp. 111–137). Springer Fachmedien Wiesbaden.

Pouikli, K., Delegkou, A., & Tsakalogianni, I. (2022). Circular economy for a climate-neutral Europe: Mapping the latest policy and legislative developments. *CCLR, 16*, 281.

Power, M. E., Chandra, S., Gleick, P., & Dietrich, W. E. (2024). Anticipating responses to climate change and planning for resilience in California's freshwater ecosystems. *Proceedings of the National Academy of Sciences, 121*(32), Article e2310075121.

Priewe, J. (2022). Growth in the ecological transition: Green, zero or de-growth? *European Journal of Economics and Economic Policies, 19*(1), 19–40.

Rahla, K. M., Mateus, R., & Bragança, L. (2021). Implementing circular economy strategies in buildings—from theory to practice. *Applied System Innovation, 4*(2), 26.

Raimi, M. O., Saliu, A. O., Babatunde, A., Okon, O. G., Taiwo, P. A., Ahmed, A. K., & Telu, M. (2022). The challenges and conservation strategies of biodiversity: the role of government and non-governmental organization for action and results on the ground. In *Biodiversity in Africa: Potentials, threats and conservation* (pp. 473–504). Springer Nature Singapore.

Rashid, S., & Malik, S. H. (2023). Transition from a linear to a circular economy. *Renewable energy in circular economy* (pp. 1–20). Springer International Publishing.

Rashid, M. I., & Shahzad, K. (2021). Food waste recycling for compost production and its economic and environmental assessment as circular economy indicators of solid waste management. *Journal of Cleaner Production, 317*, Article 128467.

Reddy, V. V., Kumar, P., Rao, A. L. N., Kumar, R., Singh, S., Asha, V., & Kareem, S. H. (2023). Waste to wealth generation: Innovative methodologies in resource utilization and minimization in circular economy. In *E3S web of conferences* (vol. 453, p. 01035). EDP Sciences.

Rossi, F., & Morone, P. (2023). North-South waste trade: Prime example of the circular economy or major environmental threat? *Circular Economy and Sustainability, 3*(4), 2159–2182.

Roy, S., Rautela, R., & Kumar, S. (2023). Towards a sustainable future: Nexus between the sustainable development goals and waste management in the built environment. *Journal of Cleaner Production, 415*, Article 137865.

Roy, A., Ganguly, N., Banerjee, S., & Pandit, S. (2024a). From linear economy to circular bioeconomy: A paradigm shift in wastewater management. *Advancements in bio-systems and technologies for wastewater treatment* (pp. 135–154). Springer International Publishing.

Roy, T., Kalambukattu, J. G., Sarkar, A., Rashmi, I., Pal, R., Singhal, V., & Kumar, S. (2024b). Climate crisis and adoption of climate-smart agriculture technologies. In *Climate crisis: Adaptive approaches and sustainability* (pp. 229–252). Springer Nature Switzerland.

Sahavacharin, A., Sompongchaiyakul, P., & Thaitakoo, D. (2022). The effects of land-based change on coastal ecosystems. *Landscape and Ecological Engineering, 18*(3), 351–366.

Sarkhoshkalat, M. M., Afkham, A., Bonyadi Manesh, M., & Sarkhosh, M. (2024). Circular economy and the recycling of e-waste. In New technologies for energy transition based on sustainable development goals: Factors contributing to global warming (pp. 319–354). Springer Nature Singapore.

Scavo, A., Fontanazza, S., Restuccia, A., Pesce, G. R., Abbate, C., & Mauromicale, G. (2022). The role of cover crops in improving soil fertility and plant nutritional status in temperate climates. A review. *Agronomy for Sustainable Development, 42*(5), 93.

Selvan, T., Panmei, L., Murasing, K. K., Guleria, V., Ramesh, K. R., Bhardwaj, D. R., & Deshmukh, H. K. (2023). Circular economy in agriculture: Unleashing the potential of integrated organic farming for food security and sustainable development. *Frontiers in Sustainable Food Systems, 7*, 1170380.

Shahhoseini, A., Heydari, S., & Pedrammehr, S. (2023). Manufacturing and assembly for the ease of product recycling: A review. *Designs, 7*(2), 42.

Shang, M., & Xie, J. (2024). Agricultural sustainable development: Soil, water resources, biodiversity, climate change, and technological innovation. *Advances in Resources Research, 4*(2), 181–204.

Shanker, R., Khan, D., Hossain, R., Islam, M. T., Locock, K., Ghose, A., & Dhodapkar, R. (2023). Plastic waste recycling: Existing Indian scenario and future opportunities. *International Journal of Environmental Science and Technology, 20*(5), 5895–5912.

Sharma, H. B., Vanapalli, K. R., Samal, B., Cheela, V. S., Dubey, B. K., & Bhattacharya, J. (2021). Circular economy approach in solid waste management system to achieve UN-SDGs: Solutions for post-COVID recovery. *Science of the Total Environment, 800*, Article 149605.

Sharma, S., Reddy, V. M., Nijhawan, G., Yadav, D. K., Raed, R., & Tyagi, L. K. (2024). Leveraging waste-to-energy technologies for sustainable development: A comprehensive review. In E3S web of conferences (vol. 529, p. 02010). EDP Sciences.

Sharma, S. (2024). Fostering green product design and innovation for a sustainable future. In *Waste management and life cycle assessment for sustainable business practice* (pp. 86–110). IGI Global.

Shell, H. R. (2014). Shoddy heap: A material history between waste and manufacture. *History and Technology, 30*(4), 374–394. https://doi.org/10.1080/07341512.2014.1003165

Singh, A. K. (2024). *Framework for integration of circular economy into business models: A study of samsung.*

Skawińska, E., & Zalewski, R. I. (2018). Circular economy as a management model in the paradigm of sustainable development. *Management, 22*(2), 217–233.

Smol, M., Mejia, A., & Howarth, M. (2023). Business opportunities in the water and wastewater sector through the ReSOLVE framework. *Water in circular economy* (pp. 225–233). Springer International Publishing.

Stathis, M. (2023). *Shaping our fate: analyzing circular economy principles in corporate sustainability strategy as a critical climate solution for the apparel industry* [doctoral dissertation, University of Oregon].

Sunkur, R., Kantamaneni, K., Bokhoree, C., & Ravan, S. (2023). Mangroves' role in supporting ecosystem-based techniques to reduce disaster risk and adapt to climate change: A review. *Journal of Sea Research*, 102449.

Sylva, K., & Iyer-Raniga, U. (2023). Partnership in the built environment for realizing the 2030 agenda: A soft systems model incorporating systems theory and the circular economy. *The role of design, construction, and real estate in advancing the sustainable development goals* (pp. 169–186). Springer International Publishing.

Tebbett, S. B., Bennett, S., & Bellwood, D. R. (2024). A functional perspective on the meaning of the term 'herbivore': Patterns versus processes in coral reef fishes. *Coral Reefs, 43*(2), 219–232.

Tedesco, M., Simioni, F. J., Sehnem, S., Soares, J. F., & Junior, L. M. C. (2022). Assessment of the circular economy in the Brazilian planted tree sector using the ReSOLVE framework. *Sustainable Production and Consumption, 31*, 397–406.

Tleuken, A., Tokazhanov, G., Jemal, K. M., Shaimakhanov, R., Sovetbek, M., & Karaca, F. (2022). Legislative, institutional, industrial and governmental involvement in circular economy in central Asia: A systematic review. *Sustainability, 14*(13), 8064.

Truong, D., Duc, K. N., & Dinh, N. C. (2024). Circular economy essentials. In *Navigating the circular age of a sustainable digital revolution* (pp. 1–26). IGI Global.

Türkeli, S., Huang, B., Ohnishi, S., & Kemp, R. (2022). Issues, interventions, and innovations in the cement industry: A comparative trajectory analysis of eco-cement transitions in the Netherlands, China, and Japan. In *Circular economy and sustainability* (pp. 545–566). Elsevier.

Twidwell, D., Scholtz, R., Donovan, V., Weir, J., & Scasta, J. D. (2022). Prescribed fires of the future: technology and fire. *Global Application of Prescribed Fire, 258.*

Uhl, B., Schall, P., & Bässler, C. (2024). Achieving structural heterogeneity and high multi-taxon biodiversity in managed forest ecosystems: a European review. *Biodiversity and Conservation,* 1–32.

Ulloa-Murillo, L. M., Villegas, L. M., Rodríguez-Ortiz, A. R., Duque-Acevedo, M., & Cortés-García, F. J. (2022). Management of the organic fraction of municipal solid waste in the context of a sustainable and circular model: Analysis of trends in Latin America and the Caribbean. *International Journal of Environmental Research and Public Health, 19*(10), 6041.

Upreti, G. (2023). Metaphysics of dominant development paradigm and its critique. In *Ecosociocentrism: The earth first paradigm for sustainable living* (pp. 189–216). Springer Nature Switzerland.

Ur Rahim, H., Qaswar, M., Uddin, M., Giannini, C., Herrera, M. L., & Rea, G. (2021). Nano-enable materials promoting sustainability and resilience in modern agriculture. *Nanomaterials, 11*(8), 2068.

van de Gronden, N. (2022). The circular economy discourse among dutch policy actors: Discussing the prospects for the circular economy in the Netherlands.

van Doorn-Hoekveld, W. J., Gilissen, H. K., Groothuijse, F. A. G., & van Rijswick, H. F. M. W. (2022). Adaptation to climate change in Dutch flood risk management: Innovative approaches and related challenges. *Utrecht Law Review, 18*(2).

Viles, E., Kalemkerian, F., Garza-Reyes, J. A., Antony, J., & Santos, J. (2022). Theorizing the principles of sustainable production in the context of circular economy and industry 4.0. *Sustainable Production and Consumption, 33*, 1043–1058.

Villasante, S., Ainsworth, G. B., Pita, P., Belgrano, A., Bennett, N., & Sumaila, U. R. (2023). The role of marine protected areas (MPAs) in providing ecosystem services to improve ocean and human health. In *Oceans and human health* (pp. 23–37). Academic Press.

Voulvoulis, N. (2022). Transitioning to a sustainable circular economy: The transformation required to decouple growth from environmental degradation. *Frontiers in Sustainability, 3*, Article 859896.

Wang, L. K., Wang, M. H. S., Cardenas, R. R., Sabiani, N. H. M., Yusoff, M. S., Hassan, S. H., & Hung, Y. T. (2021). Composting processes for disposal of municipal and agricultural solid wastes. *Solid Waste Engineering and Management:, 1*, 399–523.

Wang, J. X., Burke, H., & Zhang, A. (2022a). Overcoming barriers to circular product design. *International Journal of Production Economics, 243*, Article 108346.

Wang, T., Jian, S., Wang, J., & Yan, D. (2022b). Dynamic interaction of water–economic–social–ecological environment complex system under the framework of water resources carrying capacity. *Journal of Cleaner Production, 368*, Article 133132.

Ward, D., Melbourne-Thomas, J., Pecl, G. T., Evans, K., Green, M., McCormack, P. C., & Layton, C. (2022). Safeguarding marine life: Conservation of biodiversity and ecosystems. *Reviews in Fish Biology and Fisheries, 32*(1), 65–100.

Ward, E. J., Hunsicker, M. E., Marshall, K. N., Oken, K. L., Semmens, B. X., Field, J. C., & Tolimieri, N. (2024). Leveraging ecological indicators to improve short term forecasts of fish recruitment. *Fish and Fisheries*.

Watabe, A., & Yamabe-Ledoux, A. M. (2023). Low-carbon lifestyles beyond decarbonisation: Toward a more creative use of the carbon footprinting method. *Sustainability, 15*(5), 4681.

Wu, W., Huang, Y., Fath, B. D., Schwarzfurtner-Lutnik, K., & Harder, M. K. (2024). Using desirable urban states to understand key linkages between resilience subsystems. *Journal of Cleaner Production, 436*, Article 140678.

Xu, J., Zhang, Y., Huang, C., Zeng, L., Teng, M., Wang, P., & Xiao, W. (2022). Forest restoration shows uneven impacts on soil erosion, net primary productivity and livelihoods of local households. *Ecological Indicators, 134*, Article 108462.

Xu, X., & Sakai, A. (2024). The impact on ecosystem services from rural revitalization activity in China Shandong Province, from a human well-being perspective of local residential. *Environmental and Sustainability Indicators, 100419*.

Yadav, A., Rene, E. R., Sharma, M., Jatain, I., Mandal, M. K., & Dubey, K. K. (2022). Valorization of wastewater to recover value-added products: A comprehensive insight and perspective on different technologies. *Environmental Research, 214*, Article 113957.

Yang, F., Wen, X., Aziz, A., & Luhach, A. K. (2021). The need for local adaptation of smart infrastructure for sustainable economic management. *Environmental Impact Assessment Review, 88*, Article 106565.

Yang, M., Chen, L., Wang, J., Msigwa, G., Osman, A. I., Fawzy, S., & Yap, P. S. (2023). Circular economy strategies for combating climate change and other environmental issues. *Environmental Chemistry Letters, 21*(1), 55–80.

Yang, J. (2022). *From zero waste to material closed loop: the way towards circular economy*. Springer Nature.

Zhang, X., Tang, Y., Zhang, H., Jiang, Z., & Cai, W. (2021). Remanufacturability evaluation of end-of-life products considering technology, economy and environment: A review. *Science of the Total Environment, 764*, Article 142922.

Zhang, P., Huang, Y., Pan, S., Chen, W., Zhong, H., Xu, N., & Zhong, M. (2022). Does resilience exist in China's tourism economy? From the perspectives of resistance and recoverability. *Sustainability, 14*(17), 10641.

Zupancic, N. (2023). Systematic literature review: Inter-reletedness of innovation, resilience and sustainability-major, emerging themes and future research directions. *Circular Economy and Sustainability, 3*(3), 1157–1185.

# Circular Economy and Green Transition: Foundations, Practices, and Pathways to Sustainability

**Avanish Kumar and Dan Bahadur Pal**

**Abstract** The chapter "Circular Economy and Green Transition: An Introduction to the Concepts" offers a comprehensive overview of two pivotal models in sustainable development. It begins by defining the circular economy (CE) and green transition (GT), emphasizing their significance in modern environmental policy and economic practices. Tracing the historical context, the chapter highlights the transition from traditional linear economic models to more sustainable approaches. Investigating into the theoretical foundations, the discussion illuminates the core principles of CE is to minimise w*a*ste and reinforcing natural systems—and compares them with linear economic practices. The principles of GT, including strategies for achieving sustainable development and appropriate policy frameworks, are also analysed. The chapter further breaks down the components and strategies of CE, covering design and innovation, resource management, and sustainable business models. For GT, it examines pathways and practices in energy transition, sustainable industrial practices, and policy governance. With the help of detailed case studies, the chapter presents successful implementations and best practices and providing insights into real-world applications and their impact. Addressing the challenges and barriers to CE and GT, the chapter detects economic, technical, logistical, social, and cultural obstacles. It also aspects ahead to upcoming directions, discussing emerging trends, policy evolution, and the crucial role of multi-stakeholder collaboration. The chapter emphasizes the importance of adopting CE and GT practices to accomplish a sustainable future, offering a hopeful outlook on the continuous journey towards environmental sustainability and economic resilience.

**Keywords** Sustainable development · Circular economy (CE) · Green transition (GT)

A. Kumar · D. B. Pal (✉)
Department of Chemical Engineering, Harcourt Butler Technical University, Kanpur, Uttar Pradesh, India
e-mail: danbahadur.hbtu@gmail.com

 
P. Singh et al. (eds.), *Circular Economy and Environmental Resilience*,
https://doi.org/10.1007/978-3-031-93091-1_3

# 1 Introduction

CE, is an economic structure targeted at removing discarded and the continual use of resources. Unlike the traditional linear economy, in a CE resources, products or materials, are kept in use for long period and discarded is minimized with application in reuse, remanufacturing or recycling and sustainable condition (Reike et al., 2018). While in the Green Transition (GT) represents to the process of changing from an economy typically dependent on conventional fuels to one that is sustainable, low-carbon, and eco-friendly (Abbasi et al., 2024). So, changes are required to various sectors, including energy, transportation, agriculture, and industry field which are targeted reduce greenhouse gas emissions, maintaining resources, and protect ecosystem. The Circular Economy and Green Transition are essential in respect of climate change. By reducing waste, minimising resources, and reducing carbon emissions, these approaches lead to control the temperature rise and reduce the effects of climate change (Zeng et al., 2022) The CE in courage impart to reduce environmental degradation. Both CE and GT give attention to ecosystems and biodiversity. By minimising pollution, protecting natural resources, and promoting to preserve the environment for future generations. The concept of linear economy uses "use and through policy "believe to use to manufacture products, and then eliminate as waste once their useful life ends (Agovino and Musella, 2023). This model has been the leading economic model since the Industrial Revolution, driven by the presence of huge resources and the focus on maximizing short-term profits and consumer suitability. In the 1960s and 1970s, the publication of Rachel Carson's "*Silent Spring"* in 1962, which highlighted the hazards of pesticides (Kroll, 2001), and the first Earth Day in 1970. These movements raised public attentiveness about environmental concerns and placed the groundwork for future environmental policies. And on the same way the oil crises of the 1970s exposed the susceptibilities of economies dependent on fossil fuels, leading to increased interest in alternative energy sources and energy efficiency. Governments and industries began to explore nonconventional energy resources such as wind, solar power, as a way to minimise application of conventional energy resources like coal etc. In 1992 the development of significant international environmental agreements, such as Rio Earth Summit, which led to the creation of the UNFCCC, United Nations Framework Convention on Climate Change and in 1987 and Montreal Protocol to protect the ozone layer and the (Bradbrook, 2001). These agreements set the stage for global cooperation on environmental issues and the green transition. The Kyoto Protocol, adopted in 1997, was the first international treaty to set legally binding targets for reducing greenhouse gas emissions. While its impact was limited, it represented a significant milestone in the global effort to combat climate change and promoted the development of carbon markets and renewable energy initiatives (Iwata and Okada, 2014). In 2015, The Paris Agreement is accepted which is major milestone in the green transition. It brought together nearly all the world's nations in a commitment to limit global warming to well below 2 °C above pre-industrial levels, with efforts to limit the increase to 1.5 °C. The agreement emphasized the need for a transition to a low-carbon economy and encouraged countries to develop and implement ambitious climate action plans (Warren et al., 2022).

## *1.1 Principles and Goals of Circular Economy (CE)*

CE, promotes the continuous use of resources by extending the lifecycle of products with the application of product as reuse, reprocess etc., (Bressanelli et al., 2018). This can be attained with application of concept of recycling, remanufacturing, and recycling, ensuring that materials are kept within the use as much as possible with a long period. Seeks to restore and enhance natural ecosystems rather than depleting them. This principle of CE focuses on using renewable resources, improving soil health, and increasing biodiversity through regenerative agricultural practices and sustainable resource management (Cárceles et al., 2022). Eco-design emphasizes creating products that have minimal environmental impact throughout their lifecycle. Innovation plays a crucial role in developing new materials, technologies, and processes that support circularity. The CE aims to use resources more efficiently, decreasing the want for raw materials and lowering environmental effect. This is achieved by maximizing the value extracted from resources and minimizing waste. The concept of circulation economy could be shown as following Fig. 1.as schematic diagram of circulation economy. Creating new business models and markets focused on sustainability, the CE can drive economic growth and generate employability in different field such as reduce recycle, reuse concept of product (Horbach et al., 2015). A core goal of the CE is to minimise the environmental scratch of human activities, including reducing greenhouse gas emissions, minimizing pollution, and protecting reserves. The CE seeks to build resilience against resource scarcity and environmental challenges by creating systems that are sustainable and adaptable in the long term.

## *1.2 Concept of Green Transition (GT)*

Green technology emphasis on minimising carbon emissions by changing from fossil fuels to renewable energy sources such as wind, solar, and hydropower (Horbach et al., 2015). The reduction of carbon emission is reducing climate change and succeeding a low-carbon economy. GT accepting sustainable agricultural practices and encouraging green infrastructure, and guaranteeing the responsible use of reserves (Kumar et al., 2023).

The aim of the GT is to control climatic change due to global warming or evolution of greenhouse gas emissions and promoting climate resilience. This involves adopting policies and technologies that support a low-carbon and climate-resilient economy. GT coordinate the wealth growth and climate protection and social happiness (Apostu et al., 2023) which includes promoting green technology and reducing

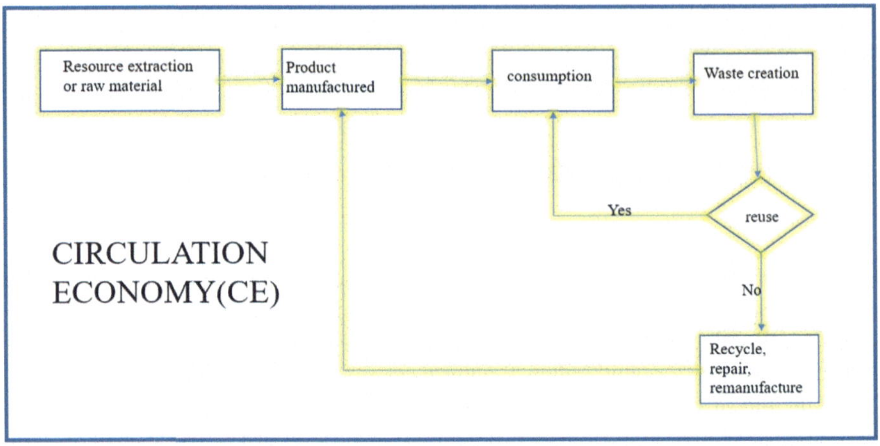


**Fig. 1** Schematic diagram of circulation economy

environmental degradation, and ensuring access to clean energy and resources for all. Reducing pollution and promoting clean energy. The GT aims to improve public health outcomes cleaner air, water, and environments contribute to healthier communities and reduce the burden of disease. GT seeks to position nations as leaders in the global effort to combat change in climate and promote sustainability (Gössling and Scott, 2018). This involves international cooperation, sharing best practices, and supporting developing countries in their green transitions. Its concept as diagram could be shown as Fig. 2.

**Fig. 2** Schematic diagram of green transition or green economy

## *1.3 Circular Economy (CE) Versus Linear Economy*

The CE emphasizes the application of renewable reserves over finite ones, reducing the impact on natural ecosystems. The sources such as solar and wind, along with sustainably sourced materials. The CE encourages practices that restore and regenerate natural systems, such as returning nutrients to the soil through composting and supporting biodiversity through sustainable land use (Kiryluk and Kostecka, 2022). Biological materials are cycled back into the environment safely, where they can decompose and regenerate natural systems. This compares with synthetic materials that do not break down and often harm ecosystems. The linear model is based on utilisation of natural reserves such as fossil fuels and minerals without consideration for regeneration (Bradshaw and Hamacher, 2012). This type of policy leads for environmental degradation, such as deforestation, soil depletion, and loss of biodiversity. Waste obtain from the linear economy often impart in pollution and the destruction of natural habitats (Luttenberger, 2020). The comparison Circular Economy vs. Linear Economy might be shown like Table 1

**Table 1** Comparison circular economy versus linear economy

| Aspect | Circular economy | Linear economy |
|---|---|---|
| Resource utilization | Emphasizes renewable resources and efficient use of materials | Relies on finite resources and extensive resource extraction |
| Product design | Designed for longevity, recyclability, and modularity | Designed for short lifespan and disposability (Zahariev et al., 2024) |
| Waste management | Minimizes waste through recycling, reuse, and closed-loop systems (Kirchherr et al., 2023) | Generates significant waste, often discarded in landfills |
| Economic model | Focuses on keeping products in use through repair, reuse, and sharing | Encourages consumption and disposal with new product purchases |
| Environmental impact | Regenerates natural systems, reduces pollution, and supports biodiversity | Depletes natural resources, pollutes ecosystems, and harms biodiversity |
| Sustainability | Promotes long-term sustainability by designing out waste and regenerating natural systems (Scheel, 2016) | Unsustainable due to continuous resource depletion and waste generation |

## 2 Initiation of Policy Frameworks on National and International Agreements

National contributions often focus on policy development, implementation, and promotion innovation within their borders, while international contributions focus on collaboration, knowledge sharing, and setting global standards. Both levels are crucial for advancing CE and GT to address global environmental challenges effectively (Alsarhan et al., 2021). National Resource Efficiency Policy (Draft), India is focusing on resource efficiency and circular economy to reduce waste and improve sustainability. The policy encourages recycling, efficient use of resources, and sustainable practices in industries like agriculture and manufacturing. Swachh Bharat Mission primarily a cleanliness drive, it also promotes waste management and recycling which key components of a circular economy (Prakash, 2014). The Circular Economy Promotion Law (2009) China has been a pioneer in implementing CE policies, with pay emphasis on increasing resource efficiency (Zhu et al., 2019). 13th Five-Year Plan, is also incorporated circular economy goals, emphasizing recycling, resource efficiency, and the development of green European Green Deal. The EU has been at the forefront of the green transition, targeting to make Europe the first climate-neutral zone by 2050 (Peeters and Misonne, 2022). The Green economy includes comprehensive strategies for circular economy, renewable energy, and sustainable agriculture. Circular Economy Action Plan, which focus on the entire lifecycle of products, from design to waste management, promoting sustainability across industries. Sustainable Development Goals (SDGs), The UN has integrated CE and GT principles into several SDGs, including those on responsible consumption and production (SDG 12), climate action (SDG 13), and life below water (SDG 14). An international treaty as Paris Agreement, focusing on minimising climate change, supports the green transition by encouraging countries to adopt renewable energy, reduce emissions, and enhance sustainability (Jacob et al., 2021). World Economic Forum (WEF), Global Lighthouse Network, The WEF recognizes manufacturing sites globally that demonstrate leadership in adopting advanced technologies for sustainability and circular economy practices. Platform for Accelerating the Circular Economy (PACE). The public–private collaboration platform aiming to accelerate the global transition to a circular economy (Gorbushina, 2023). IRENA, International Renewable Energy Agency encourages the global adoption of renewable energy, a cornerstone of the green transition (Ghorbani et al., 2023). It includes policy advice, technical assistance, and knowledge sharing to advance green energy solutions. The Paris Agreement (2015) is landmark international treaty under the United Nations Framework Convention on Climate Change (UNFCCC) that brings together almost every nation in the world to combat climate change (Protocol, 1997). The agreement sets out a global framework to avoid dangerous climate change by limiting global warming to well below 2 °C and pursuing efforts to limit it to 1.5 °C. Countries are required to set nationally determined contributions (NDCs) and update them every five years. The European Green Deal (2019), which is inclusive policy initiative by Europe to make the first climate-neutral continent by 2050. The Kyoto Protocol

(1997). The first international agreement, The Kyoto Protocol was to set mandatory targets for all developed nations minimise to generate greenhouse gases (Telesetsky, 1999). Although it has been superseded by the Paris Agreement, it was a critical milestone in global climate policy and helped pave the way for future agreements.

## 3 Strategies to Achieve Circular Economy

### 3.1 *Transition to Renewable Energy*

CE promotes renewable energy sources application such as wind, solar, hydropower, and geothermal to replace natural reserves like coal, petrol etc. and minimise generation carbon (Hassan et al., 2024). Enhancing energy efficiency in buildings, transportation, and industry to reduce energy consumption and associated emissions. Designing cities that are sustainable, resilient, and inclusive, with a focus on green infrastructure, sustainable transportation, and energy-efficient buildings. CE Promoting urban planning that reduces sprawl, conserves natural areas, and encourages the use of public transportation and non-motorized travel (Simeonova and Valk, 2016). Waste incineration can be coupled with energy recovery systems to generate electricity or heat from the combustion of waste. While this method reduces waste volume and provides energy, it must be managed carefully to minimize air pollution and ensure it complements recycling efforts. Organic waste, such as food waste and agricultural residues and food west, can be treated in anaerobic digesters to yield biogas (Pramanik et al., 2019). The shift to renewable energy sources, such as solar, wind, hydro, and geothermal, reduces dependence on fossil fuels and lowers greenhouse gas emissions. Techniques like energy audits, retrofitting buildings, and adopting energy-efficient appliances reduce energy consumption and enhance resource efficiency.

### 3.2 *Sustainable Product Design*

It is essential for circular economy to work on concept of recycle and reuse or reduces waste of product. The product should design in this way that their part should utilize again and again if it were disassembled, allowing for easier reuse or recycle of components, this will lead to the circular economy. By doing eco-design can reduce the manufacturers cost. Additionally, products designed can also lead to lower operational costs for consumers. Eco-design reduces the waste of material, reduces the costs involved in waste management (Gottberg et al., 2006). If we support green technologies that will develop new solutions for sustainability challenges and to upgrade the green transition. Eco-design emphasizes the use of fewer and more sustainable materials, minimising the reduction of natural resources and minimizing waste (Mao et al., 2018). This can include choosing recycled or renewable materials and designing products to use less energy and water. Researchers are working to develop biodegradable polymers that can degrade naturally in the environment. Minimise the effect of plastic waste material on ecosystems. The optimum number of recycled plastics in manufacturing should utilize that helps to minimise the demand for virgin plastic and also transforming waste materials into new products with optimum value of product. Researchers are working on grapheme and other nanomaterials for sustainable applications as water purification, and lightweight, sustainable construction materials. Innovations in bio-based materials include the development of alternatives to conventional materials as bio-based adhesives, resins, and composites which are more sustainable. The essential filler and Additive can be utilized for accurate production with minimal waste. The Closed-loop systems recycle waste and by-products recycle and minimise the need of raw materials and reduces waste (Kara et al., 2022). This policy can be applied in industries like textiles, electronics, and automotive manufacturing. Innovations in manufacturing processes focus on minimising carbon emissions through energy-efficient technologies, such as electric arc furnaces in steel production or low-temperature curing in the chemical industry. Companies are incorporating renewable energy resources in manufacturing processes to reduce dependency on conventional resource of energy (Gurtu et al., 2016).

## 3.3 Resource Management

The Effective management of resources are essential for achieving the aim of the Green Transition (Sharma et al., 2021). By adopting sustainable practices in utilization of water and energy and material efficiency would lead the green or circular economy. Simultaneously, waste management techniques, such as concept of reuse, recycling, and energy recovery, are important for minimizing waste, reducing pollution, and support a circular economy (Sankaran, 2019). In the context of the Green Transition, effective resource management and waste management techniques are leads to promote sustainability (Behera, 2023). These techniques help conserve resources, reduce waste, and upkeep to change over circular economy. Practices such as selective logging, reduced-impact logging, and reforestation ensure to preserve ecosystems and biodiversity (Gardner, 2010).

## 3.4 Waste Prevention and Minimization

Implementing practices that reduce waste generation at the source, such as designing products with fewer materials, improving manufacturing processes, and promoting efficient resource use, helps minimize waste Products are designed with a focus on reducing waste, using fewer materials, and increasing the potential for reuse and recycling (Reike et al., 2018). This method supports waste prevention by considering the entire product lifecycle. Encouraging the reuse of products extends their lifecycle and reduces the need for new materials. This can include practices like donating, reselling, or repurposing used items. Remanufacturing involves restoring used products to like-new condition, often upgrading them in the process (Köpman and Majava, 2024). This method is common in industries such as automotive, electronics, and heavy machinery. Recycling involves processing unused materials into useful products such as recycled materials include metals, paper, plastics, waste and other discarded products (Ponte, 2015). This technique not only recycles materials but also reduces the need for traditional mining. Anaerobic digestion is a biological process that breaks down organic waste in the absence of oxygen, producing biogas (a renewable energy source) and a nutrient-rich fertilizer. Manufacturers take back used products for recycling or remanufacturing (Awan et al., 2022). This practice encourages consumers to return products rather than discard them, supporting material recovery. Industrial symbiosis involves the exchange of waste materials or by-products between industries, where the discarded of one process becomes the feedback for another. This approach reduces waste and maximizes resource efficiency. CE, initiate to apply discarded from landfills include encouraging recycling, composting, and production of energy from waste (Xiao et al., 2020). Diverting waste reduces the environmental impact of landfills and conserves Methane an effective greenhouse gas, is generated by the decomposition of organic waste in landfills. Capturing and utilizing methane as an energy source reduces greenhouse gas emissions and generates renewable energy (Liang et al., 2022).

## *3.5 Circular Economy Practices*

Implementing policies and practices that minimise surplus generation, promote recycling, and encourage the reprocess of materials. Designing products for durability, repair ability, and recyclability to keep them in use longer and reduce the need for new resources (Ljungberg, 2007). By optimizing the design process, products can be created to produce fewer emissions during manufacturing, use, and disposal, contributing to the reduction of greenhouse gases and other pollutants. Eco-design focuses on creating durable products that last longer, reducing the need for frequent replacements. The sharing economy and collaborative consumption are economic models that emphasize the sharing, renting, and collaborative use of goods and services (Selloni and Selloni, 2017). Instead of owning products, consumers share access to goods, services, and resources, often facilitated by digital platforms. These models are integral to the Green Transition as they promote resource efficiency, reduce waste, and foster community engagement. The sharing economy allows individuals and businesses to maximize the use of underutilized assets, such as cars, homes, tools, and equipment. This reduces the need for each person to own a product, leading to more proficient application of natural resources. By sharing goods, the overall demand for new products decreases, which in turn minimise the environmental impact associated with manufacturing, transportation, and disposal (Ghisellini et al., 2016).

## 4 Case Studies and Best Practices from Various Industries

Real-world examples from industries like electronics, fashion, automotive, food and beverage, and construction demonstrate the potential of circular economy practices to reduce waste, conserve resources, and create sustainable business models. The lessons learned from these examples highlight the importance of consumer engagement, collaboration, and integration into core business strategies. For circular economy practices to scale successfully, there must be a focus on regulatory support, changing consumer behaviour, economic viability, and investment in technology and infrastructure. As these practices expand globally, they have the potential to transform industries and contribute significantly to the Green Transition.

### *4.1 Electronics Industry*

Fair phone is a Dutch company known for its modular smartphones, designed to be easily repairable and upgradable (Rafi et al., 2024). The company focuses on using ethically sourced materials, promoting fair labour practices, and extending the product life cycle. Fair phone's modular design allows change components as battery or camera without needing to replace the entire phone. The company also offers repair guides and spare parts, encouraging customer to repair rather than replace their devices. The success of Fair phone demonstrates the viability of modular design in electronics, which can significantly reduce electronic waste and extend product life. It also shows the importance of transparency in sourcing and manufacturing processes, which resonates with consumers. Modular design can be scaled across other electronic products, such as laptops and tablets. However, it requires a shift in industry standards, consumer education, and the availability of repair services.

### *4.2 Fashion Industry*

Patagonia, an outdoor clothing brand, has been a pioneer in promoting sustainability and circular economy principles (Rattalino, 2018). The company emphasizes quality, reparability, and responsible sourcing. Patagonia offers a repair program called Worn Wear, where customers can send in their damaged garments for repair. The company also promotes to sale again Patagonia products through its Worn Wear online store. Patagonia's success exemplifies that consumers are enthusiastic to pay a premium for durable, sustainable products, and that offering repair and resale options can enhance brand loyalty. It also highlights the importance of educating consumers about sustainability. Repair and resale programs can be scaled across the fashion industry, particularly in the high-end and outdoor clothing sectors. However, scalability may be challenged by fast fashion's low-cost, high-volume model, which prioritizes quick turnover and disposable clothing (Kew, 2023).

### *4.3 Automotive Industry*

Renault, a global car manufacturer, has implemented circular economy practices in its manufacturing and after-sales services (Rashid et al., 2020). The company focuses on recycling, remanufacturing, and extending the life of automotive components. Renault operates a remanufacturing plant in France where used car parts, such as engines and gearboxes, are refurbished and sold as certified pre-owned parts. The company also promotes vehicle take-back programs and material recycling. Renault's approach shows that remanufacturing and recycling can be profitable while reducing environmental impact (Ayres et al., 1997). It also demonstrates the importance of integrating circular economy practices into the core business model rather than treating them as peripheral activities. Remanufacturing and recycling can be scaled across the automotive industry, particularly as regulatory pressures increase and consumer demand for sustainable vehicles grows. However, scalability may require investment in new technologies and infrastructure.

### *4.4 Food and Beverage Industry*

Loop, an initiative by Terra Cycle, is a circular shopping platform that provides consumers with products in reusable packaging (Grace, 2019). Where the customer used the product once, the packaging is returned, cleaned, and refilled for reuse. Circle partners with major brands, such as Unilever and Nestlé, to offer a wide range of products in durable, reusable packaging. Consumers pay a deposit for the packaging, which is refunded when the packaging is returned. Loop's model demonstrates the feasibility of reusing packaging in consumer goods and highlights the importance of collaboration between brands, retailers, and logistics providers (Accorsi et al., 2020). It also shows that consumers are willing to adopt new shopping habits when offered a convenient and sustainable alternative. The Loop model can be surmounted in every urban areas where logistics networks might be implemented.

### 4.5 Construction Industry

CE principles in the construction and interior design industry must focuses on minimising waste, using recycled materials. Interface's products like carpet tiles made from recycled materials that can be easily replaced and reducing waste (Accorsi et al., 2020). The company should also offers a take-back program where used carpet tiles are collected and recycled into new products for future application. Interface's approach shows that designing for disassembly and using recycled materials can be successful apply in the construction industry. It also highlights the importance of corporate commitment to sustainability at all levels of the organization. CE concept in the construction trade can be implemented through the adoption of modular designs, the use of sustainable materials, and the implementation of take-back programs (Sáez-de-Guinoa et al., 2022). However, scalability and implementation parameter may require changes in industry design like standards building codes, and procurement practices (Abbott and Fisher, 2009).

## 5 Challenges and Future Directions

### 5.1 Economic and Financial Barriers

Shifting to circular business models often requires considerable open capital investment. This may include costs for infrastructure, new technologies, research and development, and workforce training (Rizos et al., 2016). Developing products that should follow circular economy principles, such as modular or recyclable products, can incorporate higher design and manufacturing costs. The financial benefits of circular economy practices as cost savings from resource efficiency or waste reduction, may take years too benefited. This long payback period can discourage companies from making the necessary investments, especially in industries with tight margins or uncertain market conditions (Moya and Pardo, 2013). Investors and companies may be unclear to finance in CE for initiatives due to apparent risks, such as the doubt of consumer demand, regulatory changes, or technological outmodedness. This risk aversion can reduce the adoption of CE and GT practices. Implementing circular economy practices often requires new or advanced infrastructure, such as waste management abilities, renewable energy plants, or reverse logistics systems. The cost of developing or retrofitting infrastructure to upkeep CE and GT can be substantial, especially in regions with out-of-date systems or limited resources. Accepting new technologies, such as advanced recycling techniques, energy-efficient machinery, digital platforms for tracking materials, can be expensive. Companies may also face challenges related to technology transfer, integration, and maintenance. Circular economy products and services to be economically feasible, there must be sufficient market demand (Korhonen et al., 2018). However, consumers may be unenthusiastic to pay a first-rate for sustainable products, especially if they are not fully aware of the environmental benefits or if cheaper. While the shifting to a Circular Economy and Green Transition offers significant environmental and social benefits, it also presents considerable economic and financial barricades. High initial costs, long payback periods, admittance to financing, market demand, and regulatory challenges all show a role in reducing the adoption of sustainable practices (Ghisetti et al., 2017). However, with the right support from governments, financial institutions, and consumers, these barricades can be overcome, by providing financial incentives, such as tax breaks, subsidies, and grants, to support circular economy and green transition projects. Regulatory agendas that encourage sustainable practices, such as extended producer responsibility (EPR) or carbon pricing, can also drive adoption. The innovative financing mechanisms, such as green bonds, impact investing, and blended finance attract capital to circular economy projects. Financial institutions can also create custom-made products for SMEs, such as low-interest loans or credit guarantees for circular economy ventures. Companies can use marketing movements, eco-labels, and transparency initiatives to educate consumers and build trust for circular products. Association between businesses, governments, NGOs, and academia can help overwhelmed supply chain challenges, share best practices, and develop common standards. PPP, Public–private partnerships can also take initiative in circular economy (Roos Lindgreen et al., 2022).

## 5.2 *Technical and Logistical Challenges*

Circular Economy (CE) and Green Transition (GT) involves significant technical and logistical challenges (Vassileva, 2022). These challenges can mark as technology gaps among infrastructure needs and hurdles related sustainable practices. Below is a discussion of these issues. Traditional supply chains are designed for a linear flow of materials—from production to consumption to disposal. CE requires the development of reverse logistics systems to collect, sort, refurbish, remanufacture, or recycle products after use. Setting up these systems is complex and resource-intensive, often requiring new technologies and partnerships. Managing the transfer of materials, tracking materials through various life stages, ensuring their quality, and maintaining the integrity of recycled or reused materials. This complexity can lead to inefficiencies and increased costs if not properly managed **p**roducts designed for a linear economy are often not made with disassembly, reuse, or recycling in mind. Transitioning to CE requires innovation in product design to ensure that products can be easily disassembled, repaired, or recycled (Sankaran, 2019). This often demands new design methodologies, materials, and technologies. Finding materials that are both sustainable and compatible with existing production processes can be challenging. Some sustainable materials may not perform as well or may require significant adjustments to manufacturing processes, which can complicate adoption. Effective circular economy practices rely on the ability to track products and materials throughout their lifecycle. This requires robust data management systems, including the use of digital technologies such as block chain, the Internet of Things and advanced analytics. However, implementing these technologies across complex global supply chains can be technically challenging and costly. There is a lack of standardized methods for tracking and reporting on material flows and the environmental impact of products. Without standardization, it is difficult to compare data, ensure compliance, and scale circular economy practices across industries (Roos Lindgreen et al., 2022). Current recycling technologies are often inefficient or incapable of processing certain materials, especially complex or composite materials. For example, while metals and glass are relatively easy to recycle, plastics, textiles, and electronic waste present significant challenges due to their diverse compositions. Adulteration in the waste stream can reduce the quality of recycled materials, making them less viable for reuse in manufacturing (Kerdlap et al., 2019). This problem is exacerbated by inadequate sorting technologies and systems, which struggle to separate materials accurately. Some circular economy practices, such as recycling or remanufacturing, can be energy-intensive. For example, melting down metals for recycling or processing waste materials into new products can require significant energy inputs, which may offset some of the environmental benefits if not managed with renewable energy sources. Shifting to a green transition requires not only the acceptance of renewable energy sources but also their incorporation into existing energy grids and industrial processes. This requires advancement in infrastructure, developing storage solutions, and ensuring the consistency of renewable energy supply, all of which present technical challenges (Fernandez Mendoza and Ibarra, 2022).

## *5.3 Social and Cultural Barriers*

For long period of time, societies have "use and through" mind-set has led to deeply rooted consumption patterns that given facilitation to, newness, and proprietorship. Switching consumer behaviour from this linear model to CE one that values durability, reuse, and sharing is a significant challenge (Barbu et al., 2018). Many consumers have a strong liking for new products over second-hand or renewed goods. This desire for newness is often focussed by social status, fashion drifts. Convincing consumers to squeeze products that are recycled, remanufactured can be difficult, especially when new products are readily available and inexpensive. Circular economy exercises, such as recycling, product repair, or contribution in sharing platforms, may be supposed as inconvenient or time-consuming. This perception can frighten consumers from adopting more sustainable conducts, particularly in fast-paced, urban environments where accessibility is highly valued (Kim et al., 2020). In many cultures, possessing products, especially high-value items like cars, homes, or electronics, is related with social status and success. The concept of sharing, central to CE models like Product-as-a-Service (PaaS), can crash with cultural norms that value proprietorship as a sign of wealth or achievement. Societal opposition to change can be a significant barricade to adopting new practices. People are often resistant to changing their conducts, especially when it involves adopting new technologies, moving toward consumption patterns, or participating in unfamiliar business models like circular supply chains. Consumers may be unconvinced of the quality and safety of recycled, reprocessed, or second-hand products. There can be a lack of confidence in the durability, reliability, or concert of these products, which can hamper their acceptance and adoption. Sustainable products and services are often perceived as more expensive than their traditional complements (Ottman et al., 2006). Even when this perception is not accurate, it can affect consumer choices, leading them to opt for cheaper, less sustainable opportunities. Additionally, consumers may not fully grow the long-standing value, repairable, or reusable products. There can be a gap between the values of sustainability and those of consumers. For example, consumers may prioritize convenience, low cost, or brand eminence over environmental impact. This value gap can make it difficult to stimulate ecological products and facilities in the market place Social and cultural barriers, due to lack of education and awareness, present significant challenges to the adoption of Circular Economy and Green Transition repetitions (Sunny et al., 2020). Speak to these barriers requires a multidimensional methodology that includes raising attentiveness, educating customer. By shifting societal customs and values toward sustainability, and by making sustainable liking more reachable and appealing, it is possible to overcome these barriers and indorse the widespread adoption of CE and GT practices.

## *5.4 Role of Stakeholders in Circular Economy and Green Transition*

The shifting CE to Green Transition require the active participation of multiple shareholders, including businesses, governments, and civil society. Each group carries unique strengths and viewpoints to the table, and their collective efforts are essential for overcoming the challenges related with sustainable development (van Langen et al., 2021). By working together in a coordinated and transparent manner, stakeholders can get-up-and-go the systemic changes needed to improve sustainable and resolute future. Companies can innovate by generating new business models, such as Product-as-a-Service or sharing platforms, which based on the concept of CE. By inserting CE into their operations, companies can create value not only for themselves but also for their suppliers, customers, and communities. Companies are giving importance of corporate social responsibility (CSR) and sustainability reporting (Amran et al., 2014). By measuring and publicly reportage on their environmental influence, businesses can validate their promise to sustainability and hold themselves responsible to investors. This transparency can build trust with consumers, investors, and regulators. Governments can pass essential laws and regulations that inspire sustainable practices, such as extended producer responsibility (EPR), waste reduction targets, and incentives for renewable energy acceptance. Governments can also implement carbon pricing, which incentivizes businesses to reduce their greenhouse gas emissions. Governments can give financial support for the growth of circular infrastructure, such as renewable energy projects, and public transportation systems. Public investment in R&D, can also shoot modernisation in sustainable technologies and materials (Sarpong et al., 2023). By funding and facilitating infrastructure development, governments can remove barriers to the acceptance of circular practices. Governments have the ability to lead public education and awareness drives that stimulate the repayments of the CE and GT. Through educational initiatives, governments can help change consumer behaviour, encourage sustainable intake, and build support for environmental policies. Private and public sector, play a important role in advocating for environmental sustainability and holding businesses These organizations can raise public focus about the importance of CE and GT, influence policy decisions, and organise communities to take action (Charnovitz, 2012).

## 6 Emerging Trends in Circular Economy (CE) and Green Transition (GT)

### *6.1 Circular Business Models*

Companies are increasingly adopting "product-as-a-service" models, where products are leased or rented instead of sold. This encourages companies to design durable, repairable products, reducing waste and promoting the circular economy the sharing economy, characterized by sharing or renting of goods and services, is gaining momentum (Kauffman and Naldi, 2020). Examples include car-sharing services, shared office spaces, and peer-to-peer lending platforms. Cities are becoming focal points for circular economy initiatives, with a attention on sustainable urban planning, waste management, and resource competence. Concepts like zero-waste cities, circular urban systems, and decentralized energy systems are gaining traction as cities seek to reduce their environmental footprint. The application of different nonconventional energy systems, such as solar plates, micro grids, and energy storage solutions, is supporting helping the green transition in urban areas (Walls grove et al., 2021). The issuance of green bonds, which finance projects with environmental benefits, is on the rise. These links are used to fund renewable energy projects, green buildings, and sustainable infrastructure, supporting the green transition. Environmental, Social, and Governance (ESG) dealings are increasingly stimulating investment decisions (Sultana et al., 2017). Shareholders are seeking to align their selections with sustainability goals, driving capital towards companies and projects that line up circular economy and green transition practices.

### 6.2 Sustainable Materials and Eco-Design

It is critical for monitoring and augmenting resource use, minimizing waste, and increasing the sustainability of supply chains. Block chain technology is being used to enhance directness and trace ability in supply chains (Sunny et al., 2020). By providing absolute records of product origins, materials used, and recycling processes, block chain supports circular economy practices by ensuring responsibility and fostering consumer trust in sustainable products: Innovations in materials science are foremost to the develop of biodegradable plastics, recyclable composites, and bio-based materials. These materials are designed to be easily reused, recycled, or decomposed, bring into line with circular economy principles. There is a growing emphasis on eco-design, where products are designed with their entire lifecycle in mind, from raw material extraction to end-of-life disposal. This approach aims to minimize environmental impact by promoting durability, reparability, and recyclability (Maitre-Ekern and Dalhammar, 2016). Many companies are obligating to net-zero carbon emissions by specific target dates, typically 2050 (Erb et al., 2022). These pledges are driving funds in renewable energy and carbon capture technologies. The putting into practice of carbon taxes and cap-and-trade systems, is gaining energy globally. These mechanisms incentivize concerns to minimise their carbon emission and promote the circular economy (CE) and green transition (GT) practices (Hoffman, 2007).

## 7 Conclusion

In conclusion, the circular economy (CE) and green transition (GT) represent transformative approaches to sustainability, offering alternatives to traditional linear economic models. Rooted in principles of resource efficiency, waste reduction, and innovation, CE and GT aim to create regenerative systems that balance economic growth with environmental stewardship. By integrating design innovation, sustainable practices, and renewable energy, these models provide a framework for addressing global environmental challenges. Successful case studies highlight the potential impact of these approaches, yet challenges remain, including economic, technical, and social barriers. Future advancements will require continued innovation, policy support, and a collaborative effort to overcome obstacles and drive sustainable development forward.

## References

Abbasi, K. R., Zhang, Q., Ozturk, I., Alvarado, R., & Musa, M. (2024). Energy transition, fossil fuels, and green innovations: Paving the way to achieving sustainable development goals in the United States. *Gondwana Research, 130*, 326–341.

Abbott, M. L., & Fisher, M. T. (2009). *The art of scalability: Scalable web architecture, processes, and organizations for the modern enterprise*. Pearson Education.

Accorsi, R., Baruffaldi, G., & Manzini, R. (2020). A closed-loop packaging network design model to foster infinitely reusable and recyclable containers in food industry. *Sustainable Production and Consumption, 24*, 48–61.

Agovino, M., & Musella, G. (2023). *EU waste regulation in a linear-circular economy transition: Waste management in Italy*. Springer Nature.

Alsarhan, L. M., Alayyar, A. S., Alqahtani, N. B., & Khdary, N. H. (2021). Circular carbon economy (CCE): A way to invest CO2 and protect the environment, a review. *Sustainability, 13*(21), 11625.

Amran, A., Lee, S. P., & Devi, S. S. (2014). The influence of governance structure and strategic corporate social responsibility toward sustainability reporting quality. *Business Strategy and the Environment, 23*(4), 217–235.

Apostu, S. A., Gigauri, I., Panait, M., & Martín-Cervantes, P. A. (2023). Is Europe on the way to sustainable development? Compatibility of green environment, economic growth, and circular economy issues. *International Journal of Environmental Research and Public Health, 20*(2), 1078.

Awan, U., Gölgeci, I., Makhmadshoev, D., & Mishra, N. (2022). Industry 4.0 and circular economy in an era of global value chains: What have we learned and what is still to be explored? *Journal of Cleaner Production, 371*, 133621.

Ayres, R., Ferrer, G., & Van Leynseele, T. (1997). Eco-efficiency, asset recovery and remanufacturing. *European Management Journal, 15*(5), 557–574.

Barbu, C. M., Florea, D. L., Ogarcă, R. F., & Barbu, M. C. (2018). From ownership to access: How the sharing economy is changing the consumer behavior. *Amfiteatru Economic, 20*(48), 373–387.

Behera, D. K. (2023). Promoting sustainable development through environmental policy, green technologies, and effective waste management: A comprehensive review. *Journal of Multidisciplinary Science: MIKAILALSYS, 1*(2), 179–198.

Bradbrook, A. J. (2001). The development of a protocol on energy efficiency and renewable energy to the United Nations framework convention on climate change. *New Zealand Journal of Environmental Law, 5*, 55.

Bradshaw, A. M., & Hamacher, T. (2012). Nonregenerative natural resources in a sustainable system of energy supply. *Chemsuschem, 5*(3), 550–562.

Bressanelli, G., Adrodegari, F., Perona, M., & Saccani, N. (2018). Exploring how usage-focused business models enable circular economy through digital technologies. *Sustainability, 10*(3), 639.

Cárceles Rodríguez, B., Durán-Zuazo, V. H., Soriano Rodríguez, M., García-Tejero, I. F., Gálvez Ruiz, B., & Cuadros Tavira, S. (2022). Conservation agriculture as a sustainable system for soil health: A review. *Soil Systems, 6*(4), 87.

Charnovitz, S. (2012). Accountability of non-governmental organizations in global governance. In *NGO accountability* (pp. 21–42). Routledge.

Erb, T., Perciasepe, B., Radulovic, V., & Niland, M. (2022). Corporate climate commitments: The trend towards net zero. *Handbook of climate change mitigation and adaptation* (pp. 2985–3018). Springer International Publishing.

Fernandez Mendoza, J. M., & Ibarra, D. (2022). Circular economy business models and technology management strategies in the wind industry: Sustainability potential, industrial challenges and opportunities.

Gardner, T. (2010). *Monitoring forest biodiversity: Improving conservation through ecologically-responsible management*. Routledge.

Ghisellini, P., Cialani, C., & Ulgiati, S. (2016). A review on circular economy: The expected transition to a balanced interplay of environmental and economic systems. *Journal of Cleaner Production, 114*, 11–32.

Ghisetti, C., Mancinelli, S., Mazzanti, M., & Zoli, M. (2017). Financial barriers and environmental innovations: Evidence from EU manufacturing firms. *Climate Policy, 17*(sup1), S131–S147.

Ghorbani, Y., Zhang, S. E., Nwaila, G. T., Bourdeau, J. E., & Rose, D.H., 2023. Embracing a diverse approach to a globally inclusive green energy transition: Moving beyond decarbonisation and recognising realistic carbon reduction strategies. *Journal of Cleaner Production*, 140414.

Gorbushina, A. (2023). *Platform for accelerating the circular economy.*

Gössling, S., & Scott, D. (2018). The decarbonisation impasse: Global tourism leaders' views on climate change mitigation. *Journal of Sustainable Tourism, 26*(12), 2071–2086.

Gottberg, A., Morris, J., Pollard, S., Mark-Herbert, C., & Cook, M. (2006). Producer responsibility, waste minimisation and the WEEE Directive: Case studies in eco-design from the European lighting sector. *Science of the Total Environment, 359*(1–3), 38–56.

Grace, R. (2019). In the crosshairs: Single-use, disposable packaging: New loop circular platform promotes reusable consumer product packaging. *Plastics Engineering, 75*(4), 18–23.

Gurtu, A., Searcy, C., & Jaber, M. Y. (2016). A framework for reducing global manufacturing emissions. *The Journal of Environment & Development, 25*(2), 159–190.

Hassan, Q., Viktor, P., Al-Musawi, T. J., Ali, B. M., Algburi, S., Alzoubi, H. M., Al-Jiboory, A. K., Sameen, A. Z., Salman, H. M., & Jaszczur, M. (2024). The renewable energy role in the global energy Transformations. *Renewable Energy Focus, 48*, Article 100545.

Hoffman, A. J. (2007). *Carbon strategies: How leading companies are reducing their climate change footprint*. University of Michigan Press.

Horbach, J., Rennings, K., & Sommerfeld, K. (2015, November). Circular economy and employment. In *3rd IZA Workshop: Labor market effects of environmental policies* (vol. 1, p. 39).

Iwata, H., & Okada, K. (2014). Greenhouse gas emissions and the role of the Kyoto Protocol. *Environmental Economics and Policy Studies, 16*, 325–342.

Jacob-John, J., D'Souza, C., Marjoribanks, T., & Singaraju, S. (2021). Synergistic interactions of SDGs in food supply chains: A review of responsible consumption and production. *Sustainability, 13*(16), 8809.

Kara, S., Hauschild, M., Sutherland, J., & McAloone, T. (2022). Closed-loop systems to circular economy: A pathway to environmental sustainability? *CIRP Annals, 71*(2), 505–528.

Kauffman, R. J., & Naldi, M. (2020). Research directions for sharing economy issues. *Electronic Commerce Research and Applications, 43*, Article 100973.

Kerdlap, P., Low, J. S. C., & Ramakrishna, S. (2019). Zero waste manufacturing: A framework and review of technology, research, and implementation barriers for enabling a circular economy transition in Singapore. *Resources, Conservation and Recycling, 151*, Article 104438.

Kew, S. J. (2023). *The path to net zero for the fashion industry: Five strategies for decarbonisation.* Taylor & Francis

Kim, S., Filimonau, V., & Dickinson, J. E. (2020). The technology-evoked time use rebound effect and its impact on pro-environmental consumer behaviour in tourism. *Journal of Sustainable Tourism, 28*(2), 164–184.

Kirchherr, J., Yang, N. H. N., Schulze-Spüntrup, F., Heerink, M. J., & Hartley, K. (2023). Conceptualizing the circular economy (revisited): An analysis of 221 definitions. *Resources, Conservation and Recycling, 194*, Article 107001.

Kiryluk, A., & Kostecka, J. (2022). Sustainable development in rural areas in the perspective of a decade of ecosystem restoration. *Ekonomia i Środowisko, 4*, 117–148.

Köpman, J., & Majava, J. (2024). The role of product design in advancing the circular economy of electric and electronic equipment. *Resources, Conservation & Recycling Advances*, 200207.

Korhonen, J., Honkasalo, A., & Seppälä, J. (2018). Circular economy: The concept and its limitations. *Ecological Economics, 143*, 37–46.

Kroll, G. (2001). The"Silent Springs"of Rachel Carson: Mass media andthe origins of modern environmentalism. *Public Understanding of Science, 10*(4), 403.

Kumar, P., Kaur, D., & Kaur, A. (2023, February). Green infrastructure-a roadmap towards sustainable development. In *IOP conference series: Earth and environmental science* (vol. 1110(1), p. 012060). IOP Publishing.

Liang, X., Kurniawan, T. A., Goh, H. H., Zhang, D., Dai, W., Liu, H., Goh, K. C., & Othman, M. H. D. (2022). Conversion of landfilled waste-to-electricity (WTE) for energy efficiency improvement in Shenzhen (China): A strategy to contribute to resource recovery of unused methane for generating renewable energy on-site. *Journal of Cleaner Production, 369*, Article 133078.

Ljungberg, L. Y. (2007). Materials selection and design for development of sustainable products. *Materials & Design, 28*(2), 466–479.

Luttenberger, L. R. (2020). Waste management challenges in transition to circular economy–case of Croatia. *Journal of Cleaner Production, 256*, Article 120495.

Maitre-Ekern, E., & Dalhammar, C. (2016). Regulating planned obsolescence: A review of legal approaches to increase product durability and reparability in Europe. *Review of European, Comparative & International Environmental Law, 25*(3), 378–394.

Mao, J., Li, C., Pei, Y., Xu, L., Mao, J., Li, C., Pei, Y., & Xu, L. (2018). Eco-design. In *Circular economy and sustainable development enterprises* (pp. 223–236).

Moya, J. A., & Pardo, N. (2013). The potential for improvements in energy efficiency and $CO_2$ emissions in the EU27 iron and steel industry under different payback periods. *Journal of Cleaner Production, 52*, 71–83.

Ottman, J. A., Stafford, E. R., &Hartman, C. L. (2006). Avoiding green marketing myopia: Ways to improve consumer appeal for environmentally preferable products. *Environment: Science and Policy for Sustainable Development, 48*(5), 22–36.

Peeters, M., & Misonne, D. (2022). The European Union and its rule-creating force on the European continent for moving to climate neutrality by 2050 At the latest. In *Research handbook on climate change mitigation law* (pp. 59–102). Edward Elgar Publishing.

Ponte, V. M. (2015). *Regional perspective of recuay mortuary practices: A view from the hinterlands*, Callejon de Huaylas, Peru [Master's thesis, The University of Wisconsin-Milwaukee].

Prakash, G. (2014). An article on Swachh Bharat Abhiyana: A mission to clean India. In *Edited volume on-theme 1: Government policies since 2014* (p. 112).

Pramanik, S. K., Suja, F. B., Zain, S. M., & Pramanik, B. K. (2019). The anaerobic digestion process of biogas production from food waste: Prospects and constraints. *Bioresource Technology Reports, 8*, Article 100310.

Protocol, K. (1997). United Nations framework convention on climate change. *Kyoto Protocol, Kyoto, 19*(8), 1–21.

Rafi, S., Dia, R. R., Gazi, M. A., & Islam, Q. T. (2024). *Fairphone: The phone that protects the planet*. SAGE Publications.

Rashid, A., Roci, M., & Asif, F. M. (2020). Circular manufacturing systems. In *Handbook of the circular economy* (pp. 343–357). Edward Elgar Publishing.

Rattalino, F. (2018). Circular advantage anyone? Sustainability-driven innovation and circularity at Patagonia, Inc. *Thunderbird International Business Review, 60*(5), 747–755.

Reike, D., Vermeulen, W. J., & Witjes, S. (2018). The circular economy: New or refurbished as CE 3.0?—exploring controversies in the conceptualization of the circular economy through a focus on history and resource value retention options. *Resources, Conservation and Recycling, 135*, 246–264.

Rizos, V., Behrens, A., Van der Gaast, W., Hofman, E., Ioannou, A., Kafyeke, T., Flamos, A., Rinaldi, R., Papadelis, S., Hirschnitz-Garbers, M., & Topi, C. (2016). Implementation of circular economy business models by small and medium-sized enterprises (SMEs): Barriers and enablers. *Sustainability, 8*(11), 1212.

Roos Lindgreen, E., Opferkuch, K., Walker, A. M., Salomone, R., Reyes, T., Raggi, A., Simboli, A., Vermeulen, W. J., & Caeiro, S. (2022). Exploring assessment practices of companies actively engaged with circular economy. *Business Strategy and the Environment, 31*(4), 1414–1438.

Sáez-de-Guinoa, A., Zambrana-Vasquez, D., Fernández, V., & Bartolomé, C. (2022). Circular economy in the european construction sector: A review of strategies for implementation in building renovation. *Energies, 15*(13), 4747.

Sankaran, K. (2019). Carbon emission and plastic pollution: How circular economy, blockchain, and artificial intelligence support energy transition? *Journal of Innovation Management, 7*(4), 7–13.

Sarpong, D., Boakye, D., Ofosu, G., & Botchie, D. (2023). The three pointers of research and development (R&D) for growth-boosting sustainable innovation system. *Technovation, 122*, Article 102581.
Scheel, C. (2016). Beyond sustainability. Transforming industrial zero-valued residues into increasing economic returns. *Journal of Cleaner Production, 131*, 376–386.
Selloni, D., & Selloni, D. (2017). New forms of economies: sharing economy, collaborative consumption, peer-to-peer economy. In *Codesign for public-interest services*, pp.15–26.
Sharma, H. B., Vanapalli, K. R., Samal, B., Cheela, V. S., Dubey, B. K., & Bhattacharya, J. (2021). Circular economy approach in solid waste management system to achieve UN-SDGs: Solutions for post-COVID recovery. *Science of the Total Environment, 800*, Article 149605.
Simeonova, V., & Van Der Valk, A. (2016). Environmental policy integration: Towards a communicative approach in integrating nature conservation and urban planning in Bulgaria. *Land Use Policy, 57*, 80–93.
Sultana, S., Zainal, D., & Zulkifli, N. (2017). The influence of environmental, social and governance (ESG) on investment decisions: The Bangladesh perspective. *Pertanika Journal of Social Sciences and Humanities, 25*, 155–173.
Sunny, J., Undralla, N., & Pillai, V. M. (2020). Supply chain transparency through blockchain-based traceability: An overview with demonstration. *Computers & Industrial Engineering, 150*, Article 106895.
Telesetsky, A. (1999). The Kyoto protocol. *Ecology Law Quarterly, 26*(4), 797–813.
van Langen, S. K., Vassillo, C., Ghisellini, P., Restaino, D., Passaro, R., & Ulgiati, S. (2021). Promoting circular economy transition: A study about perceptions and awareness by different stakeholders groups. *Journal of Cleaner Production, 316*, Article 128166.
Vassileva, A. G. (2022). Green Public-Private Partnerships (PPPs) as an instrument for sustainable development. *Journal of World Economy: Transformations & Transitions, 2*(5), 2.
Wallsgrove, R., Woo, J., Lee, J. H., & Akiba, L. (2021). The emerging potential of microgrids in the transition to 100% renewable energy systems. *Energies, 14*(6), 1687.
Warren, R., Andrews, O., Brown, S., Colón-González, F.J., Forstenhäusler, N., Gernaat, D.E., Goodwin, P., Harris, I., He, Y., Hope, C., & Manful, D. (2022). Quantifying risks avoided by limiting global warming to 1.5 or 2 °C above pre-industrial levels. *Climatic Change, 172*(3), 39.
Xiao, H., Li, Z., Jia, X. and Ren, J., 2020. Waste to energy in a circular economy approach for better sustainability: a comprehensive review and SWOT analysis. Waste-to-energy, pp.23–43
Zahariev, A., Mihaylova, P.A.M., Deatcu, A.L.M.C., Ciobanu, V.C.G., Vasile, V., Saraci, Y., Gokmen, A.M., Viga, S.O., Ivanovic-Djukic, E.E.G.M., Stevanovic, M.R.S. and Talic, B.K.M., 2024. Circular Economy.
Zeng, Y., Koh, L.P. and Wilcove, D.S., 2022. Gains in biodiversity conservation and ecosystem services from the expansion of the planet's protected areas. Science Advances, 8(22), p.eabl9885
Zhu, J., Fan, C., Shi, H., & Shi, L. (2019). Efforts for a circular economy in China: A comprehensive review of policies. *Journal of Industrial Ecology, 23*(1), 110–118.

# Global Environmental Degeneration: A Catalyst for Implementing Circular Practices

**Deepa Santhosh, Hammadhu HaitherAli, G. Anjali, and P. C. Sabumon**

**Abstract** Anthropogenic activities such as industrialization, urbanization, deforestation, excessive resource consumption and irresponsible waste management have severely disrupted ecosystems. The linear economic model is unsustainable, as it fails to consider environmental and social aspects of economic development. Excessive exploitation, beyond nature's regeneration capacity combined with environmental pollution beyond its' carrying capacity has negatively impacted fragile ecosystems causing severe ecological disruptions. For instance, reservoir encroachment has led to flash floods and drought, while deforestation has resulted in increased human-animal conflicts and the spread of zoonotic diseases. All these factors have contributed to environmental degradation, affecting the planet's livability, accelerating species extinction, and intensifying climate-related disasters. Addressing these challenges requires the adoption of sustainable and responsible practices that prioritize environmental health. This chapter explores the causes and impacts of environmental degradation across various industrial sectors, highlighting the limitations of the linear economic model. It introduces the Circular Economy (CE) as a viable solution and outlines pathways for a circular transition through sustainable practices across the supply chain—from resource extraction to waste management. Additionally, the chapter examines how environmental degradation influences policy, business strategies, corporate responsibility, and consumer awareness, emphasizing the urgent need for CE adoption and the responses from governments and businesses. Ultimately, it serves as a guide for policymakers, business leaders, and environmental advocates to align their efforts with the Sustainable Development Goals (SDGs).

**Keywords** Circular economy · Environmental degradation · Linear economy · Responsible consumption · Sustainable development

D. Santhosh
School of Advanced Sciences, Vellore Institute of Technology, Chennai, India

H. HaitherAli · G. Anjali
School of Civil Engineering, Vellore Institute of Technology, Chennai, India

P. C. Sabumon (✉)
Centre for Climate Change and Environment, Vellore Institute of Technology, Chennai, India
e-mail: pcsabumon@vit.ac.in

 
P. Singh et al. (eds.), *Circular Economy and Environmental Resilience*,
https://doi.org/10.1007/978-3-031-93091-1_4

# 1 Introduction

Environmental degradation has emerged as a critical challenge, demanding immediate and decisive action in resource management and sustainable development. The world generates over 2.01 billion metric tons of municipal solid waste annually, underscoring the severe consequences of unsustainable linear economic models (World Bank, 2018). Additionally, the extraction of natural resources—both renewable and non-renewable—has surged by over 235% since 1970, reaching an estimated 104 billion tons by 2024. This overexploitation, coupled with inefficient waste disposal, has led to biodiversity loss, ecosystem deterioration, and accelerated climate change. Given the current ecological overshoot, there is an urgent need for strategic interventions. This chapter explores how Circular Economy (CE) principles—focused on reducing, reusing, and recycling—can provide a transformative solution to these environmental challenges and pave the way for a more sustainable future.

Climate change and biodiversity loss are closely interconnected. Resource extraction and agricultural activities account for over 90% of pressures on biodiversity and half of climate-related impacts (IRP, 2019). Ecosystem services play a crucial role in climate change mitigation and adaptation, offering benefits such as carbon sequestration, temperature regulation, and resilience against natural disasters. However, climate change is further destabilizing ecosystems, worsening biodiversity loss (IPBES, 2018). Since more than 50% of global GDP is moderately or heavily dependent on ecosystem services, the decline of biodiversity not only disrupts natural systems but also impacts economic stability (WEF, 2020). Furthermore, climate change-driven issues—including rising temperatures, shifting precipitation patterns, ocean acidification, and extreme weather events—exacerbate environmental and economic risks (Swiss Re, 2021).

Unlike the Linear Economy (LE), which follows a take-make-dispose approach, the CE prioritizes sustainability by reducing waste, extending product lifecycles, and regenerating natural systems. CE principles help mitigate environmental degradation by lowering resource extraction pressures, promoting sustainable production, and supporting ecosystem restoration (Ellen MacArthur Foundation, 2020). By improving resource efficiency and prolonging material lifecycles, CE reduces reliance on virgin materials and fosters a more balanced supply–demand dynamic.

Policy frameworks play a critical role in driving the transition to a CE. Challenges such as technological limitations, lack of expertise, infrastructure gaps, and public resistance can be addressed through well-designed policies and collaborative initiatives. Research highlights the interconnectedness of climate, biodiversity, and economic systems, emphasizing the need for holistic policy approaches (Borges et al., 2021; Paleari, 2024).

Failure to recognize these interdependencies—or implementing fragmented solutions focused on short-term economic gains—risks unintended consequences. Shifting from a LE to a CE is essential to halting environmental degradation and ensuring long-term sustainability. This transition requires reducing resource consumption, restoring ecosystems, and mitigating biodiversity loss and climate change (Dhyani et al., 2021).

This chapter examines the root causes and consequences of environmental degradation, exploring how the linear model has contributed to climate change, biodiversity loss, deforestation, pollution, and resource depletion. It also discusses barriers to CE adoption across various sectors and outlines CE principles in detail, showcasing industry-specific strategies for sustainable transformation. Finally, the chapter analyzes how environmental, and climate crises have driven policy changes and highlights businesses and organizations leading the shift toward sustainability. This chapter provides policymakers and industry stakeholders with a comprehensive understanding of the environmental consequences of unsustainable practices. It underscores the urgency of transitioning to a CE to mitigate climate change, enhance resilience, and promote socio-economic benefits. The discussion aligns with key Sustainable Development Goals (SDGs), particularly: SDG 12 (Responsible Consumption and Production): Advocating waste reduction, resource efficiency, and sustainable production; SDG 13 (Climate Action): Addressing climate change and its direct impact on biodiversity; SDG 15 (Life on Land): Highlighting the importance of biodiversity protection and ecosystem conservation and SDG 17 (Partnerships for the Goals): Emphasizing the role of collaboration in achieving sustainability.

## 2 Understanding Environmental Degeneration

### *2.1 Definition*

As per the Oxford definition environmental degeneration refers to the "*Depletion or destruction of a potentially renewable resource such as air, water, soil, forest, or wildlife, by using it at a rate faster than it can be naturally renewed*" (Environmental Degradation, 2018). Anthropogenic activities such as industrialization, deforestation and excessive exploitation and use of natural resources are the major causes of environmental degeneration. Additionally, it is important to understand that environmental degradation is a global phenomenon that is creating impacts from polar ice caps to tropical forests.

### *2.2 Causes and Consequences*

#### 2.2.1 Resource Depletion

Resource depletion occurs when natural resources are consumed at a quicker pace than the rate of replenishment (Azwardi Andaiyani et al., 2023). According to the UNEP, pollution, biodiversity loss and climate change are mainly driven by resource exploitation (UNEP, 2019). Population growth and economic development have fueled natural resource consumption in both developed and developing countries (Nawaz et al., 2019). Fossil fuel extraction has risen from six to fifteen billion tons, biomass consumption has increased from 9 to 24 billion tons and the mineral resource depletion increased fivefold. Moreover, unsustainable practices such as land fragmentation, over-exploitation, deforestation and intensive farming further worsen environmental degradation (Maja & Ayano, 2021).

It is also important to understand that the world is facing not only resource depletion but also disparity in sharing those resources. According to the UN, the material footprint of high-economic countries is 13 times higher than the low economic and 60% higher than the upper middle-income countries. Every day, over 800 million people do not get enough food and nutrition, while 650 million are obese. Hence, sustainable development stresses not only environmental factors but also social equity. The rising demand pressure on the food supply chain is adding more stress to food production which is already facing other issues such as soil depletion, diminishing water resources, declining pollinator's population and climate change. It is found that exploitation and habitat loss due to agriculture are responsible for 80% of bird and mammal extinction. In such a case, it is alarming to realize that the demand for food will further increase by 70% by 2050 (OECD, 2019; Population Matters, 2024).

Similarly, the built sector is the largest extractor of mineral resources and contributes 37% of global emissions. It is responsible for 62% of energy consumption and 55% of GHG emission. Waste generated by the construction sector comprises 30–65% of global solid waste occupying 35% of landfills. Likewise, each industry has a significant environmental impact concerning resource consumption, pollution and waste.

### 2.2.2 Deforestation

Forests cover one-third of the land surface of earth, acting as terrestrial carbon sinks reducing greenhouse gas emissions (Kumar et al., 2022). It also offers a huge variety of ecosystem goods and services including nutrient cycling, climate regulation, biodiversity conservation, etc. (Brockerhoff et al., 2017). Deforestation at the pace of 4.7 million hectares per year leads to loss of forest cover, destruction of ecosystem services and biodiversity loss. An increase in the release of GHG and depletion of forest resources drives climate change. It also leads to loss of habitat, diminishing water quality and air pollution. Conversion of forest land for agricultural activities and extraction of wood for cooking, product manufacturing and construction are some of the causes of deforestation. Expansion of ecological footprints, change in land use and land cover are found to be the major causes of deforestation due to population rise. Croplands are mainly targeted during urban sprawl and improvement of urban–rural mobility. The urban sprawl, to accommodate the increasing population further leads to encroachment of water bodies and forest land causing environmental degradation leading to loss of catchment area, groundwater depletion, worsening of flood, spread of zoonotic diseases and human-animal conflict. It is estimated that cropland displacement has directly led to loss of 32.4 Mha of global forest land (Li et al., 2024). Hence, the demand for resources and encroachment lead to deforestation.

### 2.2.3 Pollution

The addition of any unwanted material into the environment due to anthropogenic activities that lead to undesirable alterations in the ecology and environment is called pollution (Team, 2023). With the rise in population, the intensity of anthropogenic activities increases. Increasing consumerism such as fast fashion and e-commerce is leading to mountain piles of waste (Proindia.net, 2023). The improper disposal of waste from various economic activities such as manufacturing, tourism, healthcare and municipal solid waste leads to severe degradation of soil, water and air. Globalization and industrialization further lead to the emergence of new pollutants from pharmaceutical wastes, personal care products and pesticides., which are highly hazardous to the environment and human health (Mishra et al., 2023). Heavy metals and microplastics are also acting as severe threats to biotic and abiotic environments (Ziani et al., 2023). About 22% of the global burden of illness and 23% of fatalities is due to exposure to environmental toxins (Shetty et al., 2023). Moreover, environmental pollution is not a local issue as the transboundary movement of pollutants impacts on a global scale (Ukaogo et al., 2020).

GHG emission is a major parameter to measure the environmental impact of different sectors or activities. The GHG emission of different sectors around the world is shown in Fig. 1.

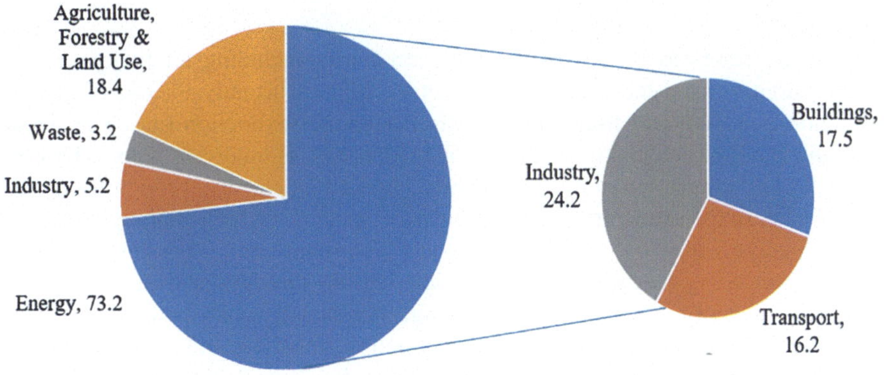


**Fig. 1** Sector-wise Greenhouse gas emissions (adapted from Hannah et al. (2020))

### 2.2.4 Waste Generation

The volume, type and complexity of waste increases with an increase in economic development. It also varies among the developed and developing countries. By 2050, the globe is expected to generate 3.4 billion tons of municipal waste (World Bank, 2018). The waste management system is still in primitive stages in third world countries, which is also used as dumping ground of e-waste and hazardous waste by developed countries. Dumping of waste without proper treatment has led to severe environmental damage and health impacts. For instance, mismanaged organic wastes act as breeding grounds for rodents, insects and reptiles leading to the spread of infections. Marine littering and land-based sources ending in ocean severely impact the marine ecosystem and its biotic environment. Mismanagement of solid waste such as uncollected organic wastes, open dumps, open burning, unscientific landfills and uncontrolled incineration leads to fire, explosion, GHG emission, air pollution, spread of infections and soil and water contamination. Its health impacts range from allergies, respiratory ailments and reproductive illnesses to cancer (Abubakar et al., 2022). Hence, it is crucial to adopt sustainable and scientific waste management to prevent resource wastage, promote CE and reduce environmental and human health impacts.

In Europe, the plastic industry demands around 50 million tons annually, generating a turnover of nearly 300 billion Euros. About 40% of this demand is for packaging, with the remaining used in various applications. Despite efforts, approximately 27% of plastic waste still ends up in landfills, leading to lost energy and materials. To address this, the Waste Framework Directive has aimed to achieve 50% domestic waste collection through system improvement as wastes act as potential raw material sources at the end of life (Baldassarre et al., 2020; IBM, 2024).

### 2.2.5 Biodiversity Loss

Biodiversity refers to the diversity of life on Earth, including all living beings, genetic variations and ecosystems (Borges et al., 2021). Currently, the rate of species extinction is a thousand times higher than the natural extinction rate. The linear economic practice is found to be the reason behind 90% of biodiversity loss (Dhyani et al., 2021). Human activities significantly impact biodiversity through urbanization, farmland expansion, introduction of invasive species and biotic homogenization. These changes lead to habitat loss, ecological disturbance and fragmentation, with deforestation being a major threat in tropical regions that harbor over 50% of the world's biodiversity. Although climate change significantly contributes to biodiversity loss, other factors, including habitat destruction, unsustainable resource use and the introduction of non-native species, also drive the sharp decline in biodiversity. This confluence of factors is contributing to what some scientists are calling the sixth mass extinction event, largely driven by human activity (Zattara & Aizen, 2021).

### 2.2.6 Climate Change

Climate change represents a defining challenge of this century. Rising global surface temperatures are fueling a greater incidence of natural disasters, particularly those stemming from hydrological and meteorological anomalies. Present levels of carbon dioxide ($CO_2$) in the atmosphere are unprecedented in the past two million years, while methane and nitrous oxide concentrations have also reached peak levels not seen in the last 800,000 years. Human activities, especially resource extraction, energy generation and agricultural practices, are major drivers of both biodiversity loss and climate change (United Nations, 2023; US EPA, 2016; Wilting et al., 2017). The Intergovernmental Panel on Climate Change (IPCC) reports that recent human-caused climate alterations are without precedent over the last two millennia and are intensifying globally. Climate change is expected to negatively impact global GDP by mid-century due to systemic economic risks like physical damage (Swiss Re, 2021). Frequent natural disasters and extreme weather events such as floods, wildfires, droughts and storms lead to severe environmental destruction.

Unsustainable practices across the sectors and unplanned developments in eco-sensitive areas lead to severe ecosystem degradation. Ecological disruption due to anthropogenic activities combined with climate crises such as ocean acidification, sea level rise and surface temperature rise further threatens the biotic and abiotic environment.

In conclusion, climate change, biodiversity loss and environmental degradation are closely interconnected to each other. Economic prosperity is feasible only with environmental well-being. Figure 2 shows the drivers and impacts of environmental degradation on the people, the planet and the economy. The sustainable transformations to prevent environmental harm and the key drivers to trigger the change are also shown in Fig. 2.

As per UNEP report, it is evident that the climate impacts, biodiversity loss, water stress and health impacts are increasing with the increase in population and GDP (UNEP, 2023). Global collaboration, intense local-level actions and knowledge sharing are essential to facilitate sustainable development and CE. Continuation of current unsustainable production and consumption practices would lead to irreversible degradation and fatal consequences.

## 3 Why Linear Practices Are Unsustainable

The LE, characterized by the "take, make, dispose" model, has been the dominant system since the Industrial Revolution. It involves extracting raw materials, producing goods and discarding them as waste, leading to significant environmental degradation, resource depletion and social inequalities. This approach is unsustainable due to its reliance on finite resources, excessive waste generation and environmental pollution. The linear model produces large amounts of waste, including wastewater and sludge, which contaminate ecosystems. Additionally, products are

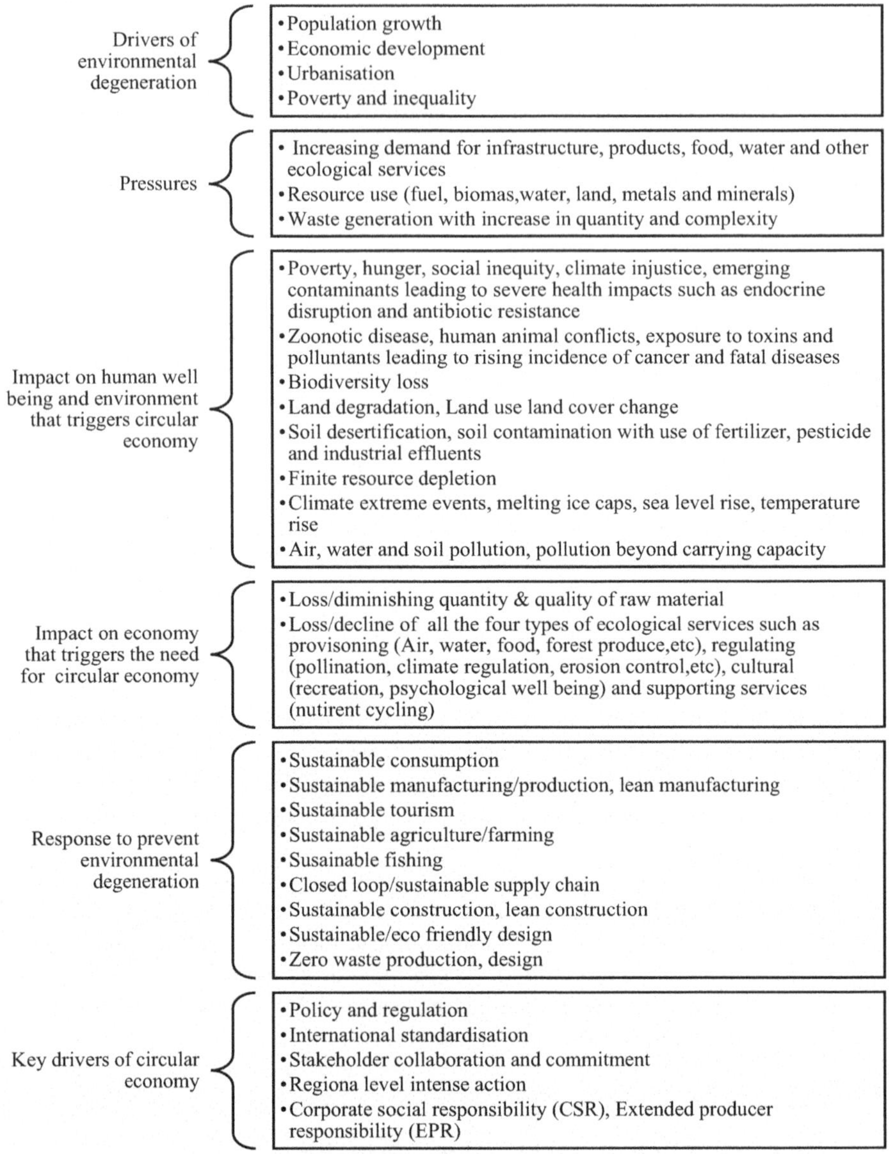


**Fig. 2** Drivers, impacts and responses to environmental degeneration (*Source* Authors)

designed with short lifespans and limited purposes, further contributing to waste. The linear model makes these problems worse as the world's need for resources, energy and food rises (Mihai & Minea, 2021). Furthermore, it is critical to limit landfilling by waste elimination and reduction, extending the usable life of goods. Landfills occupy valuable urban areas, while socioeconomic issues complicate the establishment of new ones. Hence, CE integration provides a sustainable solution to the issues created by LE through resource efficiency and waste reduction (Lederer et al., 2020).

a. **Environmental Impact**: The linear economic model, as explained in Sect. 2, is very resource-intensive and wasteful, which has negative effects on ecosystems, the environment and human health.
b. **Economic Unsustainability**: Economic instability results from rising extraction and manufacturing costs as resources become depleted. The costs of resource extraction and waste management, which are often externalised and paid by society and future generations, are not taken into account by the linear model. The short-term focus on profit maximization overlooks the long-term sustainability of economic activities. Failure to consider the environmental and social factors that lead to loss of shareholder value impacts the ability to survive competition, build resilience and achieve long-term financial success (Linnenluecke, 2022). Environmental considerations are crucial in business, as they can have mild to severe impacts, such as the loss of local raw material supplies to loss of business due to environmental compliance breaches.
c. **Social Inequity**: It is important to address the inequality in resource distribution, as developed countries have higher per capita consumption and waste generation, enjoying luxury at the cost of less developed countries, the environment and future generations (Pybus et al., 2022). The LE exacerbates social inequalities by concentrating wealth and resources in the hands of a few, often at the expense of vulnerable communities. These communities, particularly in developing regions, are disproportionately affected by environmental degradation, experiencing pollution, loss of livelihoods and more prone to health hazards from industrial activities and waste disposal. The integrity of natural ecosystems and the stability of economies, both vital to human life, are in danger from the detrimental impacts of the LE (Ghisellini & Ulgiati, 2020). Income inequality and social inequity make the CE adoption more difficult. Education acts as leverage for CE, while people sticking to traditional practices and mindsets hamper the change and adoption of CE (Neves & Marques, 2022).

## 4 The Emergence and the Principles of CE

Considering the shortcomings of the linear model, the concept of a CE has emerged as a transformative solution. A change from LE to CE is crucial to reduce pressure on the environment and to improve the security of supply of primary raw materials (Neves & Marques, 2022). The Ellen MacArthur Foundation provides the most well-known definition, stating that a CE is intentionally designed to be restorative and regenerative (Ellen MacArthur Foundation, n.d). It seeks to separate economic growth from resource consumption and environmental impact, promoting a system where materials and products are kept in use for as long as possible and waste is minimized (Geissdoerfer et al., 2017). The CE was initially associated with the 3Rs (reduce, reuse and recycle), which remain fundamental to effective waste management. As understanding of the CE has deepened, a broader set of principles has emerged, first with the 6Rs—reuse, reduce, recycle, redesign, refurbish and repurpose—and subsequently with the 10Rs: refuse, rethink, reduce, reuse, repair, refurbish, remanufacture, repurpose, recycle and recover (Reike et al., 2018).

These CE concepts offer the potential for substantial economic, environmental and social benefits. Economic benefits include optimized resource utilization, reduced costs through material and energy conservation, improved operational efficiency, financial savings, value retention, fostering innovation, stimulating economic growth and creating new business opportunities (Berry et al., 2023). Environmental benefits relate to minimizing pollution, recovering energy or materials and decreasing the need for inputs (energy, water, virgin resources), which usually leads to diminished negative environmental impacts (Morseletto, 2020). Social advantages may encompass job creation, empowerment of the local community, cohesiveness, social inclusion and equality. The principles of the CE as depicted in Fig. 3 focus on creating a sustainable economic model that reduces waste and maximizes resource use.

The CE aims to eliminate waste and pollution by ensuring materials re-enter the economy at the end of their use, rather than following the linear "take-make-waste" model. Focusing on design helps prevent waste from being created in the first place. Products are manufactured to be maintained, repurposed, repaired, refurbished and remanufactured, while biological.

materials can regenerate natural systems. By optimizing supply chains and materials, minimizing raw material extraction and continuously innovating sustainable solutions, the CE can create a system that supports both human and environmental health. The CE also emphasizes regenerating natural systems by supporting sustainable practices like agriculture, reforestation and renewable energy (Ellen MacArthur Foundation, n.d).

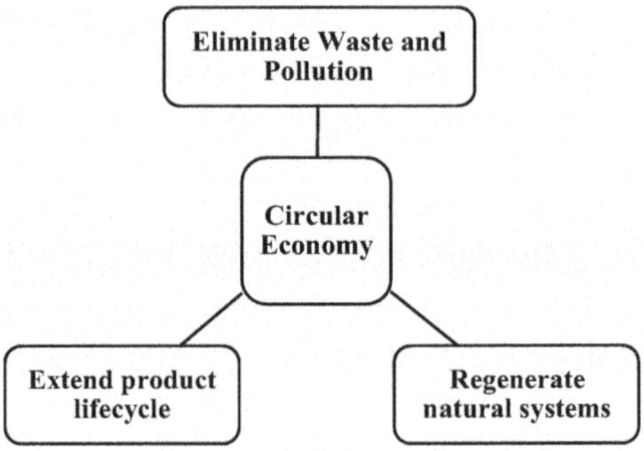


**Fig. 3** Principles of circular economy (*Source* Ellen MacArthur Foundation (n.d))

## 5 Approaches to Circular Transition to Prevent Environmental Degeneration

Responsible management of limited natural resources within the Earth's carrying capacity is essential for transforming the global economy toward a sustainable future. The upcoming section explains how circular transition is achievable through sustainable consumption, production and waste management.

### *5.1 Sustainable Consumption*

Sustainable consumption and production (SCP) refers to "the use of services and related products that meet basic needs and improve the quality of life, while minimizing the use of natural resources and toxic materials, as well as reducing waste and pollutant emissions over the product or service life cycle, without compromising the needs of future generations" (Environment, 2017).

The overall environmental footprint of human activity relies on the following three key factors—population size, technologies employed to manufacture the goods and services used and the average per capita consumption. Human consumption directly or indirectly leads to significant environmental impacts such as over 60% GHG emissions and 50–80% land, water and material usage. Hence, SCP is crucial to reduce negative environmental impact, combat poverty and support low carbon and green economies.

The primary objectives of SCP are:

a. *Separating environmental degradation from economic growth*: minimizing the resource consumption, pollution and degradation throughout the entire life cycle of goods and services while enhancing the quality of life.
b. *Applying life cycle thinking*: sustainable and efficient management of resources from the extraction of raw materials to waste disposal.
c. *Seizing opportunities for developing countries*: Offering green jobs, new markets and efficient resource management opportunities for developing countries and helping in poverty eradication. It helps developing countries to leapfrog to be more resource efficient and adopt environmentally sound technologies, thereby bypassing environmental degradation experienced by developed nations during their development phase.
d. *Consumer behavior*: Consumer behavior is changing rapidly and sometimes in unexpected ways. To make effective strategic decisions, companies need to anticipate and adapt to these shifts in consumer preferences. Recently, there has been a significant increase in Conscious Consumerism, with consumers becoming more intentional about their purchasing decisions regarding products and services (Environment, 2017).

## 5.2 Sustainable Production/Manufacturing

Sustainable manufacturing is defined as producing goods through methods that reduce environmental impact and preserve energy and natural resources while ensuring safety for workers, communities and consumers. It encompasses the full product lifecycle—pre-production, production, utilization and end of life—guided by the 6R principles (Reduce, Reuse, Recycle, Recover, Redesign, Remanufacture). Achieving sustainability requires optimizing the product, process and system to reduce waste, improve efficiency and maintain product quality. Key principles of sustainable manufacturing include adopting circular supply chains, utilizing renewable energy, sourcing ethical materials and designing sustainable products. Conservation efforts and partnerships with environmental organizations are also becoming more significant. Businesses reduce waste and increase efficiency, avoiding penalties and gaining consumer trust. Environmental factors initially considered as choice, inconvenient and costly, but now a mandatory obligation in many countries and businesses.

With the rise of Industry 4.0, manufacturing is becoming more digitized, intelligent and connected, integrating emerging technologies to enhance sustainability, flexibility and efficiency (Machado et al., 2020; Omojola et al., 2022). Technologies like Industrial IoT, predictive analytics, blockchain and AI play a vital role in improving green manufacturing by optimizing operations, improving supply chain transparency and enabling automation. Sustainable production offers financial benefits, including cost savings, brand equity, subsidies, investor interest and the ability to attract talent, all while addressing growing consumer demand for ethical practices.

In the UK, sustainable industrial ecosystems are being pursued with goals for material efficiency, low-carbon technology and waste management from 2013 to 2025, followed by new system experimentation up to 2050. By 2050, the focus will be on resource conservation, keeping materials in a productive loop and ensuring supply chain resilience. The European roadmap aims to develop sustainable value networks through CE strategies, energy-efficient factories and remanufacturing systems. Similarly, efforts in Sweden focus on reducing emissions, improving recyclability and a fossil-free economy with regulations supporting sustainable production and resource efficiency (Salvatori et al., 2019; UK government, 2021.)

## 5.3 Sustainable Design

The design phase accounts for 80% of a product's environmental impact. Consequently, many companies are prioritizing eco-design, due to increased environmental and ecological pressures. Eco-design helps reduce material consumption, energy use and emissions across all stages, including manufacturing, use, transportation, recycling and repurposing. For industries with elevated energy consumption and pollution levels, optimizing the production process is particularly critical. Eco-design, Product

Service System Design, Business Model development and Collaborative Ecosystem Design are some of the design alternatives businesses should consider achieving environmental, economic and social benefits.

Eco-design strategies are driven by internal development and sustainability objectives as well as external stakeholder pressure. Effective adoption of sustainable supply chain practices necessitates collaboration among all participants, yielding benefits throughout the entire product lifecycle. Research indicates that supply chain dynamics, including competitive pressures, evolving consumer expectations and financial incentives such as carbon taxes and subsidies, significantly influence eco-design techniques and broader sustainability initiatives.

Given its capacity to minimize waste, pollution and other adverse environmental effects even prior to consumption, design serves as a powerful tool for combating environmental degradation. Sustainable design, which supports the increasing organizational and societal focus on minimizing environmental impacts and achieving Environmental, Social and Governance (ESG) goals, offers multiple benefits. Specifically, it: (a) reduces environmental impact through the integration of renewable energy sources, sustainable materials and energy-efficient designs; (b) enhances indoor environmental quality by optimizing air quality, lighting and thermal comfort, thereby contributing to a healthier and more productive environment; and (c) facilitates lifecycle considerations (Cicconi, 2020; Kottaridou & Bofylatos, 2019; Liu et al., 2021; Munaro et al., 2022). Frequently referred to as eco-design or green design, its overarching objective is to develop sustainable solutions that address climate change and protect the environment, thereby supporting the Sustainable Development Goals (SDGs).

Eco-design, also known as Design for Environment (DfE), involves integrating environmental considerations into the product design process throughout its lifecycle (Directive 2009/125/EC). It aims to improve various aspects of a product depending on its lifecycle stage, utilizing different "Design for X" strategies:

a. *Design for Reassembly*: Focuses on redesigning existing products to extend their life cycle.
b. *Design for Manufacture and Assembly*: Enhances the production process.
c. *Design for Disassembly*: Optimizes product lifespan and recyclability through easier disassembly and repair.
d. *Design for Reuse*: Aims to maximize product lifespan by facilitating reuse.
e. *Design for Recycling*: Improves recyclability and minimizes impacts from end-of-life treatments (Sanyé-Mengual et al., 2014).

Adopting sustainable design practices also helps in achieving economic benefit, environmental compliance and stewardship. For instance, green buildings help to save energy costs and achieve global certifications. Some of the global certifications that support sustainability and CE include the following:

a. *LEED (Leadership in Energy and Environmental Design)*: Widely recognized certification for green buildings, emphasizing energy efficiency, reduced greenhouse gas emissions and improved quality of life.

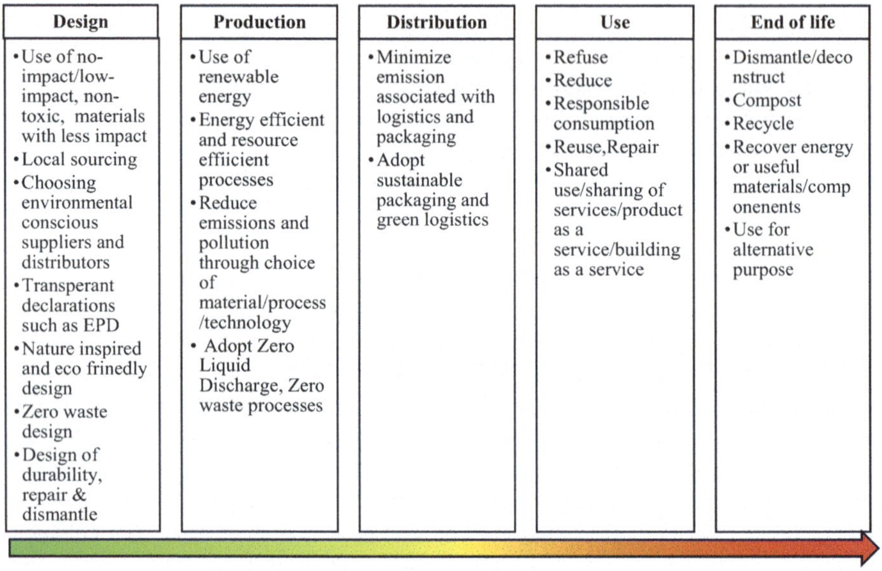


**Fig. 4** Life cycle of a product and the ability to control environmental impact (*Source* Authors)

b. *Energy Star and Cradle-to-Cradle* (C2C) are other systems that rate products or buildings based on sustainable design and energy efficiency.

Figure 4 shows the life cycle stages of the product and the sustainable practices to lower the environmental effect in each stage. The arrow below shows the ability to control the impact where the degree of control is higher during design stage and least during the end-of-life stages. Reduction in waste produced or waste elimination is always better than waste management since the latter needs cost and effort.

## 5.4 *Sustainable Agriculture*

In 2018, one-third of the world's agricultural land, totalling 4.8 billion hectares, was used for cropland. However, up to 25% of this production is lost from post-harvest to distribution. The food chain is a major source of waste and environmental impact, involving land use, animal farming, food processing, transport and consumption, which leads to organic waste, packaging waste, greenhouse gas emissions and wastewater.

Sustainable agriculture focuses on long-term productivity while conserving resources for future generations. It is based on five key principles: enhancing food production, conserving environmental resources, improving societal well-being and economic growth, supporting resilient ecosystems and communities and aligning with governmental initiatives. The main goals of sustainable agriculture are to ensure food and fibre security for the present and future, while also maintaining soil fertility, promoting biodiversity, reducing pollution, reducing the use of finite resources, supporting rural economic development and improving the quality of life for farmers (*Sustainable Agriculture*, 2022).

Key practices in sustainable agriculture include crop rotation, which enhances soil health and pest control, efficient irrigation methods like drip irrigation and cover crops that protect soil and support pollinators and no-till farming to prevent soil erosion and reduce fossil fuel emissions. Integrated Pest Management (IPM) uses biological control and natural predators to minimize chemical use, while permaculture mimics natural ecosystems to reduce waste and improve soil fertility (*Sustainable Agriculture*, 2022).

Sustainability in agriculture is measured across three pillars: environmental, societal and economic. The benefits of sustainable agriculture are wide-ranging, including lower costs, improved productivity, energy savings, fair wages, public health benefits, social equity, ecosystem protection and the conservation of soil and water. However, challenges remain, such as increased labour demands, reduced crop yields and a need for greater governmental support (Lin et al., 2024).

Technologies like satellite-based crop monitoring, smart irrigation systems and precision farming tools are helping to optimize resource use, reduce chemical inputs and monitor crop health, contributing to more sustainable, efficient and productive farming. Integration of CE principles into food production enhances sustainability and reduce environmental impacts (Payne & Kwofie, 2024).

## 5.5 *Sustainable Fishing*

Overfishing, fuelled by unregulated practices, industrial fishing techniques, rising seafood demand and ineffective regulations, severely impacts marine ecosystems and dependent economies. Coastal areas, critical for biodiversity and livelihoods, face threats from destructive gear and industrial methods, which further damage habitats and increase the bycatch. Climate change compounds these challenges, causing ocean acidification, coral bleaching and shifts in fish distribution due to rising sea temperatures. Addressing these issues requires robust sustainable fisheries management, which involves using scientific data for informed decision-making, implementing community-led management systems and adopting practices such as regulating gear, establishing no-fishing zones and enforcing quotas.

Sustainable fishing is vital for several reasons: it protects ocean biodiversity and habitats, supports coastal and indigenous communities, aids in adapting to climate change, contributes to global food security and reduces food loss and waste in the supply chain. Despite this, over 35% of the world's fisheries are overfished or overexploited, causing a global decline in fish stocks and jeopardizing future sustainability. Data collection is crucial for monitoring fish populations and ecosystem health while understanding human behaviour and employing behavior-centered design strategies can promote responsible fishing practices. By fostering cooperation among fishers and involving local communities in decision-making, sustainable fishing practices can be encouraged, benefiting both marine ecosystems and the people who depend on them (Morseletto, 2020; World Wildlife, n.d.).

Sustainable fishing not only supports the livelihoods of millions globally but also provides a major protein source for over a third of the world's population and reduces pressure on land-based agriculture. Certified sustainable fisheries, like those recognized by the Marine Stewardship Council (MSC), follow strict standards to manage fish populations responsibly, minimize environmental impacts and ensure long-term viability. These practices are crucial for preserving the health of marine ecosystems and ensuring global food resources (MSC, n.d.).

## *5.6 Sustainable Waste Management*

Sustainable waste management (SWM) practices play a vital function in protecting the environment by minimizing the negative impacts of waste on natural resources and ecosystems. These practices include the effective management of solid waste, turning waste into an opportunity for sustainable development by providing inexpensive raw materials. The goal is to conserve natural resources and to promote usage of resources efficiently, thereby reducing pollution (Stories, 2024). Several upcoming policies to regulate waste management include landfill taxes, landfill bans, incineration taxes, deposit refund schemes (DRS) and Waste Framework Directive (WFD) regulations.

Key sustainable waste management practices include recycling, source separation and organic waste management, which help conserve resources, reduce landfill usage and minimize waste. Composting organic waste not only enhances soil fertility but also reduces methane emissions. Waste-to-energy technologies change non-recyclable waste into energy, reducing reliance on fossil fuels. Extended Producer Responsibility (EPR) programs encourage eco-design and responsible disposal, holding producers responsible for the whole lifecycle of their products. Principles such as "Reduce, Reuse and Refuse" and CE approaches promote waste prevention and the design of durable, repairable and recyclable products (Abdel-Shafy & Mansour, 2018; Abubakar et al., 2022; Hong et al., 2023).

Following is some of the CE strategies to reduce or eliminate waste generation.

### 5.6.1 Product-as-a-Service

The transition to Product-as-a-Service (PaaS) is increasingly being adopted as businesses shift from selling products to offering services, fostering a more circular economy. This shift necessitates significant adjustments in business operations, value delivery models, and overall strategy. A key aspect of PaaS is reconfiguring traditional business models, with the **Business Model Canvas (BMC)** serving as a strategic tool for design and implementation. Effective product take-back systems and improved repairability play a vital role in extending product lifespans and preserving resource value, supporting both economic and environmental sustainability. However, optimizing reverse logistics and incentivizing repairs are critical to maximizing these benefits. Additionally, **Life Cycle Costing (LCC)** provides a financial framework to navigate uncertainties during the transition, helping businesses achieve long-term viability despite initial disruptions (Hidalgo-Crespo et al., 2024; Sanyé-Mengual et al., 2014).

### 5.6.2 Zero Waste (ZW)

ZW is a systematic approach aimed at eliminating waste by ensuring resources are utilized efficiently and preventing unnecessary waste generation. The core principles of ZW focus on resource conservation, responsible material use, recycling, and reuse to minimize environmental impact and safeguard public health. This approach promotes innovative strategies to reduce reliance on incineration, landfilling, and uncontrolled dumping.

Within the manufacturing sector, Zero Waste Manufacturing (ZWM) focuses on increasing material efficiency, promoting recycling and reuse, and reducing waste generation at every stage of production. While achieving absolute zero waste is challenging, significant progress can be made by designing products and systems that allow materials to be repurposed or refurbished for extended use. Implementing advanced manufacturing techniques and sustainable design practices plays a crucial role in minimizing material waste and enhancing sustainability (Kerdlap et al., 2019; Omojola et al., 2022).

### 5.6.3 Extended Producer Responsibility (EPR)

EPR is a regulatory framework designed to minimize waste, particularly plastics, by enhancing recycling efforts and reducing dependency on virgin raw materials. Based on the "**polluter pays**" principle, EPR shifts the responsibility of post-consumer waste management from local authorities to manufacturers, requiring them to oversee the collection, recycling, and proper disposal of their products. This approach discourages waste accumulation and helps prevent plastic pollution in marine ecosystems.

EPR is widely recognized as an effective waste management strategy, promoting higher recycling rates and reducing landfill dependency. For instance, in the European Union (EU), EPR policies legally hold producers, manufacturers, brand owners, and importers accountable for the lifecycle management of their products and packaging materials. While well-established in regions like Europe, Canada, Japan, and South Korea, EPR policies are still evolving in parts of Asia and Latin America. When implemented effectively, EPR can lead to multiple benefits, such as improved waste collection and recycling, reduced public sector spending on waste management, lower overall waste treatment costs, and enhanced product designs that prioritize durability, compostability, and reusability (Bhadra & Mishra, 2021; Kosior & Crescenzi, 2020; Shooshtarian et al., 2021).

### 5.6.4 Take-Back Schemes

To promote responsible product stewardship, regulations mandate that manufacturers of consumer goods take responsibility for product take-back. Recognizing the environmental harm caused by improper disposal of End-of-Life (EOL) consumer electronics, the European Commission's Waste Electrical and Electronic Equipment (WEEE) directive has established collection targets to regulate producers' collection practices. With advances in technology, many products and their components can now be reused multiple times. For instance, factory-refurbished iPhone X models and similar devices, which retain all their functions, have established a strong secondary market in developing countries (Xintong et al., 2021; Xu et al., 2021).

### 5.6.5 Closed-Loop Supply Chains

Sustainability concerns have pushed many industries to adopt closed-loop supply chains (CLSCs), creating a more complex process with various risks due to the circular or multi-circular structure and multiple reverse flows of goods, parts and materials. Furthermore, integration between partners and digitalization levels in CLSCs remain low, despite the significant societal benefits they offer. A sustainable supply chain involves managing the flow of materials, information and funds, as well as fostering collaboration between firms along the supply chain, while addressing the three pillars of sustainable development: environmental, social and economic. Companies can leverage sustainable programs as strategic tools not only for environmental benefits but also to enhance brand image, generate revenue, improve customer service and reduce production costs (Cannella et al., 2016; Simonetto et al., 2022).

One type of sustainable supply chain is the Closed Loop Supply Chain (CLSC) which manages the recycling and recovery of end-of-life products. In a CLSC, manufacturers handle reverse logistics, either directly recovering returned goods or through third-party channels. Following appropriate processing, returned products are resold through both primary and secondary markets. A Closed-Loop Supply Chain (CLSC) extends the conventional supply chain framework by integrating reverse channels that manage product returns, recycling, recovery, remanufacturing and resale processes (Cannella et al., 2016).

Linear structure of most supply chains focuses solely on maximizing economic returns, often overlooking the environmental and social impacts associated with sourcing and processing raw materials, product distribution and the post-consumer stage. Common issues include selecting raw material suppliers without considering environmental and social sustainability criteria and manufacturing and distribution processes that harm the environment and society.

Products are often designed without considering circular strategies like remanufacturing, repair, reuse or recycling resulting in shorter product lifecycles. Many of these products ultimately end up as waste, contributing to environmental pollution and related social issues, such as health hazards, pollution and resource scarcity. Furthermore, the processes used to produce these products can also negatively impact the environment and society, depending on the resources consumed, internal treatments applied and the technologies implemented.

These concerns have led to a rethinking of supply chain designs across industries, incorporating not only economic factors but also social and environmental aspects to promote sustainable futures. As a result, there is a growing interest in advancing efforts toward achieving economic, environmental and social sustainability, with major organizations and institutions worldwide setting targets for sustainable development (Wang et al., 2022). The environmental impact caused by various industries and appropriate sustainable solutions for preventing the environmental degeneration is shown in Table 1.

## 6 The Urgency of the Shift

The shift from a linear to a CE is not just an option; it is a necessity for the survival and well-being of our planet. The urgency of this shift cannot be overstated. If current trends in resource use, waste generation and environmental degradation continue, the consequences will be catastrophic for ecosystems, economies and human societies. By implementing circular practices, we can generate a more sustainable, strong and equitable future. This paradigm shift represents a fundamental change in how we view resources, waste and the economy, moving us towards a model that prioritizes long-term environmental health, economic stability and social well-being. The advantages of CE transformation would be beneficial for the people, planet and businesses (Cezarina et al., 2021).

## 7 Accelerating Environmental Degradation and Policy Responses

The accelerating pace of environmental degradation has reached a critical juncture, compelling corporations, governments and individuals to re-evaluate existing development paradigms and policies. The pressing issues of pollution, resource depletion and climate change have spurred the development of innovative sustainable practices across various sectors and prompted the implementation of urgent regulatory measures.

Research indicates that establishing robust regulatory frameworks to monitor, promote and manage the transition towards a CE is crucial for regional and national

**Table 1** Environmental impact and sustainable solution for various sectors

| Sector | Environmental impact | Solution to prevent environmental degeneration | References |
|---|---|---|---|
| Textile industry | • Harmful chemicals, dyes and heavy metals from untreated wastewater can contaminate water bodies, impacting aquatic ecosystems and human communities<br>• High water consumption during fibers, dyeing and finishing process<br>• Harmful pollutants contribute to air quality degradation and human health issues<br>• Generate a large amount of solid waste, including off-cuts from manufacturing, discarded clothing and synthetic fibers that do not biodegrade. Huge quantities of textiles waste end up in landfills<br>• Synthetic fibers like polyester, nylon and acrylic shed microplastics during washing enters waterways and impact marine life<br>• Hazardous chemicals such as formaldehyde, chlorine bleach and heavy metals, harm both workers and the environment | • Use of Sustainable Materials such as organic and renewable fibers, recycled materials such as polyester made from plastic bottles and recycled cotton to reduce waste<br>• Implementing water-saving technologies and energy-efficient machinery to minimize consumption<br>• Adopt non-toxic dyes and waterless dyeing methods (e.g., supercritical $CO_2$)<br>• Focus on textile recycling, durable design and upcycling to reduce waste and promote a circular economy<br>• Reduce transportation emissions by sourcing and producing locally<br>• Adopting innovative processes like digital printing and supercritical $CO_2$ dyeing, which significantly reduce water use | Luo et al. (2021) |

(continued)

**Table 1** (continued)

| Sector | Environmental impact | Solution to prevent environmental degeneration | References |
|---|---|---|---|
| Agriculture | • Soil erosion, nutrient depletion and loss of fertility reduce land productivity over time<br>• Runoff from pesticides, fertilizers and animal waste contaminates water bodies, causing eutrophication and harming aquatic ecosystems<br>• Greenhouse gas emissions from livestock, rice paddies and synthetic fertilizers contribute to climate change<br>• Large-scale agriculture drives deforestation, leading to habitat loss and biodiversity decline<br>• Excessive use of pesticides and fertilizers pollutes ecosystems and harms pollinators, beneficial insects and human health | • Implement crop rotation, agroforestry and conservation tillage to maintain soil health, prevent erosion and improve land productivity<br>• Use drip irrigation and rainwater harvesting to reduce water waste and promote more efficient water use<br>• Support reforestation and agroforestry to restore ecosystems, increase biodiversity and reduce deforestation<br>• Encourage the use of biological pest control and integrated pest management to minimize harmful chemical use | Koul et al. (2022) |
| Oil and gas | • Releases $CO_2$, methane and VOCs, contributing to climate change and smog<br>• Spills and leaks contaminate water bodies, harming aquatic life and water quality<br>• Exploration and drilling disrupt natural habitats, affecting wildlife habitat<br>• Spills and waste disposal contaminate soil, impacting land use and agriculture<br>• Burning fossil fuels releases $CO_2$ and other gases, driving global warming<br>• Accidents cause severe environmental damage to marine and coastal ecosystems<br>• The industry uses significant energy, often from fossil fuels, increasing environmental impact<br>• Produces hazardous waste that requires careful management and disposal | • Use advanced technologies to reduce air pollution, control GHG emissions, leak detection, equipment maintenance and environmental monitoring<br>• Invest in better spill prevention and rapid response systems to minimize water pollution<br>• Minimize habitat disruption<br>• Address soil contamination with effective clean-up techniques and proper waste management<br>• Shift towards infinite energy sources to reduce dependence on fossil fuels<br>• Implement waste reduction strategies, recycle materials and ensure safe disposal of hazardous waste<br>• Optimize energy use in extraction, processing and transportation | Masudin et al. (2024) |

(continued)

**Table 1** (continued)

| Sector | Environmental impact | Solution to prevent environmental degeneration | References |
|---|---|---|---|
| Mining | • Mining activities cause deforestation, land degradation and loss of biodiversity<br>• Runoff with toxic chemicals and heavy metals contaminate soil and water<br>• Excavation and stripping of vegetation lead to soil erosion, impacting land stability and fertility<br>• Dust and emissions of chemicals like sulfur dioxide ($SO_2$) and particulate matter<br>• Overexploitation leads to the exhaustion of finite resources | • Use sustainable methods that minimize land disturbance and implement reclamation plans to restore habitats. Implement proper handling and treatment of mine tailings<br>• Use water recycling systems and treatment processes to reduce pollution<br>• Employ dust suppression techniques and control emissions<br>• Optimize resource extraction and invest in technologies that improve efficiency and reduce waste | Onifade et al. (2024) |
| Transportation sector | • Construction of roads and infrastructure leads to land alteration, habitat fragmentation and loss of green spaces<br>• Vehicle production generates waste, including scrap metal, plastics and hazardous materials<br>• Fossil fuel-powered vehicles emit significant $CO_2$ and other greenhouse gases, contributing to climate change<br>• Vehicles emit nitrogen oxides (NOx), particulate matter and volatile organic compounds (VOCs), degrading air quality and impacting human health | • Switch to electric and hybrid vehicles<br>• Enhance the fuel efficiency<br>• Increase investment in and use of public transit systems<br>• Use advanced emissions control technologies<br>• Improve recycling processes for vehicle parts and materials<br>• Design and build transportation infrastructure with minimal environmental impact, incorporating green practices and technologies | Heidari et al. (2023) |

(continued)

**Table 1** (continued)

| Sector | Environmental impact | Solution to prevent environmental degeneration | References |
|---|---|---|---|
| Construction sector | • Heavy reliance on natural resources like sand, stone, cement, timber and metals leading to depletion of resources and cause environmental degradation, including habitat loss and ecosystem damage<br>• Large amounts of waste generated during construction, demolition and renovation projects<br>• Unplanned or poorly designed construction contributes to urban sprawl | • Implement sustainable design practices, minimize off-cuts and surplus materials through careful planning and design<br>• Salvage and reuse materials from demolition sites<br>• opt for deconstruction instead of demolition<br>• Incorporate design for disassembly principles in new projects, enabling easy removal, recovery and reuse of materials at the end of a building's life cycle<br>• Decarbonize built sector | UNEP (2023) |
| Electronics industry | • Extraction of rare earth metals leads to habitat destruction, deforestation and water contamination. Mining and refining require significant energy, increasing carbon emissions. Electronics manufacturing (e.g., semiconductor fabrication) is energy-intensive, contributing to a large carbon footprint<br>• The global nature of the supply chain adds additional environmental costs through transportation and energy use. Electronic waste (e-waste) is one of the staunchest-growing waste streams worldwide<br>• Improper disposal leads to harmful chemicals (e.g., lead, mercury and cadmium) leaching into environment, affecting ecosystems and human health | • Implement stricter regulations and practices to reduce habitat destruction, control water pollution and minimize land degradation during mining operations<br>• Design electronics for longevity, repairability and recyclability. Products that are easier to repair and upgrade will extend their life cycle, lowering the demand for new resources and reducing production-related emissions | |

(continued)

**Table 1** (continued)

| Sector | Environmental impact | Solution to prevent environmental degeneration | References |
|---|---|---|---|
| Pharmaceutical industry | • Discharge of residues and chemicals contaminate soil, water, aquatic ecosystems and affect wildlife<br>• Industry generates large quantities of hazardous waste, including unused medications, packaging materials and production by-products | • Develop and enforce systems for proper disposal and recycling<br>• Use environmentally friendly chemicals and processes<br>• Enhance wastewater treatment processes to remove pharmaceutical residues<br>• Reduce resource consumption and improve energy efficiency in pharmaceutical production<br>• Increase public awareness and provide guidelines for the safe disposal of medications to prevent environmental contamination | Milanesi et al. (2020) |
| Tourism industry | • Tourism growth often leads to urban sprawl and the displacement of local communities, altering traditional landscapes and cultures<br>• Tourists generate large amounts of solid waste, including plastic, which is often poorly managed in less developed areas, resulting in land and ocean pollution<br>• Cruise ships and coastal tourism contribute to water pollution by dumping sewage and waste into oceans and waterways, harming marine life<br>• Air travel, cruise ships and vehicles used by tourists contribute to GHG emissions | • Green certifications for hotels and resorts, promoting local and sustainable sourcing of materials and reducing plastic use<br>• Travel focused on experiencing and preserving natural areas with minimal environmental impact, often benefiting conservation efforts<br>• Some airlines and travel companies offer carbon offset programs to reduce the carbon footprint of travel | Séraphin and Chaney (2023) |

(continued)

**Table 1** (continued)

| Sector | Environmental impact | Solution to prevent environmental degeneration | References |
|---|---|---|---|
| Chemical industry | • Chemical spills and improper disposal of hazardous waste led to soil contamination, affecting land usability and agriculture. Discharge of toxic chemicals and untreated wastewater contaminates rivers, lakes and groundwater, harming aquatic life | • Implement cleaner technologies<br>• Enhance recycling and safe disposal methods for hazardous and chemical waste<br>• Adopt Green chemistry that minimize environmental and health risks<br>• Enhance pollution controls in manufacturing processes<br>• Adhere to environmental regulations and standards to minimize environmental impact | De Faria et al. (2021) |
| Aviation | • Aircraft emit significant amounts of $CO_2$ and other greenhouse gases<br>• Aircraft noise impacts communities near airports, leading to health issues such as sleep disturbance and stress<br>• Airport construction and expansion lead to habitat destruction and increased energy use | • Use biofuels and other alternative energy sources to reduce greenhouse gas emissions<br>• Invest in more efficient aircraft designs and technologies to reduce fuel consumption and emissions<br>• Optimize flight routes and air traffic control to reduce unnecessary fuel use and emissions<br>• Use renewable energy, improve waste management and incorporate green building standards in airport operations and development | Guimarans et al. (2019) |
| Fishing industry | • Depletes fish stocks, disrupting marine ecosystems and threatening species<br>• Non-target species are caught unintentionally, leading to declines in their populations<br>• Fishing practices like trawling damage seabed and destroy critical marine habitats<br>• Pollutants, such as plastics and chemicals, enter oceans from fishing operations and shipping | • Use methods like catch limits, quotas and selective fishing gear to prevent overfishing and bycatch<br>• Establish marine protected areas and use less destructive fishing techniques to safeguard habitats<br>• Improve waste management practices and minimize pollutants from fishing operations and shipping<br>• Support sustainable seafood choices and certifications (e.g., MSC, ASC) to reduce demand for overexploited species<br>• Enhance research on marine ecosystems and fish populations to better manage resources and assess the impacts of fishing practice | Moshood et al. (2022), Msc.org (n.d) |

authorities. This is particularly important in addressing environmental concerns such as greenhouse gas emissions and climate change, where individual and business engagement may be limited. Evidence suggests that stringent environmental laws, coupled with the adoption of CE principles, can effectively reduce carbon emissions (Tiwari et al., 2024).

Governments play a vital role in strengthening corporate environmental management, facilitating the adoption of cleaner energy sources and enhancing the stringency of environmental regulations. Their policies and legislative frameworks are essential for furthering the CE. These changes need the development of comprehensive and consistent strategies for consumer behaviour, company operations and governmental legislation. All the stakeholders must work together to guarantee the effective implementation of CE principles. Governments are essential in raising public awareness integrate environmental policies, energy transitions and supply chains to reduce environmental harm. To integrate CE frameworks factors such as sustainability standards, auditing and monitoring, collaboration and data sharing platforms, financial support, technology and transparent supply chains are required. For which, current legal structures and environmental laws must reorient to support the factors (Tiwari et al., 2024).

Despite the introduction of several global green initiatives such as the Kyoto Protocol, the Paris Agreement and the Sustainable Development Goals (SDGs) environmental challenges persist (Li & Li, 2024). Hence, the UN has worked to minimize developmental losses by bringing nations together in the battle for environmental sustainability. The need for a sustainable future was emphasized during COP27 conference by narrowing the focus on four important areas: International cooperation, adoption of green technology, funding for green projects and environmental mitigation. These themes aim to protect environment from damage for both present and future generations (Alayza et al., 2022).

Governments started realising the importance to address environmental deterioration by laws and regulations. Therefore, national and regional policies concentrate on tackling waste management, pollution and resource conservation while international accords such as the Paris Agreement have set objectives for decreasing greenhouse gas emissions. These regulations often provide financial incentives to companies that implement sustainable practices, enforce fines for noncompliance and fund the advancement of green technology research and development.

In addition, the CE ought to be inclusive, guaranteeing that local people, marginalised groups and stakeholders participate fully in decision-making procedures. Businesses as well as society may gain from social capital and the management of natural resources, such as ecological resources and communal cohesion, which can help reduce waste and $CO_2$ emissions. Reducing $CO_2$ emissions and environmental damage via CE practices is essential under the Sustainable Development Goals (SDGs), especially Goal 12 (responsible consumption and production) and Goal 3 (excellent health and well-being) (Tiwari et al., 2024).

## 7.1 *Trigger for Policy and Regulation*

The growing environmental crisis has pushed governments worldwide to implement policies aimed at mitigating ecological degradation. The concept of a CE is actively advocated by the EU and various national governments, such as those of China, Japan, the UK, France, Canada, the Netherlands, Sweden and Finland. Numerous global businesses also support this shift. According to a recent projection by the European Commission, adopting CE practices could bring annual economic benefits of 600 billion euros to the EU designing sector alone.

(a) **Netherlands**: The Dutch government seeks to achieve full circularity by 2050, with a target of reaching 50% circularity by 2030. This plan involves cutting the use of primary raw materials such as minerals, fossil fuels and metals by 50% by 2030. The Netherlands has introduced transition agendas in five priority sectors: construction, plastics, consumer goods, biomass and food and manufacturing. These agendas emphasize resource optimization, a shift to renewable and recycled resources and the creation of new markets and business models. The Netherlands Circular Accelerator, a joint initiative involving the Ministry of Infrastructure and Water Management, plays a key role in advancing this policy through concrete actions. Amsterdam is also noted for integrating socioeconomic factors with the CE and adopting 'Doughnut Economics.' Starting in 2023, the Dutch government will ensure that all contracts are circular.

(b) **China**: China has been a pioneer in the CE, starting with its 11th Five-Year Plan in 2006 and the CE Promotion Law in 2008. The CE is a key component of China's vision for a harmonious society, emphasizing cleaner manufacturing and trade ecology principles.

(c) **Japan**: Japan, an early adopter of CE practices, has introduced several policies since the early 2000s aimed at reducing, reusing and recycling. Its CE Vision 2020 encourages industries to adopt circular business models and improve resource efficiency. The country is known for its high recycling rates and unremitting industrial innovation, despite its policies not always being explicitly linked to the CE.

(d) **Chile**: Following its hosting of COP25, Chile has made significant progress in environmental initiatives. Its Roadmap for a Circular Chile by 2040, published in June 2021, seeks to create 100,000 green jobs by 2030.

(e) **World Economic Forum (WEF)**: The WEF is engaged in worldwide initiatives to accelerate the shift to a CE, serving as a platform for collaboration between public and private sectors. Notable initiatives include the Platform for Accelerating the CE, which supports 18 projects worldwide and major value chain action partnerships focusing on sectors like plastics, electronics, batteries and fashion. The WEF also launched Scale 360° to foster innovation in circular technologies across various industries and the CE for Net Zero Industry Transition initiative to promote decarbonization in hard-to-abate sectors such as steel, cement, chemicals and aluminium (*Leading Countries in Embracing CE*, 2022).

## 7.2 *Trigger for Businesses*

The shift towards circularity is not only driven by governments but also by innovative business practices. Companies around the world are embracing CE principles to reduce waste and minimize their environmental footprint. For instance, Patagonia, an outdoor apparel company, has been a leader in promoting circularity since 1986 through its Worn Wear program, which repairs and resells used products. IKEA, a multinational furniture retailer, has set a goal to achieve full circularity in its business operations by 2030. As part of this commitment, they have launched a buyback program designed to incentivize customers to return used furniture for resale or recycling. Similarly, Unilever, a leading global consumer goods company, is prioritizing the incorporation of recycled materials in its packaging and working closely with suppliers to minimize waste throughout its supply chain. Unilever has also pledged to reach net-zero emissions in its operations by 2030 and across its entire value chain by 2039. These examples illustrate the growing trend of businesses taking responsibility for their environmental impact, leveraging innovation to promote sustainability while ensuring profitability. The increasing importance of ESG (Environmental, Social and Governance) factors, CSR (Corporate Social Responsibility) initiatives and heightened consumer awareness have all contributed to companies' decisions to integrate CE practices into their core business models.

(A) **Corporate Responsibility**

Corporate responsibility is becoming an increasingly vital element in addressing environmental challenges. Public and governmental expectations are rising, with companies facing pressure to actively reduce their carbon footprints and adopt sustainable business models. Numerous major corporations, including H&M and Accenture, have incorporated CE principles into their strategies. For example, H&M has partnered with Remondis to collect used garments for resale or recycling, effectively reducing waste within the fashion industry. Concurrently, Accenture is dedicated to fostering sustainable growth by collaborating with organizations that facilitate circular supply chains.

The move towards corporate environmental responsibility is increasingly driven by initiatives like the World Economic Forum's (WEF) CE for Net Zero Industry Transition. This initiative supports decarbonization by fostering collaborations among industries and key stakeholders to create circular solutions for challenging sectors such as steel, cement and chemicals (Bancilhon, 2021).

(B) **Consumer Awareness**

Consumer awareness plays critical role in enabling the shift towards a CE. Nowadays people become more aware about the environmental footprint of their buying choices and started demanding sustainable products and services. In response, companies are pushed to provide more eco-friendly alternatives such as including reusable packaging and environmental footprint declarations. For example, Loop—a reusable packaging platform by Tesco enable customers to buy popular products in sustainable packaging.

Few other companies like Patagonia's repair and rental services align their service with CE principles by promoting consumers to buy-less and encouraging them to extend the lifespan of their products through repair and rental services. This shift in consumer preferences is essential for driving systemic change, as it places pressure on businesses and governments to prioritize sustainability in their practices and policies (Deloitte, 2022).

(C) **Pioneers**

Despite the challenges, several companies have implemented CE and harnessed its benefits. Following are a few such companies:

(a) An Australian firm called **'Close the Loop'** recycles old printer cartridges and soft plastics into durable road surfaces. Their process incorporates recycled materials, extending road life by up to 65% and keeping waste out of landfills.
(b) A Canadian company called **'Enerkem'** converts non-recyclable trash into biofuels and chemicals using a carbon extraction process. Their technology helps cities like Edmonton recycle 90% of their waste, reducing landfill use significantly.
(c) **Schneider Electric**, the most renowned French multinational company, with expertise in energy management and mechanization, integrates reprocessed materials into its products, extends product lifespans and implements take-back schemes. Circular activities now contribute to 12% of its revenue and help conserve resources.
(d) **Lehigh Technologies**, Atlanta, translates old tyres and rubber waste into micronized rubber powder used in various applications. Their products have been used in 500 million new tyres, earning them the CE SME Award.
(e) **Miniwiz** focuses on upcycling and turning waste materials into new, sustainable products. Their Trashpresso machine recycles plastic bottles into building materials, using solar power and no water.
(f) The Dutch company **DyeCoo** has developed a waterless dyeing process for textiles using supercritical carbon dioxide. This method eliminates the need for water and chemicals, reduces drying time and costs less, with major brands like Nike and IKEA as partners.
(g) **TriCiclos** is a Chilean company that operates the largest recycling network in South America, diverting 33,000 metric tons of material from landfills and reducing carbon emissions by over 140,000 metric tons (World Economic Forum, 2019).

## 8 Conclusion

Population growth, economic development, resource consumption and waste generation are interconnected leading to severe environmental degeneration. As discussed in the chapter, linear economy leads to deforestation, climate change acceleration, biodiversity loss, higher pollution and waste generation. As climate change is already there and we are facing the consequences of our actions, it is high time we need to rethink, reorient and collaborate to prevent environmental degradation. Further socio-economic factors impact exposure level and vulnerability to environmental and climate risks, as socially and economically disadvantaged groups (such as children, women and low-income countries) are more prone to pollution and hazards. As policy acts as trigger for any change, it is vital to develop policy and regulation to make the shift to CE happen. Several countries and local regulations have banned landfills and imposed stringent measures on waste and effluent disposal. However, lack of awareness, technology, infrastructure, funding and societal and cultural challenges prevent low-income countries from implementing CE. Consumer awareness is necessary to pull the industrial transformation. Developed countries must adopt very stringent measures to reduce per capita consumption and focus on behavioural change achieve responsible consumption. As environmental consequences are global, world organization must take effective strategies to gain stakeholder collaboration to prevent environmental harm, combat climate change and improve resilience. Additionally, it is important to ensure social equity in resource distribution and waste disposal.

In conclusion, the severe consequences of environmental degradation, such as the frequent occurrence of extreme climate events, massive species extinction, emerging contaminants, rising health threats such as COVID, cancer and toxic contaminants and food and water demand stress, along with issues from mismanagement of waste and the increasing demand for raw materials, have triggered the shift towards the CE. To achieve significant impact, the CE must be adopted widely across regions and industries. The CE must be integrated into core business strategies and data on raw material acquisition to end-of-life scenarios must be transparent to consumers to facilitate responsible choices. Furthermore, international cooperation is essential for harmonizing standards, sharing best practices and managing cross-border waste.

Future research can analyze the effectiveness of various policies and strategies, such as incentives and penalties, in achieving the CE. There is significant potential for integrating technology into the material supply chain, which would support real-time data collection and decision-making from designing to end-of-life. As the majority of current work focuses on recycling, regulatory, economic and societal triggers are further required to shift toward rethinking and redesigning for zero waste. Therefore, a strong emphasis on design thinking and systems thinking is needed, considering the life cycle of products and their utilization.

## References

Abdel-Shafy, H. I., & Mansour, M. S. M. (2018). Solid waste issue: Sources, composition, disposal, recycling and valorization. *Egyptian Journal of Petroleum, 27*(4), 1275–1290. https://doi.org/10.1016/j.ejpe.2018.07.003

Abubakar, I. R., Maniruzzaman, K. M., Dano, U. L., AlShihri, F. S., AlShammari, M. S., Ahmed, S. M. S., Al-Gehlani, W. A. G., & Alrawaf, T. I. (2022). Environmental sustainability impacts of solid waste management practices in the global south. *International Journal of Environmental Research and Public Health, 19*(19), 12717. https://doi.org/10.3390/ijerph191912717

Alayza, N., Bhandari, P., Burns, D., Cogswell, N., Zoysa, K. de, Finch, M., Fransen, T., González, M. L., Krishnan, N., Langer, P., Larsen, G., Srouji, J., Warszawski, N., & Waskow, D. (2022). COP27: Key takeaways and what's next. https://www.wri.org/insights/cop27-key-outcomes-un-climate-talks-sharm-el-sheikh

Azwardi Andaiyani, S., Igamo, A. M., & Wijaya, W. A. (2023). The environmental impacts of natural resources depletion. In D. F. Swasto, D. H. Rahmi, Y. Rahmawati, I. Hidayati, J. Al-Faraby, & A. Widita (Eds.), *Proceedings of the 6th International Conference on Indonesian Architecture and Planning (ICIAP 2022)* (pp. 705–714). Springer Nature. https://doi.org/10.1007/978-981-99-1403-6_47

Baldassarre, B., Keskin, D., Diehl, J. C., Bocken, N., & Calabretta, G. (2020). Implementing sustainable design theory in business practice: A call to action. *Journal of Cleaner Production, 273*, Article 123113. https://doi.org/10.1016/j.jclepro.2020.123113

Bancilhon, C. (2021). What business needs to know about the EU corporate sustainability reporting directive | Blog | sustainable business network and consultancy | BSR. Retrieved September 6, 2024, from https://www.bsr.org/en/blog/what-business-needs-to-know-eu-corporate-sustainability-reporting-directive

Berry, B., Haverkamp, J., Isenhour, C., Bilec, M. M., & Lowden, S. S. (2023). Is convergence around the circular economy necessary? Exploring the productivity of divergence in US circular economy discourse and practice. *Circular Economy and Sustainability, 3*(3), 1597–1622. https://doi.org/10.1007/s43615-022-00199-1

Bhadra, U., & Mishra, P. P. (2021). Extended producer responsibility in India: Evidence from Recykal, Hyderabad. *Journal of Urban Management, 10*(4), 430–439. https://doi.org/10.1016/j.jum.2021.07.003

Borges, P. A. V., Gabriel, R., & Fattorini, S. (2021). Biodiversity erosion: Causes and consequences. In W. Leal Filho, A. M. Azul, L. Brandli, A. Lange Salvia, & T. Wall (Eds.), *Life on land* (pp. 81–90). Springer International Publishing. https://doi.org/10.1007/978-3-319-95981-8_78

Brockerhoff, E. G., Barbaro, L., Castagneyrol, B., Forrester, D. I., Gardiner, B., González-Olabarria, J. R., Lyver, P. O., Meurisse, N., Oxbrough, A., Taki, H., Thompson, I. D., van der Plas, F., & Jactel, H. (2017). Forest biodiversity, ecosystem functioning and the provision of ecosystem services. *Biodiversity and Conservation, 26*(13), 3005–3035. https://doi.org/10.1007/s10531-017-1453-2

Cannella, S., Bruccoleri, M., & Framinan, J. M. (2016). Closed-loop supply chains: What reverse logistics factors influence performance? *International Journal of Production Economics, 175*, 35–49. https://doi.org/10.1016/j.ijpe.2016.01.012

Cezarina, C., Viorel, P., & Afteni, M. (2021). Study on the transition from the linear economy to the circular economy. *Annals of Dunarea de Jos University of Galati, 39*. https://doi.org/10.35219/tmb.2021.1.08

Cicconi, P. (2020). Eco-design and eco-materials: An interactive and collaborative approach. *Sustainable Materials and Technologies, 23*, Article e00135. https://doi.org/10.1016/j.susmat.2019.e00135

Deloitte. (2022). Conscious consumerism: Understanding the dilemmas of today's consumer. Deloitte Netherlands. Retrieved September 10, 2024, from https://www2.deloitte.com/nl/nl/pages/consumer/articles/conscious-consumerism.html

De Faria, D. R., De Medeiros, J. L., & Araújo, O. Q. (2021). Sustainability assessment for the chemical industry: Onwards to integrated system analysis. *Journal of Cleaner Production, 278*, Article 123966. https://doi.org/10.1016/j.jclepro.2020.123966

Dhyani, S., Murthy, I. K., Kadaverugu, R., Dasgupta, R., Kumar, M., & Gadpayle, K. A. (2021). Agroforestry to achieve global climate adaptation and mitigation targets: Are South Asian countries sufficiently prepared? *Forests, 12*(3), 303. https://doi.org/10.3390/f12030303

Ellen MacArthur Foundation. (2020). Design and the circular economy: Deep dive. Ellen MacArthur Foundation. Retrieved September 12, 2024, from https://www.ellenmacarthurfoundation.org/design-and-the-circular-economy-deep-dive

Ellen Mac Arthur Foundation. (n.d). Circular economy introduction. Retrieved August 29, 2024, from https://www.ellenmacarthurfoundation.org/topics/circular-economy-introduction/overview

Environment, U. N. (2017, October 2). Sustainable consumption and production policies | UNEP—UN Environment Programme. https://www.unep.org/explore-topics/resource-efficiency/what-we-do/sustainable-consumption-and-production-policies

Environmental Degradation. (2018). Oxford reference. https://doi.org/10.1093/oi/authority.20110803095753548

Geissdoerfer, M., Savaget, P., Bocken, N. M. P., & Hultink, E. J. (2017). The circular economy—a new sustainability paradigm? *Journal of Cleaner Production, 143*, 757–768. https://doi.org/10.1016/j.jclepro.2016.12.048

Ghisellini, P., & Ulgiati, S. (2020). Circular economy transition in Italy. Achievements, perspectives and constraints. *Journal of Cleaner Production, 243*, 118360. https://doi.org/10.1016/j.jclepro.2019.118360

Global Trends of Material Use—Materialflows.net. (2021, July 19). https://www.materialflows.net/global-trends-of-material-use/

Global Waste to Grow by 70 Percent by 2050 Unless Urgent Action is Taken: World Bank Report. (2018). World Bank. Retrieved September 12, 2024, from https://www.worldbank.org/en/news/press-release/2018/09/20/global-waste-to-grow-by-70-percent-by-2050-unless-urgent-action-is-taken-world-bank-report

Guimarans, D., Arias, P., Tomasella, M., & Wu, C. (2019). A review of sustainability in aviation. *Elsevier eBooks*, 91–121. https://doi.org/10.1016/b978-0-12-814242-4.00004-1

Heidari, I., Eshlaghy, A. T., & Hoseini, S. M. S. (2023). Sustainable transportation: Definitions, dimensions and indicators—case study of importance-performance analysis for the city of Tehran. *Heliyon, 9*(10), Article e20457. https://doi.org/10.1016/j.heliyon.2023.e20457

Hidalgo-Crespo, J., Riel, A., Duberg, J. V., Bunodiere, A., & Golinska-Dawson, P. (2024). An exploratory study for product-as-a-service (PaaS) offers development for electrical and electronic equipment. *Procedia CIRP, 122*, 521–526. https://doi.org/10.1016/j.procir.2024.01.076

Hong, Y., Yan, W., & Ge, Q. (2023). Designing sustainable logistics networks for classified municipal solid wastes collection and transferring with multi-compartment vehicles. *Sustainable Cities and Society, 99*, Article 104921. https://doi.org/10.1016/j.scs.2023.104921

IBM. (n.d.). What is sustainable design? | IBM. (2024, August 13). https://www.ibm.com/think/topics/sustainable-design

International Resource Panel (IRP). (2019). Global resources outlook 2019: Natural resources for the future we want. United Nations Environment Programme.

IPBES. (2018). Summary for policymakers of the assessment report on land degradation and restoration. IPBES Secretariat.

Kerdlap, P., Low, J. S. C., & Ramakrishna, S. (2019). Zero waste manufacturing: A framework and review of technology, research and implementation barriers for enabling a circular economy transition in Singapore. *Resources, Conservation and Recycling, 151*, Article 104438. https://doi.org/10.1016/j.resconrec.2019.104438

Kosior, E., & Crescenzi, I. (2020). Chapter 16—solutions to the plastic waste problem on land and in the oceans. In T. M. Letcher (Ed.), *Plastic waste and recycling* (pp. 415–446). Academic Press. https://doi.org/10.1016/B978-0-12-817880-5.00016-5

Kottaridou, A., & Bofylatos, S. (2019, November 15). Design out waste methodology for circular economy.

Koul, B., Yakoob, M., & Shah, M. P. (2022). Agricultural waste management strategies for environmental sustainability. *Environmental Research, 206*, Article 112285. https://doi.org/10.1016/j.envres.2021.112285

Kumar, R., Kumar, A., & Saikia, P. (2022). Deforestation and forests degradation impacts on the environment. In V. P. Singh, S. Yadav, K. K. Yadav, & R. N. Yadava (Eds.), *Environmental degradation: Challenges and strategies for mitigation* (pp. 19–46). Springer International Publishing. https://doi.org/10.1007/978-3-030-95542-7_2

Leading Countries in Embracing Circular Economy. (2022). Retrieved September 12, 2024, from https://www.circle-economy.com/blog/pioneers-of-the-future-the-countries-leading-the-way-with-circular-economy-policy

Lederer, J., Gassner, A., Kleemann, F., & Fellner, J. (2020). Potentials for a circular economy of mineral construction materials and demolition waste in urban areas: A case study from Vienna. *Resources, Conservation and Recycling, 161*, Article 104942. https://doi.org/10.1016/j.resconrec.2020.104942

Li, F., Wu, S., Liu, H., & Yan, D. (2024). Biodiversity loss through cropland displacement for urban expansion in China. *Science of the Total Environment, 907*, Article 167988. https://doi.org/10.1016/j.scitotenv.2023.167988

Li, J., & Li, G. (2024). Natural resources utilization, digital technology and green development expenditures can reduce environmental stress: A case study of emerging economies. *Resources Policy, 95*, Article 105147. https://doi.org/10.1016/j.resourpol.2024.105147

Lin, K.-T., Pan, S.-Y., Yuan, M.-H., Guo, H.-Y., & Huang, Y.-C. (2024). Quantifying and monetarizing cropland ecosystem services towards sustainable soil management. *Ecological Indicators, 159*, Article 111751. https://doi.org/10.1016/j.ecolind.2024.111751

Linnenluecke, M. K. (2022). Environmental, social and governance (ESG) performance in the context of multinational business research. *Multinational Business Review, 30*(1), 1–16. https://doi.org/10.1108/MBR-11-2021-0148

Liu, J., Geng, Y., Chen, B., & Xia, X. (2021). The effect of a supplier's eco-design on the economic benefits of a supply chain and associated coordination. *International Journal of Environmental Research and Public Health, 18*(24), 13357. https://doi.org/10.3390/ijerph182413357

Luo, Y., Song, K., Ding, X., & Wu, X. (2021). Environmental sustainability of textiles and apparel: A review of evaluation methods. *Environmental Impact Assessment Review, 86*, Article 106497. https://doi.org/10.1016/j.eiar.2020.106497

Machado, C. G., Winroth, M. P., & Ribeiro da Silva, E. H. D. (2020). Sustainable manufacturing in Industry 4.0: An emerging research agenda. *International Journal of Production Research, 58*(5), 1462–1484. https://doi.org/10.1080/00207543.2019.1652777

Maja, M. M., & Ayano, S. F. (2021). The impact of population growth on natural resources and farmers' capacity to adapt to climate change in low-income countries. *Earth Systems and Environment, 5*(2), 271–283. https://doi.org/10.1007/s41748-021-00209-6

Masudin, I., Tsamarah, N., Restuputri, D. P., Trireksani, T., & Djajadikerta, H. G. (2024). The impact of safety climate on human-technology interaction and sustainable development: Evidence from Indonesian oil and gas industry. *Journal of Cleaner Production, 434*, Article 140211. https://doi.org/10.1016/j.jclepro.2023.140211

Mihai, F.-C., & Minea, I. (2021). Sustainable alternative routes versus linear economy and resources degradation in eastern romania. *Sustainability, 13*(19), Article 19. https://doi.org/10.3390/su131910574

Milanesi, M., Runfola, A., & Guercini, S. (2020). Pharmaceutical industry riding the wave of sustainability: Review and opportunities for future research. *Journal of Cleaner Production, 261*, Article 121204. https://doi.org/10.1016/j.jclepro.2020.121204

Mishra, R. K., Mentha, S. S., Misra, Y., & Dwivedi, N. (2023). Emerging pollutants of severe environmental concern in water and wastewater: A comprehensive review on current developments and future research. *Water-Energy Nexus, 6*, 74–95. https://doi.org/10.1016/j.wen.2023.08.002

Morseletto, P. (2020). A new framework for policy evaluation: Targets, marine litter, Italy and the Marine Strategy Framework Directive. *Marine Policy, 117*, Article 103956. https://doi.org/10.1016/j.marpol.2020.103956

Moshood, T. D., Nawanir, G., Mahmud, F., Mohamad, F., Ahmad, M. H., & AbdulGhani, A. (2022). Sustainability of biodegradable plastics: New problem or solution to solve the global plastic pollution? *Current Research in Green and Sustainable Chemistry, 5*, Article 100273. https://doi.org/10.1016/j.crgsc.2022.100273

MSC. (n.d.). What is sustainable fishing. MSC International—English. Retrieved September 9, 2024, from https://www.msc.org/what-we-are-ng/our-approach/what-is-sustainable-fishing

Munaro, M. R., Tavares, S. F., & Bragança, L. (2022). The ecodesign methodologies to achieve buildings' deconstruction: A review and framework. *Sustainable Production and Consumption, 30*, 566–583. https://doi.org/10.1016/j.spc.2021.12.032

Nawaz, M. A., Azam, A., & Bhatti, M. A. (2019). Natural resources depletion and economic growth: Evidence from ASEAN countries. *Pakistan Journal of Economic Studies (PJES), 2*(2), Article 2.

Neves, S. A., & Marques, A. C. (2022). Drivers and barriers in the transition from a linear economy to a circular economy. *Journal of Cleaner Production, 341*, Article 130865. https://doi.org/10.1016/j.jclepro.2022.130865

OECD. (2019). Global material resources outlook to 2060: Economic drivers and environmental consequences. OECD. https://doi.org/10.1787/9789264307452-en

Omojola, A., Kallon, D., & Bello, K. (2022). Resource recycling with the aim of achieving zero-waste manufacturing. *Sustainability, 14*, 4503. https://doi.org/10.3390/su14084503

Onifade, M., Zvarivadza, T., Adebisi, J. A., Said, K. O., Dayo-Olupona, O., Lawal, A. I., & Khandelwal, M. (2024). Advancing toward sustainability: The emergence of green mining technologies and practices. *Deleted Journal*. https://doi.org/10.1016/j.gsme.2024.05.005

Paleari, S. (2024). The EU policy on climate change, biodiversity and circular economy: Moving towards a Nexus approach. *Environmental Science & Policy, 151*, Article 103603. https://doi.org/10.1016/j.envsci.2023.103603

Payne, A., & Kwofie, E. M. (2024). Unleashing circular economy potential in agriculture: Integrating social impact assessment with the ReSOLVE framework as a tool for sustainable development. Sustainable Development, n/a(n/a). https://doi.org/10.1002/sd.2952

Population Matters. (2024). The facts: Resources and consumption. Population Matters. Retrieved September 12, 2024, from https://populationmatters.org/the-facts-resources-consumption/

Pybus, K., Power, M., Pickett, K. E., & Wilkinson, R. (2022). Income inequality, status consumption and status anxiety: An exploratory review of implications for sustainability and directions for future research. *Social Sciences & Humanities Open, 6*(1), Article 100353. https://doi.org/10.1016/j.ssaho.2022.100353

Reike, D., Vermeulen, W. J. V., & Witjes, S. (2018). The circular economy: New or Refurbished as CE 3.0?—exploring controversies in the conceptualization of the circular economy through a focus on history and resource value retention options. *Resources, Conservation and Recycling, 135*, 246–264. https://doi.org/10.1016/j.resconrec.2017.08.027

Sanyé-Mengual, E., Lozano, R. G., Farreny, R., Oliver-Solà, J., Gasol, C. M., & Rieradevall, J. (2014). Introduction to the eco-design methodology and the role of product carbon footprint. In S. S. Muthu (Ed.), *Assessment of carbon footprint in different industrial sectors* (Vol. 1, pp. 1–24). Springer. https://doi.org/10.1007/978-981-4560-41-2_1

Salvatori, G., Holstein, F., & Böhme, K. (2019). Circular economy strategies and roadmaps in Europe: Identifying synergies and the potential for cooperation and alliance building—study. *The European Economic and Social Committee.*

Séraphin, H., & Chaney, D. (2023). A research agenda for the sustainability of the tourism industry: A childism perspective on overtourism. *Journal of Cleaner Production, 414*, Article 137556. https://doi.org/10.1016/j.jclepro.2023.137556

Shetty, S. S., Deepthi, D., Harshitha, S., Sonkusare, S., Naik, P. B., & Madhyastha, H. (2023). Environmental pollutants and their effects on human health. *Heliyon, 9*(9). https://doi.org/10.1016/j.heliyon.2023.e19496

Shooshtarian, S., Maqsood, T., Wong, P. S., Khalfan, M., & Yang, R. J. (2021). Extended producer responsibility in the Australian construction industry. *Sustainability, 13*(2), Article 2. https://doi.org/10.3390/su13020620

Simonetto, M., Sgarbossa, F., Battini, D., & Govindan, K. (2022). Closed loop supply chains 4.0: From risks to benefits through advanced technologies. A literature review and research agenda. *International Journal of Production Economics, 253*, 108582. https://doi.org/10.1016/j.ijpe.2022.108582

Stories, W. D. (2024, February 1). What is sustainable waste management?—waste mission. https://wastemission.com/blog/what-is-sustainable-waste-management/

Sustainable Agriculture: Benefits & Management Tips. (2022, March 30). https://eos.com/blog/sustainable-agriculture/

Swiss Re Institute. (2021). The economics of climate change: The climate economics index. Swiss Re.

Team, N. I. C. (2023, November 15). Environment pollution: Classification, causes and its types. https://www.nextias.com/blog/environment-pollution

The Collision of Fast Fashion and E-Commerce: Unraveling the Impact on Clothing Waste. (2023). Expert EPR, ESG, sustainability, neutrality and plastic recycling consulting services | ProIndia. Retrieved September 10, 2024, from https://proindia.net/the-collision-of-fast-fashion-and-e-commerce-unraveling-the-impact-on-clothing-waste/

Tiwari, S., Si Mohammed, K., Mentel, G., Majewski, S., & Shahzadi, I. (2024). Role of circular economy, energy transition, environmental policy stringency and supply chain pressure on $CO_2$ emissions in emerging economies. *Geoscience Frontiers, 15*(3), Article 101682. https://doi.org/10.1016/j.gsf.2023.101682

Ukaogo, P. O., Ewuzie, U., & Onwuka, C. V. (2020). Environmental pollution: Causes, effects and the remedies. In *Microorganisms for sustainable environment and health* (pp. 419–429). Elsevier. https://doi.org/10.1016/B978-0-12-819001-2.00021-8

United Nation. (2023). Causes and effects of climate change. United Nations; United Nations. Retrieved September 12, 2024, from https://www.un.org/en/climatechange/science/causes-effects-climate-change

United Nations Environment Programme. (2019). UN calls for urgent rethink as resource use skyrockets. https://www.unep.org/news-and-stories/press-release/un-calls-urgent-rethink-resource-use-skyrockets

United Nations Environment Programme (UNEP). (2023). Bend the trend: Pathways to a liveable planet as resource use spikes—summary for policymakers. United Nations Environment Programme.

US EPA, O. (2016, June 27). Climate change indicators: Atmospheric concentrations of greenhouse gases [Reports and Assessments]. https://www.epa.gov/climate-indicators/climate-change-indicators-atmospheric-concentrations-greenhouse-gases

Wang, L., Deng, T., Shen, Z.-J.M., Hu, H., & Qi, Y. (2022). Digital twin-driven smart supply chain. *Frontiers of Engineering Management, 9*(1), 56–70. https://doi.org/10.1007/s42524-021-0186-9

UK Government. (2021, January). Waste management plan for England. Retrieved from https://assets.publishing.service.gov.uk/media/60103f71d3bf7f05bc42d294/waste-management-plan-for-england-2021.pdf

What a Waste: An Updated Look into the Future of Solid Waste Management. (2018). World Bank. Retrieved April 22, 2024, from https://www.worldbank.org/en/news/immersive-story/2018/09/20/what-a-waste-an-updated-look-into-the-future-of-solid-waste-management

World Economic Forum. (2019, February 26). These 11 companies are leading the way to a circular economy. World Economic Forum. https://www.weforum.org/agenda/2019/02/companies-leading-way-to-circular-economy/

World Wildlife. (n.d.). What is overfishing? Facts, effects and overfishing solutions. Retrieved September 12, 2024, from https://www.worldwildlife.org/threats/overfishing

Wilting, H. C., Schipper, A. M., Bakkenes, M., Meijer, J. R., & Huijbregts, M. A. J. (2017). Quantifying biodiversity losses due to human consumption: A global-scale footprint analysis. *Environmental Science & Technology, 51*(6), 3298–3306. https://doi.org/10.1021/acs.est.6b05296

World Economic Forum (WEF). (2020). The global risks report 2020. World Economic Forum.

World Economic Forum. (2022). Landmark study shows the circular economy can halt biodiversity loss. https://www.weforum.org/stories/2022/06/study-shows-circular-economy-can-halt-biodiversityloss/

Xintong, C., Bangyi, L., Zonghuo, L., Goh, M., & Shanting, W. (2021). Take-back regulation policy on closed loop supply chains: Single or double targets? *Journal of Cleaner Production, 283*, Article 124576. https://doi.org/10.1016/j.jclepro.2020.124576

Xu, J., Ng, C. T., & Cheng, T. C. E. (2021). Remanufacturing strategies under product take-back regulation. *International Journal of Production Economics, 235*, Article 108091. https://doi.org/10.1016/j.ijpe.2021.108091

Zattara, E. E., & Aizen, M. A. (2021). Worldwide occurrence records suggest a global decline in bee species richness. *One Earth, 4*(1), 114–123. https://doi.org/10.1016/j.oneear.2020.12.005

Ziani, K., Ioniță-Mîndrican, C.-B., Mititelu, M., Neacşu, S. M., Negrei, C., Moroşan, E., Drăgănescu, D., & Preda, O.-T. (2023). Microplastics: A real global threat for environment and food safety: A state of the art review. *Nutrients, 15*(3), 617. https://doi.org/10.3390/nu15030617

# Reimagining Sustainability: The Role of Circular Economy in Building Environmental Resilience

**Beata Zofia Filipiak and Denys Lutyi**

**Abstract** Overall, interest in the subject area is growing, and the differences between sustainability and circular economy require demonstration. In particular, interest in the circular economy has grown rapi over the past couple yearsdly. The chapter will aim to demonstrate how the concept of circular economy affects environmental resilience. The chapter will address the issue of the impact of the circular economy on the new sustainability paradigm. The issue of creating business models in the circular economy and creating the circular mechanisms of value chain transposition will be discussed. The authors will present the similarities and differences in approaches to resilience and sustainability in the circular economy. The environmental concern that affects environmental resilience and its connections with the circular economy will be discussed. The whole discussion will end with the establishment of transitions to climate-resilient development. The chapter offers a new look at the circular economy from the view of sustainability, creating new business models determined by a circular view of the economy and the value that allows for creating environmental resilience.

**Keywords** Circular economy · Environmental resilience · Circular mechanisms · Business models · Circular value

## 1 Introduction

Climate change affects all parts of the world, economic systems, the environment and humans. In some regions, extreme climate phenomena are becoming more frequent and causing irreversible changes in many areas of the economy. They constitute a very serious threat recognized as high risk, which affects many different aspects of social life and affects almost all sectors of the economy. It can be pointed out that

B. Z. Filipiak (✉) · D. Lutyi
University of Szczecin, Szczecin, Poland
e-mail: beata.filipiak@usz.edu.pl

D. Lutyi
e-mail: denys.liutyi@phd.usz.edu.pl



P. Singh et al. (eds.), *Circular Economy and Environmental Resilience*,
https://doi.org/10.1007/978-3-031-93091-1_5

not only laws and prohibitions are needed, but also the creation of mechanisms and good examples of actions that will effectively counteract climate change.

In the context of growing problems related to climate cha nge, and in particular excessive consumption of raw materials and waste production, the circular economy model is gaining importance, offering an alternative to the traditional linear model of the economy (Lubowiecki-Vikuk et al., 2021; Elgarahy et al., 2023). This method of managing natural resources aims to extend the life cycle of products, minimize waste and reuse materials, it is beneficial to the environment and humans. Circular Economy (CE) is a business models (B) that minimizes the use of raw materials and the generation of waste (Commoner, 1971; Hysa et al., 2020; Ogunmakinde et al., 2022; Mukherjee et al., 2023; Sanchez-García et al., 2024). Its goal is to reduce greenhouse gas emissions and the level of energy use. Such action helps to create a sustainable, low-emission and competitive economy implementing the principles of sustainable development based on financing that takes into account ESG factors.

There has been a discussion on the CE in the literature for many years, the paradigm is evolving, and various factors of key importance for CE are being analyzed (as shown below). However, there is a lack of a coherent view of how to navigate the changing principles and horizons of CE and how to achieve lasting Environmental Resilience.

The goals of this chapter are focused on indicating how navigating circular horizons towards sustainability. Moreover, of the chapter will be to demonstrate how the concept of circular economy affects environmental resilience. Therefore, four cognitive directions have been established (as auxiliary, intermediate goals). The first one aims to present the evolution of the Circular Economy (CE) paradigm based on the principles of 6R and the reasons for its strengthening. Of importance for the paradigm itself is to indicate how to shape new value based on the ideas of CE and key BI archetypes for the EC. Another auxiliary goal focuses on learning about the similarities and differences of approaches in the CE as key elements to build resilience and sustainability. The last secondary (intermediate) objective is to determine how environmental concern affects towards environmental resilience.

### *1.1 The Circular Economy as a New Sustainability Paradigm*

In the face of the climate crisis, limited resources, and environmental pollution, there is an urgent need to reformulate the basic principles of management in order to ensure the survival of human civilization and provide an acceptable quality of life. The current economic model was based on the "take, make, use, and throw away" scheme and wrongly assumed infinite natural resources and unlimited possibilities for waste storage. The circular economy (CE) opposes this model, pointing out that it is based on the thoughtless consumption of goods (take-make-waste). The CE is seen as a solution that allows for improving the quality of life while limiting the use of resources, without giving up on meeting social needs, but meeting them in an innovative way. The CE is understood as an economic system that replaces the concept of "end of life" of a product with reuse, recycling and recovery of materials in the production, distribution and consumption process, both at the micro-, meso- and macro- levels. Its goal is to achieve sustainable development, i.e. to simultaneously influence good environmental quality, economic prosperity and social equality, for the benefit of current and future generations (Kirchherr et al., 2017).

The approach and understanding of CE has evolved to incorporate different features and contributions from a variety of concepts that share the idea of closed loops. The following key concepts can be indicated in the literature, which had the most relevant theoretical influences on the development of the CE concept: laws of ecology (Commoner, 1971); regenerative design (Lyle, 1994); industrial ecology (Graedel & Allenby, 1995); cradle-to-cradle (McDonough & Braungart, 2002); biomimicry (Benyus, 2002); looped and performance economy (Stahel, 2010), the blue economy (Pauli, 2010); the impact on the natural environment (Rashid et al., 2013); circular product design (Bakker et al., 2014); closed loop value and supply chains (Wells & Seitz, 2005; Guide & Van Wassenhove, 2009; Govindan et al., 2015); the impact on an integrated view of sustainability (Muniz & Cruz, 2015); circular BI (Bocken et al., 2016); the impact on the CE by new technologies (Hysa et al., 2020; Sanchez-García et al. 2024); the links between CE and ESG (Ramakrishna & Ramasubramanian, 2024).

The CE emerges not only as one option to the linear system, but as an fundamental paradigm shift, representing a holistic approach to the problem of how to understand climate change affecting sustainable development and what goals to pursue in connection with it (Mukherjee et al., 2023; Sanchez-García et al., 2024). The CE focuses on providing a viable strategy to counteract the climate change that impact broadly understood ESG, in particular the "E" factor by minimizing environmental pollution, waste and impacting healthier lifestyles and environmentally friendly production, as well as strengthening actions for a future marked by prosperity, equity, and ecological balance (Abbate et al., 2023; Elgarahy et al., 2023; De Pascale et al., 2023). The literature points to the alignment of circular economic standards with the sustainable development goals (SDGs). In this respect for the CE, the following are important: responsible consumption and production, sustainable energy (renewable energy sources, green and blue energy), clean water supply and climate (Ogunmakinde et al., 2022; Khajuria et al., 2022; Ortiz-de-Montellano et al., 2023). Emerging innovations and new technologies (sustainable) play a special role in creating change, facilitating, and enhancing the principles of circularity (Khan et al., 2022; Farahbakhsh et al., 2023). The impact of innovation and new technologies is also significant, as they facilitate the circular flow of materials, and generate new employment opportunities in emerging sectors such as recycling, sustainable product design and responsible management (Sehnem et al., 2022; Jesus & Jugend, 2023). In turn, technological advances towards CE imply the application of circular rules and principles across various strata of society and their activities, that can contribute to economies, industries, and social structures, especially causing consideration of ESG factors, activities to preserve biodiversity and striving to minimize social inequalities (Böckel et al., 2021; Rusch et al., 2023).

The analysis of the indicated trends allows for the identification of principles, key mechanisms and distinguishing features of CE from the point of view of various perspectives (Table 1).

The CE has developed three basic principles of economy, called the three “R”: reduce, reuse, recycle. These principles have been expanded as 6 R and mean:

1. Refuse—don’t buy things that contribute to increasing the amount of waste.
2. Reduce—limit the amount of things you use, buy only what you really need.
3. Reuse—reuse, exchange and pass on unnecessary things/production factors, consider whether they can serve another function.
4. Repair—don’t throw away slightly damaged things right away, try to repair damaged things.
5. Recycle—segregate waste and contribute to the processing of secondary raw materials, e.g. glass bottles.
6. Rethink/Rot—don’t waste, try to use leftovers that you can’t use in another way, try to compost.

The indicated principles are often considered to be the principles that build the new paradigm of sustainability (Anderson, 2023). Table 2 indicates the directions of transition and inclusion of the CE in the paradigm of sustaianble development.

The CE, being in opposition to the linear approach, is included in the paradigm of sustainable development. The CE strengthens the degree of implementation of those SDG goals that are oriented towards resources, but also by creating new value it contributes to strengthening relationships with different stakeholders and affects the mitigation of the impact of ESG risk. Moreover, most conceptual frameworks dealing with circular BI are constructed with the goal of either linking sustainability science (Pieroni et al., 2019).

### *1.2 Business Models in the Circular Economy as a Determinant of Circular Mechanisms of Value Chain Transposition*

Important evidence for the inclusion of the CE idea as an element of the sustainable development paradigm is provided by the concept of building BI based on the concept of circular BI (Kirchherr & van Santen, 2019). The approach itself has evolved from a concept based on the functioning of closed material loops BI (Mentink, 2014), then as one that links the creation of economic value with the narrowing, delaying (reducing the speed of progress) or finalizing of resource loops (Antikainen & Valkokari, 2016; Bocken et al., 2016; Lewandowski, 2016), to one that takes into account an innovation perspective. The latter focuses on the dynamic process of redirecting BI into the direction of more sustainable or circular ways of working (Linder & Williander, 2017; Geissdoerfer et al., 2018; Pieroni et al., 2019; Gillabel et al., 2021).

The BI model is a conceptual tool to describe the way business is done based on the principles of how an entity creates delivers and captures value (Osterwalder et al.,

**Table 1** The key principles, key mechanisms and distinguishing features of CE from the point of view of various perspectives

| Perspective | Key mechanisms | Distinguishing features |
|---|---|---|
| Business perspective | Internalization of costs | Local models for delivering products and services reduce the transfer of costs and responsibility for the product and its impact on the environment |
| | | Internalization of external costs through economic policy instruments and new BI, "designing out waste", "your waste - my raw material" |
| | | Intelligent use of human labor, creating jobs in the regional economy, lifelong learning |
| | Business mechanizm | Functionality/performance as a source of value creation |
| | | Location of production and use tend to be more linked |
| | | User needs/wants drive the role of a product |
| | | Internal incentive to incorporate end-of-life phase in BI |
| | Resource use | Material and resource sufficiency and efficiency) |
| | | Decoupling economic growth from the consumption of limited resources |
| | Product | Functionality as a source of creating value |
| | | Creating benefits from combining locally-provided production and services |
| | | Extended producer responsibility |
| | | Creating products in response to market and consumer needs |
| | | Extending lifespan, sharing, increasing attractiveness (upgrading) |
| | | Design becomes the most important process that covers the entire product life cycle, taking into account the possibility of its repair and modernization (e.g. modular design, ecodesign) |
| | | Dematerialization, simplification, accessibility |
| Consumer perspective | Consumer mechanizm | Customer satisfaction is an important driver |
| | | Local-first attitude |
| | | Accessibility is the norm |
| | | End-of-use incentives incorporated |
| | Market attitudes of customer | Focusing on usability, functions, and needs that the product/service satisfies |
| | | Active participant in the process of creating goods and services that respond to real needs |
| | | Prosumer |

(continued)

**Table 1** (continued)

| Perspective | Key mechanisms | Distinguishing features |
|---|---|---|
| Policy perspective | Policy mechanizm | More focus on facilitating skilled workforce |
| | | Less risk for outsourcing jobs |
| | | Facilitate safe and healthy services with regulation |
| | | Facilitation of end-of-life management |
| | Rules determining market behavior | Stocks used in accordance with the 3R towards 6R principle and maintaining their value (not only quantity but also quality) |
| | | The slower resources flow in the economy, the better (reuse, cascading use) |
| | | The market supports the maintenance of the value of products in circulation by changing the orientation towards services and extending the useful life |
| | | Availability is a basic principle, it is possible to use without having ownership rights (the rights to use are acquired, not ownership of the thing) |

*Source* Own elaboration based on Andersen (2007), EEA Report (2017), Stahel (2019), Michalak et al. (2020)

2005). Value in sustainable BI is often the basis for defining the models themselves (Ziolo et al., 2021). It can be understood as (Bocken et al., 2014; Osterwalder & Pigneur, 2010; Zioło et al., 2022):

- the value proposition: the service or product offering and the target;
- the value creation and delivery: the services or products specific features and the delivery (supply) channels;
- the value capture: specificity (e.g. seasonality, variability) and cost structure, and sources of revenue.

Table 3 presents the general ideological assumptions for the Canvas model for an entity building its BI based on ecological innovations within the circular economy.

Circular value of circular BI focuses on value preservation and circular resource flows. Circular value propositions also allow entities to function more sustainably by facilitating increased their utilization and resource efficiency, and consumers to live more sustainably. As indicated in the literature, circular value propositions allow (Selvefors et al., 2024):

(a) respond to the needs of society everyday in a way that aligns with circular principles;
(b) enable circular user practices throughout the entire process of meeting needs;
(c) are attractive for users to adopt due to the implementation of climate goals based on circular resource use;
(d) help realize ESG, economic and climate benefits for users and entities.

**Table 2** The directions of transition and inclusion of the CE in the paradigm of sustainable development

| General direction | Type of relationship (degree of inclusion) | Short description |
|---|---|---|
| Goals | Necessary and strong but not sufficient conditional relation | Flexible and diverse goals based on the agent and their interests, aligned with the 17-SDGs, fitting within a closed-loop circular economy (CE) framework, ideally eliminating all resource input and leakage from the system |
| Perceptions of responsibilities | Degree relation | Responsibilities are shared, but not clearly defined for the paradigm of sustainable development, but private business and regulators/policymakers join the paradigm with a specific declaration/law/good practices in the scope of CE |
| Society | Active inclusion; necessary and gradual inclusivity | Collaborative, nexus oriented respect for rights (but the CE refers to resources, which means inclusion) |
| Environment (for the CE resources) | Strong conditional relation | Ransitioning from a single-criterion approach (focused on maximizing income) to a multi-criteria approach (considering selected characteristics), while managing limited resources; adopting a regenerative strategy for resources (preserving resource generation patterns), being adaptable in achieving desired outcomes, and ensuring equitable distribution of resources |
| Trade-off | Cost-benefit/trade-off relation | Having costs and benefits in regard to sustainability, which can also lead to negative outcome; fostering certain aspects of sustainability but lacking others |
| Beneficial (economic development and fair distribution of benefits resulting from it) | Beneficial relationship; subset relation (structured and unstructured) | Beneficial in terms of sustainability; one among several solutions for fostering a sustainable system; BI that fit into the general assumptions of sustainable BI, but retain their specificity |
| Financing | The degree of transformation measured by financial expenditure | Green finance, blue finance, including ESG as an important substantive decision criterion |
| Value | Creating new value based on the use of CE towards sustainability | Generating new value through the application of circular economy (CE) sustainability within a relational model by: sharing the CE narrative, collaboratively building knowledge for political decision-making, advancing industry, establishing local CE ecosystems, and enhancing BI; CE operations can contribute to sustainability by creating multidimensional value with and for stakeholders and by recognizing sustainability as a value in stakeholder relationships; sustainability value stands out as a value that is (only) created when societal-level sustainability aims are advanced |

*Source* Own elaboration based on Voulvoulis et al. (2022), Ortiz-de-Montellano et al. (2023), Tapaninaho and Heikkinen (2022)

**Table 3** Circular economy BI canvas

| **Partners** | **Key Activities** | **Value proposition** | **Customer relations** | **Customer segments** |
|---|---|---|---|---|
| As in traditional, linear BI. There is a requirement to create a new partnership for reverse logistics, or to include ecological criteria. The requirement for a new partnership will also be implemented in the area of financing (green, blue or financing taking into account ESG factors) | Production/Service Provision. Obtaining environmental certificates (credentials) for business (products/services) and circular economy credentials. Additional activities related to returns, recycling, energy efficiency, sustainable water and wastewater management, etc. | Economic value and ESG value. It results from increased efficiency of raw material consumption, reduced energy consumption of materials, sustainable water and waste management, reduced waste, reuse of waste, longer use of products, "extending life" for used products | As in traditional, linear BI. Using the image as a socially and environmentally responsible entity operating in accordance with the assumptions of the circular economy. Promoting new values based on ESG, circular economy and sustainability | Individual customers who are part of the idea and promote activities within the circular economy and sustainability. Entities in the green/blue supply chain. Entities purchasing energy service company services (based on sustainable energy). Beneficiaries of knowledge-based economy and sustainability training |
| | Key resources<br>As in traditional, linear BI. Differentiation is based on ecological, sustainable and circular economy resource management | | Distribution channels. Environmentally friendly. Reducing carbon footprint. Preferred channels with the least impact on the environment | |
| **Cost structure**<br>Costs generated by innovative pro-environmental activities (in line with the idea of a circular economy and sustainability towards counteracting climate change) | | | **Revenue structure**<br>Sale/lending of products/services certified as environmentally friendly, produced in accordance with the principles of a circular economy | |

*Source* Own elaboration based on Braun et al. (2021), Ziolo et al., (2021), Pollard et al. (2023)

In the literature on the subject, one can find a number of classifications of BI for the circular economy. Their changes occurred along with changes related to climate, concentration of activities for the development of stable rules for the circular economy, development of positions for the basic identification of principles, key mechanisms and distinguishing features of CE and specification of the importance of innovation for the CE. Table 4 presents the key archetypes of the typology of BI for the circular economy.

Due to the complexity of activities, economies, and development toward sustainability, it seems justified to state that there is no single "ideal" BI for CE. Each entity adopting a model to achieve its goals should first conduct an in-depth analysis of the conditions and only then select and implement an appropriate circular BI toward sustainability.

### *1.3 Resilience and Sustainability: Similarities and Differences of Approaches in the Circular Economy*

The concept of circular economy (CE) is built upon two fundamental principles—environmental resilience and sustainability. Often regarded as inseparable attributes of many circular economic systems, these two complex terms share similar goals, yet trying to achieve them using distinct approaches.

According to the widely accepted definition, environmental resilience refers to the ability of society, business or economies to respond effectively to the shocks of various natures—both internal and external—while minimizing welfare losses (Hallegatte, 2014). In the context of EC, this definition remains fully applicable, though bringing up additional aspects. Beyond the traditional understanding of the resilience as a framework of principles that allows the circular systems to sustain in face of challenges, it can also be viewed as a capacity to preserve the integrity of existing processes and resource loops while maintaining resource efficiency (Kennedy & Linnenluecke, 2022).

Sustainability, on the other hand, exhibits a more foundational vision over a long-term economic development of both society and economy. It focuses primarily on building systems capable to meet the needs of the present while preserving similar opportunities and resources for the future (Boar et al., 2020). In the literature on the circular economy (Geissdoerfer et al., 2017; Nikolaou & Tsagarakis, 2021), sustainability is often described as a paramount principle for creating and operating businesses that is aimed to ensure that use of resources, waste management and environmental impacts remain in line with long-term social and economic goals.

In fact, the very concept of CE dictates that both society as a whole, and businesses in particular, should strive to be resource-efficient (Hina et al., 2023), adaptive (Ferasso et al., 2020) and regenerative (Hopkinson et al., 2018) over the long-term. In this context, the terms of environmental resilience and sustainability become even more essential, as understanding how these frameworks complement and contrast

**Table 4** Key archetypes of the typology of BI for the EC

| Specificity—typology of BI for born-circular entities | Typology of BI due to the increasing degree of circularity | Typology of BI based on innovation |
|---|---|---|
| 1. **Circular inputs—**this model aims to use renewable, recycled or sustainable materials—named as 'inputs'. These are largely used in the production process to partially or even totally eliminate waste and pollution. This means lower costs for their production input because it comes from excess materials and recycled materials. They are based on recovery, quality and increased control of material flow | 1. **Resource models and design models**: (a) Models focus on the recovery and reuse of raw and processed materials and components; (b) the model is based on the design of a product process which is based on the minimum use of virgin materials. The idea is that many materials as possible are substitutable, and there is the least possible negative impact in terms of polluting emissions, depletion or social and ecological destruction | 1. **Circular start-ups**—refer to the creation of new BI that incorporate circular economy strategies outside of an existing company, the new company has its own brand, employees and resources, in its activities can be supported by nonindependent institutions, such as incubators or accelerators. The basic principles are: cycling, extending, intensifying, and/or dematerializing resource loop |
| 2. **Sharing economy concept—**the concept behind the sharing economy is that businesses share industrial assets with other companies, particularly idle assets. Thanks to the sharing, there is a reduction in purchasing costs, which helps cut costs. Thus, there is a higher utilisation of often expensive assets | 2. **lifetime extension models, platform (sharing) models, and product-as-a-service models**. This models are based on the assumption that products should be used longer, smarter, and better | 2. **circular business model diversification**—new BI operate within the existing organization using their resources and network. The currentBI of the parent organization stays in place, the new circular BI are either integrated into the organization as new businesses or spun-off as subsidiaries. The characteristic feature are joint BI innovation projects with other organizations |

(continued)

**Table 4** (continued)

| Specificity—typology of BI for born-circular entities | Typology of BI due to the increasing degree of circularity | Typology of BI based on innovation |
|---|---|---|
| 3. **Product-as-a-service**—customers purchase products as a service for a limited time specified contractually. The provider still has ownership of the products involved, upgrades and treats these products accordingly, increasing their lifecycle. The effect is to maximize the use of products and responsible disposal at the end of life | 3. **models indicating a growing degree of responsibility** (end-of-life models, and lifecycle models). They are based on the assumption that in the end-of-life phase there is the retrieval of raw/processed materials, components and products. They include producer responsibility for the entire lifecycle of a product, component or even raw/processed materials | 3. **circular BI transformation**—assumes the modification of an existing BI. It is assumed initially that both the conventional and circular models will function side by side, so as to arrive at the resulting BI incorporates CE strategies over time |
| 4. **Product use extension**—products are designed for repairability, upgradeability, reusability, ease of disassembly, reconditioning, and recyclability of all components. The approach ensures a continuing income stream throughout the product's usage cycles | | 4. **circular BI acquisition**—the model is based on merger and acquisition, that target BI that incorporate CE strategies. It comprises the identification, acquisition, and integration of new circularBI, and the process of integration may vary |
| 5. **Resource recovery**—model focuses on the end stages of the usage cycle, and involves the recovery of built-in materials, energy and resources from products that are no longer functional. The model assumes a direct economic interest in the extraction of all their products recoverable value and the design process focuses on making value recovery easy and effective. The business is based on stimulating the return of the products | | |

*Source* Own elaboration based on Lacy et al. (2020), Jonker et al. (2022), Geissdoerfer et al. (2020), Gillabel et al. (2021), Geissdoerfer et al. (2023)

with each other may be crucial for aligning existing business practices with the goals of CE.

There is still no consensus among researchers on the interconnection between resilience and sustainability. In their paper, Marchese et al. (2018) grouped the literature on environmental resilience and sustainability based on how the hierarchy between these two terms is built. The first framework that authors identified focuses

**Table 5** Key differences between resilience and sustainability in the concept of circular economy

| Resilience | Sustainability |
|---|---|
| 1. *Time perspective: long-term versus short-term* | |
| Resilience is often viewed as an immediate approach, meaning primarilily swift reaction to shocks and crises, while at the same time—controlling the damages, minimizing any disruptions and ensuring recovery of the system in the short-term perspective | Sustainability is rather perceived as long-term approach in the circular economic systems. It emphasizes developing and adopting such business practices that will be not only efficient "here and now", but also viable and stable in longer perspective |
| 2. *Approach perspective: reactivity versus proactivity* | |
| Resilience can be regarded as a reactive approach, particularly because of the fact that the central to the concept of the resilience as part of the CE systems still remains the ability to withstand and adapt to the disturbances, as well as effectively mitigate their negative impacts—rather than preventing them | Based on the fact the sustainability strives to build systems that preserve existing resources and opportunities for the future generations, it can equally be said that such systems are designed to be problem-avoiding, which in this context means a purely proactive approach |

*Source* Own elaboration based on Bansal and DesJardine (2014), Marchese et al. (2018), Suárez-Eiroa et al. (2021)

on studies that prioritize sustainability as the primary objective while resilience is considered as a mere aspect and a proxy of this broader concept (Jarzebski et al., 2016). In contrast, the second framework distinguishes studies that regard sustainability as a contributing factor to the resilience (Ahi & Searcy, 2013). Finally, the third framework considers these two terms as separate approaches that can either complement each other or complete with one another (Marchese et al., 2018).

In the context of CE, the key difference between resilience and sustainability can be described from the two major perspectives: time and approach (Table 5).

Despite obvious differences in the approaches, resilience and sustainability remain directly interconnected and interdependent within the concept of CE. Short-term and reactive nature of the environmental resilience is often complemented by long-term strategical vision of sustainability, which in combined allows to align existing business practices with the goals of the CE that strives to transform the world into environmentally-friendly and socially responsible place.

## 1.4 Environmental Concern Affect Towards Environmental Resilience—Links with Circular Economy

Public concerns about climate changes, environmental pollution and non-sustainable business practices has grown in recent years (Gifford & Nilsson, 2014). In this context, the concept of CE could become a key in addressing these environmental concerns as it offers a sustainable BI that is capable to ensure stability over the long-term (Haleem et al., 2021) while remaining more resilient and adaptive to environmental shocks than traditional BI. Given that resilience is one of fundamental principles of CE, the interconnections between these three terms need to be clearly explained.

Environmental concern can be defined as the awareness of the wide public in topics related to the environment, ecology and other related terms (Cruz & Manata, 2020). According to the findings of Deng et al. (2023) and earlier research of Gu et al. (2023), the majority of studies on environmental concern can be categorized into two groups depending on the subject of the concern: government or people. The first group emphasizes primarily governmental responsibilities, exploring so-called "top-down" approach, while the second focuses on wider societal perspective, representing "bottom-up" approach. Driven by the recent scientific findings and growing attention in the media, the people's heightened anxiety regarding climate changes and ecological issues made both businesses and policy-makers to pay greater attention to the concept of CE (Geissdoerfer et al., 2017) as one of the most effective ways to achieve long-term stability and sustainability for the society as a whole.

Consumer perception and attitude play a crucial role in forming demand for any products or services in the market, particularly sustainable ones (Lubowiecki-Vikuk et al., 2021). As businesses strive to satisfy the growing demand for these products and services, they are obliged to align their existing practices and operations with the principles and goals of sustainability. The fact that CE is perceived as a framework of principles that allow to achieve economic growth while staying aligned with principles of sustainability (Bigliardi et al., 2022; Kirchherr et al., 2017) makes it a perfect way to illustrates how wider environmental concerns could affect the promotion of principles of sustainability and environmental resilience.

Being one of cornerstones of the CE, resilience, in fact, is being affected in two significant ways. As circular practices evolve and become more widespread, the environmental resilience—as the associated principle of the CE—is gaining wider adoption as well. On other hand, the direct demand for environmental resilience—as an attribute of sustainable products and services—is gaining momentum too. Rising awareness and, what's more important, anxiety of the society regarding ongoing environmental issues pressure businesses to ensure that not only their practices and approaches, but also products themselves will be capable to withstand, adapt and respond swiftly and effectively to any shocks that may compromises principles of "reuse/remanufacture/recycle/recover" or negatively affect the integrity of resource loops and other processes aligned with the principles of sustainability.

This confirms the strong influence of public environmental concerns on environmental resilience based on its interplay and interdependence with the concept CE. While challenges in advancing principles of CE still persist (Takacs-Santa, 2007), particularly from the traditional businesses (Velenturf & Purnell, 2021), wider people's awareness and concern about ecological issues may appear to be a positive factor in promoting sustainability.

## *1.5 Transitions to Climate Resilient Development*

Climate-resilient development (CRD) is based on the approach of implementing climate action, including water protection, eliminating greenhouse gas emissions, especially $CO_2$, pro-environmental waste management and activities and risk reduction instruments, to support sustainable development for all (Trisos et al., 2022). The process of implementing CRD assumptions varies across countries and continents. There are common assumptions involving considering (in national policies or entire groups such as the European Union) a broader range of sustainable development priorities, policies and practices, as well as enabling societal choices to accelerate and deepen their implementation making climate action and sustainable development interdependent (Simpson et al., 2023). It is indicated in the literature that 'pathways' of actions, as an analytical tool, enable sequences of actions and increase the efficiency of processes (activities) within individual 'pathways' of actions for progressive implementation (Lin et al., 2017; Werners et al., 2021a, 2021b). They allow to deal with uncertainty by embedding flexibility within planning.

The result (outcomes) is future-oriented, action-oriented and requires monitoring and assessing whether existing solutions are effective, what actions should be taken to increase effectiveness and when more effective pathways for achieving predefined goals should be used (Lawrence & Haasnoot, 2017; Lin et al., 2017; Werners et al., 2021a, 2021b).

One of the most important 'pathways' of action is the path indicated by the World Bank Group as The Adaptation Principles (Hallegatte et al., 2020). They presented six universal 'Principles of Adaptation and Resilience' and 26 concrete actions that can be used to expand effective strategies and activities to strengthen sustainability through implementing the idea of CE. The indicated basic groups of activities, which are called the 6 'paths', are as follow (Hallegatte et al., 2020): (1) Build resilient and inclusive development; base with speedly; (2) Make it easier for people and businesses to fulfill their role; (3) Review and update land use plans and secure critical infrastructure; (4) Support people and companies in recovering more swiftly and efficiently; (5) Manage consequences and processes at the macro scale; (6) Focus on needs-based prioritization, implement across various sectors, and track progress.

The US administration has adopted a slightly different approach, indicating the need to take responsibility for climate risk accountability and adopt a framework of principles, which include (U.S., 2021): Mobilizing both public and private funds to facilitate the shift to a net-zero economy; Shielding climate-vulnerable and disadvantaged frontline communities; Mitigating financial risks for the government and the communities it serves; Ensuring the financial system is protected against climate-related financial risks; Showcasing global leadership. Actions towards transition to CRD are presented in Fig. 1.

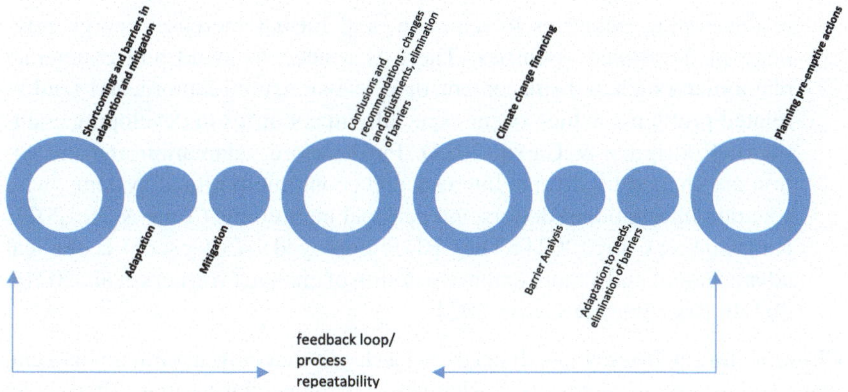


**Fig. 1** Actions towards transition to CRD. *Source* Own elaboration

Actions towards transition to CRD should focus on:

(1) **Shortcomings and barriers in adaptation and mitigation**:

(a) **Adaptation**—The IPCC Working Group II AR6 published in 2002 and UNEP's Adaptation GAP Report 2022 provide a starting point for findings on shortcomings and barriers in adaptation. There are also significant limitations in these efforts.These reports highlight that the adaptation measures are both insufficient and delayed. Progressis required and therefore additional shortcomings should be removed, such as: the lack of clarity on what defines effective and successful adaptations in current practices; adaptation resources are frequently allocated to support existing development activities and priorities, overlooking the root causes of vulnerability; adaptation efforts continue to prioritize incremental changes over the transformational measures needed to address future climate change impacts; adaptation costs are rising faster than the available funds, with a current gap likely five to ten times greater than current international adaptation finance flows; increasing evidence shows that financial providers are not strategically directing adaptation assistance to the most vulnerable countries and population groups; large-scale, non-climate, and compounding factors continue to undermine adaptation outcomes (Pörtner et al., 2022; UNEP, 2022).

(b) **Mitigation**—The climate change mitigation is passed on to developing countries, increasing the risk of lock-in and reducing their options to develop. Recent studies argue mitigation projects can have cascading negative consequences in developing countries (Bos & Gupta, 2019; Werners et al., 2021a, 2021b). On the other hand, it is argued that existing climate mitigation agendas under the Paris Agreement tend to maintain the status quo and perpetuate colonial inequalities through the appropriation of land in developing countries to maintain, and further increase, energy privilege in developed countries. There is a need to avoid and counteract phenomena such as displacement, dispossession, elite capture, and gender-related problems, which occur especially in countries in developing countries) (Rodríguez & Carril, 2024). Furthermore, adaptation and mitigation are shaped by diverse interests and economic-political systems, with separate international donors and national institutions (Yang & He, 2021; Rodríguez & Carril, 2024). The CRD is employed to foster socio-ecological advancement and a more extensive notion of change (Werners et al., 2021a, 2021b; Rodríguez & Carril, 2024).

(2) **Climate change financing**—It needs to be shown that current climate financial flows and estimated needs are inadequate and require increasing. There is an urgent necessity to transition to green and blue financing, making it more accessible, particularly from a regulatory standpoint and application procedures. A major role is attributed to changing the BI of the basic players on the financial market, especially the bank. Also important are stock exchanges and the coordination of the entire financial system towards CRD and sustainability rules (esspecialy ESG) by central banks (Ziolo et al., 2021).
(3) **Planning pre-emptive actions**—It is recommended that developing countries seek methods and paths to align a global emphasis on mitigation with its national and subnational effects, and simultaneously balance mitigation and adaptation at these levels (Rodríguez & Carril, 2024).

The approaches and voices in the discussion indicated above do not exhaust the broad discussion on the transition and transformation towards the CE and the implementation of the sustainability paradigm. The real problems and threats shown in the implementation of the CE in the sustainability paradigm will require financial resources and fast, adequate actions. New challenges will certainly also appear, which will require efficiency, adequate paths, tools and financing.

## 2 Conclusions

In this chapter, we presented how to navigate circular horizons toward sustainability, in what direction the CE paradigm is evolving, to achieve environmental resilience of the economy. The emphasis on transformation towards EC and sustainability requires numerous interactions, co-transformation with the environment and reorientation in BI. Therefore, CE transformation and building the environmental resilience of the economy at the same time has to consider the mutual dependence between socio-economic activities and environmental resilience principles in order to counter the negative effects of climate change. Based on an analysis of good practices, and empirical studies presented in the literature, we identify key conclusions that constitute a prioritization in building CE with simultaneous ecological integrity and a focus on efficiency improvements.

Our findings allow us to formulate the following conclusions:

- the actions taken so far are too slow, it is necessary to accelerate the implementation of the ideas and principles of the CE paradigm in all areas of the economy due to low climate resilience;
- it requires strengthening the tools and wider application of good practices;
- BI should not only take into account actions for sustainability but should also be part of CE and create new value;
- the new value should strive to strengthen climate-resilient development;
- the creation of climate-resilient development should be firmly anchored in legal regulations, good practices, BI, but also in sustainable financing;
- inter-sectoral cooperation between entities and financial institutions should be developed in order to create strong bonds conducive to consolidating the climate resilience of the economy.

## References


Abbate, S., Centobelli, P., Cerchione, R., Giardino, G., & Passaro, R. (2023). Coming out the egg: Assessing the benefits of circular economy strategies in agri-food industry. *Journal of Cleaner Production, 385*, Article 135665. https://doi.org/10.1016/j.jclepro.2022.135665

Ahi, P., & Searcy, C. (2013). A comparative literature analysis of definitions of green and sustainable supply chain management. *Journal of Cleaner Production, 52*, 329–341. https://doi.org/10.1016/j.jclepro.2013.02.018

Andersen, M. S. (2007). An introductory note on the environmental economics of the circular economy. *Sustainability Science, 2*(1), 133–140.

Anderson, K. (2023). The six rs of sustainability: What are they? *Greenly.* https://greenly.earth/en-gb/blog/company-guide/the-six-rs-of-sustainability-what-are-they

Antikainen, M., & Valkokari, K. (2016). A framework for sustainable circular business model innovation'. *Technology Innovation Management Review, 6*, 5–12. https://doi.org/10.22215/timreview/1000

Bakker, C. A., den Hollander, M. C., van Hinte, E., & Zljlstra, Y. (2014). *Products that last—product design for circular business models.* TU Delft Library.

Bansal, P., & DesJardine, M. R. (2014). Business sustainability: It is about time. *Strategic Organization, 12*(1), 70–78. https://doi.org/10.1177/1476127013520265

Benyus, J. M. (2002). *Biomimicry.* Harper Perennial.

Böckel, A., Nuzum, A. K., & Weissbrod, I. (2021). Blockchain for the circular economy: analysis of the research-practice gap. *Sustainable Production and Consumption, 25*, 525–539. https://doi.org/10.1016/j.spc.2020.12.006

Boar, A., Bastida, R., & Marimon, F. (2020). A systematic literature review: Relationships between the sharing economy, sustainability, and sustainable development goals. *Sustainability, 12*(17), 6744. https://doi.org/10.3390/su12176744

Bigliardi, B., Filippelli, S., & Quinto, I. (2022). Environmentally-conscious behaviours in the circular economy: An analysis of consumers' green purchase intentions for refurbished smartphones. *Journal of Cleaner Production, 378*, Article 134379. https://doi.org/10.1016/j.jclepro.2022.134379

Bocken, N. M. P., Short, S. W., Rana, P., & Evans, S. (2014). A literature and practice review to develop sustainable business model archetypes'. *Journal of Cleaner Production, 65*, 42–56. https://doi.org/10.1016/j.jclepro.2013.11.039

Bocken, N. M. P., de Pauw, I., Bakker, C., & van der Grinten, B. (2016). Product design and business model strategies for a circular economy. *Journal of Industrial and Production Engineering, 33*, 308–320.

Bos, K., & Gupta, J. (2019). Stranded assets and stranded resources: implications for climate change mitigation and global sustainable development. *Energy Research & Social Science, 56*, 101215.

Braun, A.-T., Schöllhammer, O., & Rosenkranz, B. (2021). Adaptation of the business model canvas template to develop business models for the circular economy. *Procedia CIRP, 99*, 698–702. https://doi.org/10.1016/j.procir.2021.03.093

Commoner, B. (1971). *The closing circle: Nature, man, and technology*. Random House.

Cruz, S. M., & Manata, B. (2020). Measurement of environmental concern: A review and analysis. *Frontiers in Psychology, 11*, 363. https://doi.org/10.3389/fpsyg.2020.00363

De Pascale, A., Di Vita, G., Giannetto, C., Loppolo, G., Lanfranchi, M., Limosani, M., & Szopik-Depczynska, K. (2023). The circular economy implementation at the European Union level. Past, present and future. *Journal of Cleaner Production*, 138658. https://doi.org/10.1016/j.jclepro.2023.138658

Deng, Y., Dong, K., Zhang, X., & Taghizadeh Hesary, F. (2023). How does environmental concern affect corporate resilience? Evidence from China's energy companies. *SSNR* (7 June 2023). https://doi.org/10.2139/ssrn.4471746

EEA Report. (2017). Circular by design—products in the circular economy. Luksemburg: *European Environment Agency*. https://circulareconomy.europa.eu/platform/sites/default/files/circular_by_design_-_products_in_the_circular_economy.pdf

Elgarahy, A. M., Priya, A. K., Mostafa, H. Y., Zaki, E. G., Elsaeed, S. M., Muruganandam, M., & Elwakeel, K. Z. (2023). Toward a circular economy: Investigating the effectiveness of different plastic waste management strategies: A comprehensive review. *Journal of Environmental Chemical Engineering*, 110993 https://doi.org/10.1016/j.jece.2023.110993

Farahbakhsh, S., Snellinx, S., Mertens, A., Belderbos, E., Bourgeois, L., & van Meensel, J. (2023). What's stopping the waste-treatment industry from adopting emerging circular technologies? An agent-based model revealing drivers and barriers. *Resources, Conservation and Recycling, 190*, 106792. https://doi.org/10.1016/j.resconrec.2022.106792

Ferasso, M., Beliaeva, T., Kraus, S., Clauss, T., & Ribeiro-Soriano, D. (2020). Circular economy business models: The state of research and avenues ahead. *Business Strategy and the Environment, 29*(8), 3006–3024. https://doi.org/10.1002/bse.2554

Geissdoerfer, M., Savaget, P., Bocken, N. M. P., & Hultink, E. J. (2017). The circular economy—a new sustainability paradigm? *Journal of Cleaner Production, 143*, 757–768. https://doi.org/10.1016/j.jclepro.2016.12.048

Geissdoerfer, M., Pieroni, M. P. P., Pigosso, D. C. A., & Soufani, K. (2020). Circular business models: A review. *Journal of Cleaner Production, 277*, 123741. https://doi.org/10.1016/j.jclepro.2020.123741

Geissdoerfer, M., Vladimirova, D., & Evans, S. (2018). Sustainable business model innovation: A review. *Journal of Cleaner Production, 198*, 401–416. https://doi.org/10.1016/j.jclepro.2018.06.240

Geissdoerfer, M., Santa-Maria, T., Kirchherr, J., & Pelzeter, C. (2023). Drivers and barriers for circular business model innovation. *Business Strategy and the Environment, 32*(6), 3814–3832. https://doi.org/10.1002/bse.3339

Gifford, R., & Nilsson, A. (2014). Personal and social factors that influence pro-environmental concern and behaviour: A review. *International Journal of Psychology*, January, n/a-n/a. https://doi.org/10.1002/ijop.12034

Gillabel, J., Manshoven, S., Grossi, F., Mortensen, L. F., & Coscieme, L. (2021). Business models in a circular economy. Eionet report—ETC/WMGE 2021/2, Mol: *European Topic Centre on Waste and Materials in a Green Economy.*

Govindan, K., Soleimani, H., & Kannan, D. (2015). Reverse logistics and closedloop supply chain: A comprehensive review to explore the future. *European Journal of Operational Research, 240*, 603–626.

Graedel, T. E., & Allenby, B. R. (1995).*Industrial ecology*. Prentice Hall, Englewood Cliffs.

Gu, K., Zhou, M., Luo, W., Liu, Y., Dou, P., Huang, C., & Di, Q. (2023). Long-term exposure to fine particulate matter and cardiovascular disease: The mitigation role of environmental concerns. *Science of The Total Environment, 875*, 162656. https://doi.org/10.1016/j.scitotenv.2023.162656.

Guide, V. D. R. J., Wassenhove, & Van, L. N. (2009). Or forum—the evolution of closed-loop supply chain research. *Operational Research, 57*, 10–18.

Hallegatte, S., Rentschler, J., & Rozenberg, J. (2020). *Adaptation principles: A guide for designing strategies for climate change adaptation and resilience*. World Bank. http://hdl.handle.net/10986/34780

Hallegatte, S. (2014). Economic resilience: Definition and measurement. *The World Bank. Policy Research Working Paper* 6852.

Haleem, A., Khan, S., Luthra, S., Varshney, H., Alam, M., & Khan, M. I. (2021). Supplier evaluation in the context of circular economy: A forward step for resilient business and environment concern. *Business Strategy and the Environment, 30*(4), 2119–2146. https://doi.org/10.1002/bse.2736

Hina, M., Chauhan, C., Sharma, R., & Dhir, A. (2023). Circular economy business models as pillars of sustainability: Where are we now, and where are we heading? *Business Strategy and the Environment, 32*(8), 6182–6209. https://doi.org/10.1002/bse.3480

Hopkinson, P., Zils, M., Hawkins, P., & Roper, S. (2018). Managing a complex global circular economy business model: Opportunities and challenges. *California Management Review, 60*(3), 71–94. https://doi.org/10.1177/0008125618764692

Hysa, E., Kruja, A., Rehman, N. U., & Laurenti, R. (2020). Circular economy innovation and environmental sustainability impact on economic growth: An integrated model for sustainable development. *Sustainability, 12*, 4831. https://doi.org/10.3390/su12124831

Pörtner, H.-O., Roberts, D. C., Poloczanska, E. S., Mintenbeck, K., Tignor, M., Alegría, A., Craig, M., Langsdorf, S., Löschke, S., Möller, V., & Okem, A. (2022). Summary for policymakers. IPCC. In Pörtner, H.-O., Roberts, D. C., Tignor, M., Poloczanska, E. S., Mintenbeck, K., Alegría, A., Craig, M., Langsdorf, S., Löschke, S., Möller, V., Okem, A., & Rama B. (Eds.), *Climate Change 2022: Impacts, adaptation, and vulnerability. Contribution of working group II to the sixth assessment report of the intergovernmental panel on climate change*. Cambridge University Press. https://doi.org/10.1017/9781009325844

Jesus, G. M. K., & Jugend, D. (2023). How can open innovation contribute to circular economy adoption? Insights from a literature review. *European Journal of Innovation Management, 26*(1), 65–98. https://doi.org/10.1108/ejim-01-2021-0022

Jarzebski, M. P., Tumilba, V., & Yamamoto, H. (2016). Application of a tri-capital community resilience framework for assessing the social–ecological system sustainability of community-based forest management in the Philippines. *Sustainability Science, 11*(2), 307–320. https://doi.org/10.1007/s11625-015-0323-7

Jonker, J., Faber, N., & Haaker, T. (2022). Quick scan circular business models. Inspiration for organising value retention in loops. *Ministry of Economic Affairs and Climate Policy The Hague*, The Netherlands. https://circulareconomy.europa.eu/platform/sites/default/files/quick-scan-circular-businessmodels_ebook.pdf

Kennedy, S., & Linnenluecke, M. K. (2022). Circular economy and resilience: A research agenda. *Business Strategy and the Environment, 31*(6), 2754–2765. https://doi.org/10.1002/bse.3004

Khajuria, A., Atienza, V. A., Chavanich, S., Henning, W., Islam, I., Kral, U., & Li, J. (2022). Accelerating circular economy solutions to achieve the 2030 agenda for sustainable development goals. *Circular Economy, 1*(1), 100001. https://doi.org/10.1016/j.cec.2022.100001

Khan, S. A. R., Piprani, A. Z., & Yu, Z. (2022). Digital technology and circular economy practices: Future of supply chains. *Operations Management Research, 15*(3–4), 676–688. https://doi.org/10.1007/s12063-021-00247-3

Kirchherr, J., Reike, D., & Hekkert, M. (2017). Conceptualizing the circular economy: An analysis of 114 definitions. *Resources, Conservation and Recycling, 127*, 221–232.

Kirchherr, J., & van Santen, R. (2019). Research on the circular economy: A critique of the field. *Resources, Conservation and Recycling, 151*, 104480. https://doi.org/10.1016/j.resconrec.2019.104480

Lacy, P., Long, J., & Spinder, W. (2020). *The circular economy handbook, realizing the circular advantage*. Palgrave Macmillan.

Lawrence, J., & Haasnoot, M. (2017). What it took to catalyse uptake of dynamic adaptive pathways planning to address climate change uncertainty. *Environmental Science & Policy, 68*, 47–57.

Lewandowski, M. (2016). Designing the business models for circular economy—towards the conceptual framework. *Sustainability, 8*(1), 43. https://doi.org/10.3390/su8010043

Lin, B. B., Capon, T., Langston, A., Taylor, B., Wise, R., Williams, R., & Lazarow, N. (2017). Adaptation pathways in coastal case studies: Lessons learned and future directions. *Coastal Management, 45*(5), 384–405.

Linder, M., & Williander, M. (2017). Circular business model innovation: Inherent uncertainties. *Business Strategy and the Environment, 26*(2), 182–196. https://doi.org/10.1002/bse.1906

Lubowiecki-Vikuk, A., Dąbrowska, A., & Machnik, A. (2021). Responsible consumer and lifestyle: Sustainability insights. *Sustainable Production and Consumption, 25*, 91–101. https://doi.org/10.1016/j.spc.2020.08.007

Lyle, J. T. (1994). *Regenerative design for sustainable development*. Wiley.

Marchese, D., Reynolds, E., Bates, M. E., Morgan, H., Spierre Clark, S., & Linkov, I. (2018). Resilience and sustainability: Similarities and differences in environmental management applications. *Science of the Total Environment, 613–614*, 1275–1283. https://doi.org/10.1016/j.scitotenv.2017.09.086

McDonough, W., & Braungart, M. (2002). *Cradle to cradle: Remaking the way we make things* (1st ed.). North Point Press.

Mentink, B. (2014). *Circular business model innovation: A process framework and a tool for business model innovation in a circular economy*. MSc Thesis Industrial Ecology. Delft University of Technology & Leiden University.

Michalak, J. L., Stralberg, D., Cartwright, J. M., & Lawler, J. J. (2020). Combining physical and species-based approaches improves refugia identification. *Frontiers in Ecology and the Environment, 18*, 254–260.

Mukherjee, S., Nagariya, R., Baral, M. M., Patel, B. S., Chittipaka, V., Rao, K. S., & Rao, U. A. (2023). Blockchain-based circular economy for achieving environmental sustainability in the Indian electronic MSMEs. *Management of Environmental Quality: An International Journal, 34*(4), 997–1017. https://doi.org/10.1108/MEQ-03-2022-0045

Muniz, R., & Cruz, M. J. (2015). Making nature valuable, not profitable: Are payments for ecosystem services suitable for degrowth? *Sustainability, 7*, 10895–10921.

Nikolaou, I. E., & Tsagarakis, K. P. (2021). An introduction to circular economy and sustainability: Some existing lessons and future directions. *Sustainable Production and Consumption, 28*, 600–609. https://doi.org/10.1016/j.spc.2021.06.017

Ogunmakinde, O. E., Egbelakin, T., & Sher, W. (2022). Contributions of the circular economy to the UN sustainable development goals through sustainable construction. *Resources, Conservation and Recycling, 178*, 106023. https://doi.org/10.1016/j.resconrec.2021.106023

Ortiz-de-Montellano, C. G. S., Samani, P., & van der Meer, Y. (2023). How can the circular economy support the advancement of the Sustainable Development Goals (SDGs)? A comprehensive analysis. *Sustainable Production and Consumption, 40*, 352–362. https://doi.org/10.1016/j.spc.2023.07.003

Osterwalder, A., & Pigneur, Y. (2010). B*usiness model generation: A handbook for visionaries, game changers, and challengers*. Wiley.

Osterwalder, A., Pigneur, Y., & Tucci, C. L. (2005). Clarifying business models: Origins, present, and future of the concept. *Communications of the Association for Information Systems*, 16. https://doi.org/10.17705/1CAIS.01601

Pauli, G. A. (2010). *The blue economy: 10 years, 100 innovations, 100 million jobs*. Paradigm Publications, Taos, NM.

Pieroni, M. P. P., McAloone, T. C., & Pigosso, D. C. A. (2019). Business model innovation for circular economy and sustainability: A review of approaches. *Journal of Cleaner Production*, 215, 198–216. https://doi.org/10.1016/j.jclepro.2019.01.036

Pollard, J., Osmani, M., Grubnic, S., Díaz, A. I., Grobe, K., Kaba, A., Ünlüer, Ö., & Panchal, R. (2023). Implementing a circular economy business model canvas in the electrical and electronic manufacturing sector: A case study approach. *Sustainable Production and Consumption, 36*, 17–31. https://doi.org/10.1016/j.spc.2022.12.009

Takacs-Santa, A. (2007). Barriers to environmental concern. *Human Ecology Review, 14*, 26–38.

Ramakrishna, S., Ramasubramanian, B. (2024). ESG and circular economy. In *Handbook of materials circular economy*. Springer. https://doi.org/10.1007/978-981-97-0589-4_8

Rashid, A., Asif, F. M., Krajnik, P., & Nicolescu, C. M. (2013). Resource conservative manufacturing: An essential change in business and technology paradigm for sustainable manufacturing. *Journal of Cleaner Production, 57*, 166–177.

Rodríguez, R. A. S., & Carril, L. R. F. (2024). Climate-resilient development in developing countries. *Current Opinion in Environmental Sustainability, 66*, 101391. https://doi.org/10.1016/j.cosust.2023.101391

Rusch, M., Schöggl, J. P., & Baumgartner, R. J. (2023). Application of digital technologies for sustainable product management in a circular economy: A review. *Business Strategy and the Environment, 32*(3), 1159–1174. https://doi.org/10.1002/bse.3099

Sánchez-García, E., Martínez-Falcó, J., Marco-Lajara, B., & Manresa-Marhuenda, E. (2024). Revolutionizing the circular economy through new technologies: A new era of sustainable progres. *Environmental Technology & Innovation, 33*, 103509. https://doi.org/10.1016/j.eti.2023.103509

Sehnem, S., de Queiroz, A. A. F. S., Pereira, S. C. F., dos Santos Correia, G., & Kuzma, E. (2022). Circular economy and innovation: A look from the perspective of organizational capabilities. *Business Strategy and the Environment, 31*(1), 236–250. https://doi.org/10.1002/bse.2884

Selvefors, A., Renström, S., Whalen, K. A., Fallahi, S., Leivas, M., Nordenö, H., & Fransson, A. (2024). User-centered circular value propositions—approaches in practice and research. *Resources, Conservation and Recycling, 207*, 107628. https://doi.org/10.1016/j.resconrec.2024.107628

Simpson, N. P., Simpson, K. J., Ferreira, A. T., Constable, A., Glavovic, B., Hallstrøm Eriksen, S. E., Ley, D., Solecki, W., Rodríguez, R. S., & Stringer, L. C. (2023). Climate-resilient development planning for cities: Progress from Cape Town. *npj Urban Sustain, 3*, 10. https://doi.org/10.1038/s42949-023-00089-x

Stahel, W. R. (2010). *The performance economy* (2nd ed.). Palgrave Macmillan, Basingstoke.

Stahel, W. R. (2019). *The circular economy: A user's guide*. Routledge.

Suárez-Eiroa, B., Fernández, E., & Méndez, G. (2021). Integration of the circular economy paradigm under the just and safe operating space narrative: Twelve operational principles based on circularity, sustainability, and resilience. *Journal of Cleaner Production, 322*, Article 129071. https://doi.org/10.1016/j.jclepro.2021.129071

Tapaninaho, R., & Heikkinen, A. (2022). Value creation in circular economy business for sustainability: A stakeholder relationship perspective. *Business Strategy and the Environment, 31*(6), 2728–2740. https://doi.org/10.1002/bse.3002

Trisos, C., Adelekan, I., Totin, E., Ayanlade, A., Efitre, J., Gemeda, A., Kalaba, K., Lennard, C., Masao, C., Mgaya, Y., Ngaruiya, G., Olago, D., Simpson, N., Zakieldeen, R., & Thorn, J. (2022). Africa. In *Climate change 2022: Impacts, adaptation and vulnerability. Contribution of working group II to the sixth assessment report of the intergovernmental panel on climate change* (pp. 1171–1274). Cambridge University Press. https://doi.org/10.1017/9781009325844.011

U.S. (2021). *A roadmap to build a climate-resilient economy, U.S. climate-related financial risk executive order*, 14030. The White House.

Velenturf, A. P. M., & Purnell, P. (2021). Principles for a sustainable circular economy. *Sustainable Production and Consumption, 27*, 1437–1457. https://doi.org/10.1016/j.spc.2021.02.018

UNEP. (2022). *The adaptation gap report. Too little, too slow.* United Nations Environment Programme.

Voulvoulis, N., Giakoumis, T., Hunt, C., Kioupi, V., Petrou, N., Souliotis, I., Vaghela, C., & Wan Rosely, W. I. H. binti. (2022). Systems thinking as a paradigm shift for sustainability transformation. *Global Environmental Change, 75*, 102544. https://doi.org/10.1016/j.gloenvcha.2022.102544

Wells, P., & Seitz, M. (2005). Business models and closed-loop supply chains: A typology. *Supply Chain Management, 10*, 249–251.

Werners, S., Sparkes, E., Totin, E., Abel, N., Bhadwal, S., Butler, J., Douxchamps, S., James, E., Methner, N., Siebneck, J., Stringe, R. L., Vincen, T. K., Wise, R., & Tebboth, M. (2021a). Advancing climate resilient development pathways since the IPCC's fifth assessment report. *Environmental Science & Policy, 126*, 168–176. https://doi.org/10.1016/j.envsci.2021.09.017

Werners, E. S., Sparkes, E., Totin, E., Abel, N., Bhadwal, S., Butler, J. R. A., Douxchamps, S., James, H., Methner, N., Siebeneck, J., Stringer, L. C., Vincent, K., Wise, R. M., & Tebboth, M. G. L. (2021b). Advancing climate resilient development pathways since the IPCC's fifth assessment report. *Environmental Science & Policy, 126*, 168–176. https://doi.org/10.1016/j.envsci.2021.09.017

Yang, B., & He, J. (2021). Global land grabbing: A critical review of case studies across the world. *Land, 10*, 324. https://doi.org/10.3390/land10030324

Ziolo, M., Filipiak, B. Z., & Tundys, B. (2021). *Sustainability in bank and corporate business models. The link between ESG risk assessment and corporate sustainability* (pp. 118–124). Palgrave Macmilan.

Zioło, M., Bąk, I., Filipiak, B. Z., & Spoz, A. (2022). In search of a financial model for a sustainable economy. *Technological and Economic Development of Economy, 28*(4), 920–947. https://doi.org/10.3846/tede.2022.16632

# Circular Harmony: Role of Circular Economy in Enhancing Environment Resilience

**Assunta Di Vaio, Anum Zaffar, and Daniel Balsalobre-Lorente**

**Abstract** The circular economy (CE) tackles global issues like pollution, waste, biodiversity loss, and climate change by decreasing waste, reusing materials, and recycling products to create a closed-loop system. It requires creative commercial strategies. By separating economic progress from environmental degradation, CE practices promote resilience by extending the lifespan of products and materials. The foundation of environmental resilience is enhancing ecosystems' capacity to endure shocks by reducing resource extraction and waste production. This chapter discusses the importance of circular harmony, emphasizing CE and environmental resilience by adopting decarbonization practices. The emergence of the CE concept offers numerous economic sectors, new potential for resource utilization and a new phase of development. The study directs the real issue of ecological development harmony under the framework of the CE to develop sustainable business models. Sustainable business models based on CE are more ecologically conscious and utilize the appropriate resources to increase firms' performance. This chapter further focuses on decarbonization practices to understand the role of CE in decarbonizing processes and the challenges of maintaining harmony between environmental and

A. Di Vaio (✉) · A. Zaffar
University of Naples Parthenope, Naples, Italy
e-mail: susy.divaio@uniparthenope.it

A. Zaffar
e-mail: anum.zaffar001@studenti.uniparthenope.it

D. Balsalobre-Lorente
Department of Applied Economics I, University of Castilla-La Mancha, Cuenca, Spain

UNEC Research Methods Application Center, Azerbaijan State University of Economics (UNEC), Baku, Azerbaijan

Faculty of Economics and Management, Department of Management, Czech University of Life Sciences Prague, Prague, Czech Republic

Western Caspian University, Economic Research Center (WCERC), Baku, Azerbaijan

D. Balsalobre-Lorente
e-mail: daniel.balsalobre@uclm.es

 
P. Singh et al. (eds.), *Circular Economy and Environmental Resilience*,
https://doi.org/10.1007/978-3-031-93091-1_6

social well-being. To understand these issues, supported by the institutional, legitimacy, resilience and stakeholder theories, this chapter seeks to develop a new theoretical framework addressing reshaping the organizational culture to achieve sustainable development goals (SDGs) by adopting decarbonization practices through green technology and renewable energy resources.

**Keywords** Circular Economy (CE) · Circular harmony · Decarbonization practices · Environment resilience · Sustainable Development Goals (SDGs)

## 1 Introduction

The idea of a "Circular Economy (CE)" arises as an attempt to address environmental issues such as reduced resource availability, biodiversity degradation, and climate variability through deliberate redesign of processes and products, material recycling, and the use of clean energy sources in the operational processes of firms (Lakshmi et al., 2024). The latter's detrimental effects endanger the integrity of natural ecosystems and the stability of economies, which are vital to human survival (Geng et al., 2012; Park & Chertow, 2014). The novel idea and development model of the CE has drawn increasing attention globally over the past ten years to offer a superior substitute for the prevalent "take, make, and dispose" paradigm of economic development. In this connection, numerous CE and sustainable development guidelines have been passed in many different geographical areas. The European Union (EU) has set up guidelines for the development of CE in its member states through the publication "Towards a circular economy: A zero-waste program for Europe" and its addendum (European Commission [EC], 2014). The concerning plastic materials (EC, 2018a), energy extraction from waste (EC, 2017), basic materials (EC, 2018b), sustainable power sources, energy productivity, and eco-design (EC, 2015a) are some of the continuing initiatives of the UN. In order to advance economic and social progress, the EC adopted guidelines in 2015 that aimed to move Europe closer to CE while also generating new jobs (EC, 2015a). In order to transition from a linear economy to a CE by the end of 2030, the EU set goals to reduce landfills by 10% overall and recycle 75% of packaging waste and 65% of municipal trash (EC, 2015b). Customers, stakeholders, and regional and national governments are encouraged by the EU Commission to take constructive action (EC, 2015c). Actions to address environmental issues have been dominated by international agreements, typically mediated by the United Nations. The most influential voice for climate change policy is still the Intergovernmental Panel on Climate Change, and efforts to stop biodiversity loss are also led by the Intergovernmental Science-Policy Platform on Biodiversity and Ecosystem Services.

Moreover, environmental resilience has been seen as a necessary condition for sustainability; therefore, it is required to investigate the transformation of those systems that threaten planetary resilience (Folke et al., 2010). Biggs et al. put up a framework for enhancing the socio-ecological systems' resilience to coexist with a natural system with biophysical boundaries. Depending on the application, resilience can be approached from various angles. It can be defined as the system's capacity to adjust and constructively change in response to a disturbance from the standpoint of the social-ecological system (Folke et al., 2010; Walker et al., 2004). The condition "positively" meant a new, better, and more sustainable system structure (Masnavi et al., 2019). From a narrative perspective, resilience refers to the structural components that deal with change and provide insight into the potential for transitioning to more sustainable and circular systems (Suárez-Eiroa et al., 2021). However, as far as our knowledge, CE and resilience thinking have not yet been explicitly combined. Therefore, to advance CE, we proposed a new research agenda that links EC with environment resilience, emphasizing circular harmony with the adoption of decarbonization practices.

Following a literature review, this chapter establishes a linkage between CE and circular harmony to achieve environmental resilience and lead towards sustainable development goals (SDGs). The analysis is based on institutional, legitimacy, resilience and stakeholder theories. Furthermore, the chapter also addresses the environmental issues caused by the operational processes of firms and the adoption of decarbonization practices via green technology and renewable resources to tackle those challenges balancing economic, social and environmental pillars of sustainability. The impact of people's activities after becoming aware of environmental hazards on accomplishing the CE's goals has not yet received significant attention in CE research (United in Science, 2022). The chapter focuses on decarbonization practices to understand the role of CE in decarbonizing processes and the challenges of maintaining harmony between environmental and social well-being. The study focuses on the real issue of ecological development harmony under the framework of the CE. This research aims to establish guidelines for assessing the harmony of environmental development in the development of CE. When areas' ecological growth aligns with the framework of the CE, there is an opportunity to establish a specific equilibrium of environmental conditions that further provides information on how the CE is developing to receive greater economic efficiency and ecological procedures (Weigend Rodríguez et al., 2020). The necessary steps to counter the challenges and utilize the benefits of the resources without distorting the environment should be the new goal for everyone.

This chapter is based on an analysis of the literature. A literature review enables us to compile current research on a specific topic (Leventis et al., 2024). Systematic literature reviews (SLRs) are a useful technique in accounting and management studies for thoroughly gathering the body of knowledge on subjects that have received little attention from academics (Burritt et al., 2023; Damschroder et al., 2022; Hardies et al., 2024). By considering new conceptual frameworks and research methodologies, SLRs further facilitate an inclusive interpretation of earlier scientific findings (Di Vaio et al., ). Motivated by this approach, this chapter identifies crucial scholarly work on Circular Harmony and Environmental Resilience through an organized literature review that employs important eligibility criteria. This chapter stresses the adoption of decarbonization practices via green technologies and renewable resources and how these adoptions impact the transition from CE to circular harmony to achieve environmental resilience and lead towards SDGs. Specific keywords were used to retrieve data from the Scopus database published from 1990 to the third quarter of 2024. We used the following keywords to design our search strings: "Climate Change and Sustainab* Development and Circular Economy or CE and Circular Harmony"; "Greenhouse Gas or Gas Emission* or GHG Emission* or Carbon Emission* or CO2 Emission* and Environmental Impact and Circular Economy or CE or Circular Harmony"; "Recycl* or Reuse or Waste Management and Sustainability Theory and Circular Economy or CE or Circular Harmony"; "Renewable Resource* or Green Technolog* or Decarbonization or Decarbonisation and Sustainable Development Goal* or SDG* and Legitimacy Theory and Circular Economy or CE or Circular Harmony and Resilience Theory or Environmental Resilience"; "Circular Economy or CE and Environmental Issues or Environmental Crises and Institutional Theory and Social Well-Being and Net-Zero and Business Model*".

The division of succeeding sections of this chapter is as follows: Sect. 2 explains the concept of circular harmony concerning the performance of firms. Section 3 puts forward the view of CE to achieve environmental resilience. Section 4 describes the linkage of circular harmony with an emphasis on institutional, legitimacy, resilience and stakeholder theories to shift from CE to circular harmony, balancing environmental resilience and the social well-being of humans. Section 5 describes adopting decarbonization practices to achieve circular harmony towards SDGs. Finally, Sect. 6 depicts the conclusions, containing the study's key limitations and directions for future research.

## 2 Circular Harmony: A Conceptualization Within the Performance Paradigm

Climate change adaptation is adjusting the actual and expected climate and its impact by encompassing the range of ecosystems-based approaches such as erosion control, drought or flood prevention to increase the resilience to climate change. In this connection, CE is considered an innovative business model that can harmonize sustainable development and society (Di Vaio et al., 2024b, 2024c; Lett, 2014). The traditional linear economy model caused economic inefficiency, resource depletion, and environmental deterioration (Kennedy et al., 2024). According to Bjørnbet et al. (2021), CE is becoming more attention-seeking as people become more conscious of resource scarcity and economic activity's environmental effects. The concept of CE has evolved since the 1960s. Boulding (1966) was the first author to significantly impact the CE with his groundbreaking essay "The Economics of the Coming Spaceship Earth". More than 53 years ago, this author wrote about the two opposing forms (linear and circular) of extraction, production, and consumption, which are still up for debate today. CE was also affected by Stahel (1982). He talked about extending the use life of goods to move towards a sustainable society and introduced a "performance economy" based on a spiral-loops system that "minimizes matter, energy-flow, and environmental deterioration without restricting economic growth or social and technological progress" (Stahel, 1982: p. 74). The idea of "industrial ecology," as Frosch and Gallopoulos (1989) proposed, is the third factor that influenced CE. The authors made a compelling case for the traditional industrial model back in 1989, which has not changed much since then and is primarily focused on maximizing short-term gains for producers and consumers rather than the economy as a whole. An "industrial ecology" was put up by Frosch and Gallopoulos (1989) as a comprehensive strategy that looks for a long-term equilibrium between environmental demands and economic gains. The idea that "waste equals food" suggests that leftovers from one industrial process can be used as raw materials for another, which is a revolutionary idea put forth by these authors. This proposal lessens firms' environmental impact. In order to "replace the present linear system of throughput flows with cyclical flows at sources, consumption centres, and sinks," Lyle (1996) invented the fourth influence, "regenerative design" (p. 10). Additionally, according to Lyle (p. 10), this suggested regenerative system "provides continuous replacement, through its own functional processes, of the energy and materials used in its operation". This author suggested using less fossil fuel and more renewable energy, maximizing the usage of resources through reuse, and minimizing waste volumes so that the environment can re-assimilate them without causing irrevocable harm. Therefore, CE seeks to minimize water and energy use while maximizing resource utilization product durability, and reducing carbon emissions, plastics, and organic waste (Lakshmi et al., 2024). As these challenges intensify, the need for CE provides a sustainable solution by promoting reuse, repair, refurbishment, and recycling. Nevertheless, the economic advantages of CE practices are not sufficiently supported by actual data (Lakshmi et al., 2024). By examining the impact of CE on economic growth and finding quantifiable advantages, challenges, and viable implementation strategies, this study seeks to bridge the gap existing in the current literature. Charonis (2012) defines CE as a system intended to be regenerative and restorative. He says CE is not an "alternative to growth discourse" but rather an "alternative growth discourse". The CE practices support building and restoring a healthy society, ecosystem and human health while carefully handling the resources to create value so that a shortage of raw materials does not impede economic progress. CE is a tool for accomplishing a long list of SDGs (Lakshmi et al., 2024). In the past, waste management was thought of as merely a means of disposing of waste products by incineration or landfilling. This is still the most common method of disposal in the world, which has a significant negative impact on the environment and results in a considerable loss of precious resources. A new perspective on garbage has recently emerged, acknowledging waste management as resource recovery and environmental impact prevention. With the rise of new types of operators and processes, such as the so-called "scavengers" and "decomposers," which refer to firms that can extract resources from waste by utilizing cutting-edge recovery technologies, waste management thus becomes a significant sub-sector of the CE (Ghisellini et al., 2016).

The economic, environmental, technological, and social facets of an economy, industry, or specific firm's operational processes must be balanced and considered simultaneously, as well as how these factors interact to achieve sustainable development (Ghisellini et al., 2016). The CE's basic logic, which is primarily based on political and environmental considerations as well as commercial and economic ones, helps to harmonize all the components (Birat, 2015). In addition, CE encourages more environmentally responsible and appropriate resource use to establish a greener economy, which will be characterized by a new business model and creative job opportunities (Stahel, 2014), as well as by increased well-being and noticeable effects on equity within and between generations in terms of resource use and access (MacArthur, 2013). Most of the time, CE has been thought of as a method for better waste management. Such a narrow perspective could cause CE to fail because some recycling, reuse, or recovery options might not be suitable in a particular situation but might work in another. In addition, some conversion options based on biotechnology and green chemistry might be significantly more costly and impactful than the conventional technology addressed, prioritizing prevention over treatment (Ghisellini et al., 2016). Overall, the challenge of preventative and regenerative eco-industrial development is not a "more of the same" approach that calls for increased use of "green" technologies. Green technologies refer to using technology to create positive or reduce negative environmental impacts (Du & Li, 2019). According to Loeser (2013), "green" describes environmentally friendly technologies and procedures with fewer negative environmental effects, such as reduced pollution, lower input resource requirements, and material reuse throughout their life cycles.

Moreover, adopting digital technologies increases the firms' performance in terms of CE (Di Vaio et al., 2023a, 2023b). Hence, the adoption of decarbonization practices calls for a much more thorough and comprehensive look at the design of radical alternative solutions, considering the entire life cycle of any process, as well as the interaction between the process and the environment and the economy in which it is embedded (Geng et al., 2014). This way, the regeneration is not just about recovering materials or energy but also about improving the entire living and economic model compared to the previous business-as-usual economy and resource management (Ghisellini et al., 2016). Even though the CE has acquired popularity in the corporate, political, and scientific spheres, there is ongoing debate regarding its theoretical underpinnings and key role within the sustainable development paradigm (Prieto-Sandoval et al., 2018; Schöggl et al., 2020). Although it is generally acknowledged that CE can help move toward sustainable development (Geissdoerfer et al., 2017; Schroeder et al., 2019), it is also known that the concept's limitations in its current form (Korhonen et al., 2018; Millar et al., 2019), namely the adoption of closing-the-loop production patterns with a strong focus on increasing resource efficiency, could make CE counterproductive in the long run (Suárez-Eiroa et al., 2021). It has been argued that the CE paradigm encompasses a broader range of strategies that support the transition of businesses to sustainable development (de Jesus et al., 2018), such as implementing industry 4.0 plans (Awan et al., 2021; de Sousa Jabbour et al., 2018), digitizing businesses (Antikainen et al., 2018), substituting inputs and increasing the share of renewable and recycled resources (Sousa-Zomer et al., 2018), minimizing or eliminating the use of hazardous substances (Chen et al., 2020; Silvestri et al., 2021), better transportation, carbon–neutral energy sources, more effective resource utilization (Horton & Horton, 2019) adopting other production-consumption patterns, and using big data to enable sustainable production and consumption (Jabbour et al., 2019) to lessen the negative impact of people on the environment. Although human activities and firms' operations cause problems, green technology must be a part of the solution, even though it is assumed that digitalization also contributes to the issue. However, a strong relationship between how technology may support a new relationship with the environment—one that is harmonious with it while still achieving the goals and innovations that are fundamental to mankind and its accomplishments (Horton & Horton, 2019).

The concept of resilience is defined as the ability of social-ecological systems (and their constituent entities, including businesses) "to absorb disturbance and reorganize while changing to retain essentially still the same function, structure, identity, and feedback" (Walker et al., 2004, para 7). According to Holling et al. (2002), resilience is commonly seen as a multi-scale term to evaluate interacting, nested entities (like enterprises) and systems (like supply chains, industrial systems, and social-ecological systems). The research on resilience has concentrated on examining resilience at various levels of analysis (e.g. Berkes & Ross, 2016; Linnenluecke, 2017). For example, determining "the inherent characteristics of those firms that can respond more quickly, recover faster, or develop more unusual ways of doing business under duress than others" is a common goal of researchers who study firm-level (or organizational) resilience (Linnenluecke, 2017, p. 4). However, multidisciplinary research is increasingly looking at resilience at several levels to investigate how changes in resilience at one level can affect resilience at other levels due to increased worries about issues like climate change, biodiversity loss, and sustainable development. It is implicitly assumed that business practices in the CE increase the resilience of social-ecological systems or the resilience of nested, interacting systems at several levels. In this regard, significant worries are raised by the deteriorating state of the natural environment regarding how contemporary industrial systems are undermining social-ecological systems' resilience, or ecosystems' and human society's capacity to withstand shocks and disruptions that could result in non-linear and transformative change (Folke et al., 2010). Unrestrained waste and emissions, land-use changes, and resource extraction drive social-ecological systems to surpass adaptive thresholds and transition into new operating regimes significantly less conducive to economic activity and human existence (Steffen et al., 2015, 2018). A different approach to industrial system organization, the CE aims to keep social-ecological systems within bounds that are conducive to human life by minimizing waste, pollution, and raw material exploitation (Kennedy & Linnenluecke, 2022).

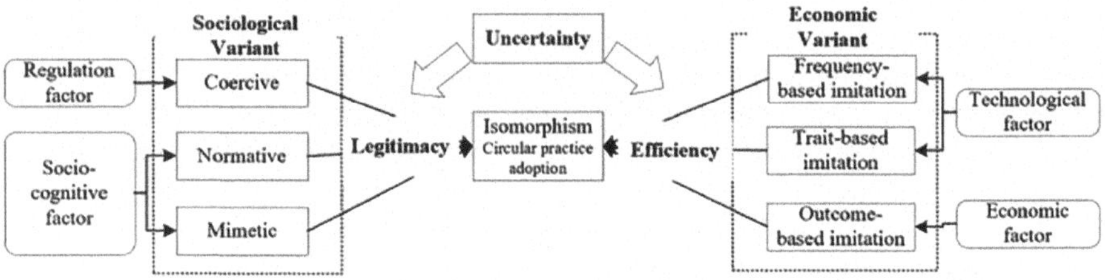


**Fig. 1** The sociological and economic variants of institutional theory. *Source* Do et al. (2022)

## 3 Contribution of the Theories to the Circular Harmony

### *3.1 Institutional Theory for Circular Harmony*

In the existing research, several ideas have been used to investigate the uptake and spread of circular behaviours (Do et al., 2022). Examples include the resource-based view (Jakhar et al., 2019) and stakeholder theory (Jabbour et al., 2020). However, because of the ability to explain and categorize the elements that cause firms to be isomorphic—identical in acts and inactions—institutional theory has grown in popularity in the CE literature (DiMaggio & Powell, 1983). According to Turkulainen et al. (2017), there are two types of institutional theory: sociological and economic. The former explains why firms adopt legal practices, such as sustainable practices (Liu et al., 2018), and its roots are in Meyer and Rowan (1977). Based on Haunschild and Miner (1997), the latter describes what motivates firms to use comparable effective strategies, such as cutting expenses or raising output. Considering all factors, the institutional theory combines social and economic variations, as shown in Fig. 1. The institutional theory is most appropriate to serve as our theoretical lens since CE makes it easier to create harmony between environmental preservation (legitimacy) and economic development (efficiency) (Murray et al., 2017).

### *3.2 Legitimacy Theory for Circular Harmony*

The firms must satisfy their stakeholders with their environmental performance to gain or maintain environmental legitimacy, which is essential for corporate survival and sustainability (Qin et al., 2019). Environmental legitimacy is defined as "the generalized perception or assumption that a firm's environmental performance is desirable, proper, or appropriate" (Bansal & Clelland, 2004, p. 94). In other words, a firm's environmental legitimacy is guaranteed when its environmental performance meets the expectations of its stakeholders. The need for corporate environmental legitimacy and accountability is growing since firms function in society through social contracts (Shocker & Sethi, 1973). A social contract's provisions are thought to give firms resources, labour, and legal footing. The firms must secure legitimacy by pursuing socially acceptable aims while being accountable to their stakeholders for their environmental resilience (Qin et al., 2019).

### *3.3 Stakeholder Theory for Circular Harmony*

It is acknowledged globally that a range of stakeholders, including communities, governments and consumers, have been putting more and more pressure on firms to embrace CE practices to create harmony for environmental resilience (Jabbour et al., 2020). In this regard, prior research has demonstrated that many stakeholders, including governments (Ghinoi et al., 2020), supply chain participants (Gupta et al., 2019), and consumers (Jaca et al., 2018), have either supported or impeded firms' efforts to implement CE. Nevertheless, little research has been done on the connections between stakeholder pressures and the adoption of CE (Jabbour et al., 2020). In addition to stakeholder pressure, the adoption of CE has been aided by the growing trend of product sharing and production digitization and the pursuit of improved operational performance through lower natural resource use (Esposito et al., 2018).

## 4 Circular Harmony for Sustainable Development Goals (SDGs)

The ability to define a certain balance of environmental conditions that further provides development of CE directions for the reception of greater efficiency of economic and ecological processes is made possible by the harmony of ecological development in the conditions of the CE formation (Kryvda et al., 2022). A sustainable society can only be developed if the economic goals are fully harmonized with environmental and social objectives that lead towards SDGs (Cao et al., 2023). For this reason, the United Nations stressed harmony with nature to promote sustainable practices in achieving SDGs (Zu, 2023). A key concept to achieve SDGs is sustainable production, which uses natural resources, energy-efficient technologies, and pollution-free, safe, and healthy systems and technologies to produce goods and services for consumers, society, and workers (Chang et al., 2024). Chang et al. (2024) further elaborated that the characteristics of the products—such as their durability, repairability, ease of recycling, composability, and biodegradability—should also be considered in sustainable production.

Additionally, production typically seeks to satisfy consumer market demands. Using services and products to meet basic needs and improve quality of life is sustainable consumption from a consumption perspective. This approach minimizes the use of natural resources and toxic materials, as well as the discharge of waste and pollutants throughout the life cycle of services or products, so as not to jeopardize the needs of future generations (Cooper, 2005). Therefore, the people must coexist peacefully regarding natural resources within the planet's limits to create a viable future. For this purpose, knowledge, technology, education, policies, and international agreements must all contribute to achieving SDGs by creating a sense of community and harmony with the natural world (Horton & Horton, 2019).

## *4.1 The Financial Performance and Environmental Concerns*

The European Parliament introduced a new work program of the EC on 29 January 2020. The Commission declared its intention to create a new climate change adaptation strategy under the priority, "a European Green Deal." On 24 February 2021, the Commission approved the new EU climate change adaptation strategy, which aims to increase the ability of the EU and the rest of the world to adapt to and mitigate the effects of climate change while also advancing and accelerating economic and social initiatives (European Parliament, 2024). The Commission intends to increase and expand assistance to partner nations and local governments to expand the strategy's global reach. The strategy's main components include funding climate adaptation, taking action to fill knowledge gaps, and developing nature-based solutions. Through the potential expansion of the European Insurance and Occupational Pensions Authority and the extension of insurance coverage to reduce the costs of climate impacts, the Commission emphasizes the significance of the insurance industry in knowledge collection (European Parliament, 2024). The transformation route depicted in Fig. 2 is based on knowledge, reason, creativity, and innovation, which drives changes in government policy, corporate regulation, and civil society through several procedures, such as instruction, campaigns, citizen assemblies, and communication. These modifications result in enlightenment for sustainability for all life on Earth, humankind, and harmony with the environment (Horton & Horton, 2019).

Every firm's socioeconomic structure has been significantly altered by environmental performance. Thus, preserving environmental stability between the value creation process and sustainability standards is a requirement of contemporary business strategies (Barnabè & Nazir, 2022). Considering that contemporary methods influence the innovative firms' environment, organizations and regulatory agencies are implementing new procedures to implement sustainability strategies and create new guidelines for sustainability. Beyond financial success, firms must better understand the sustainability of their processes and frameworks for environmental impacts (Barnabè & Nazir, 2021). Considering that sustainable business practices are predicated on a triple bottom line, they must embrace a solid dedication to those strategic actions for enduring benefits (Nazir & Capocchi, 2024).

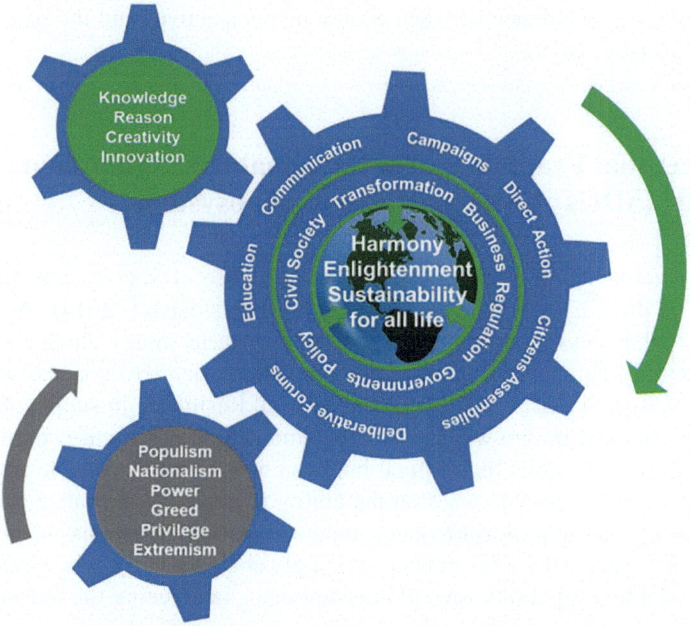


**Fig. 2** Harmony enlightenment sustainability for all life. *Source* Horton and Horton (2019)

## *4.2 The Financial Performance and Social Well-Being*

CE is important to solve firms, human and technological issues. Three fundamental ideas underpin CE: using nature as a mentor, a model for resolving human matters, and a yardstick for evaluating human inventions (Suárez-Eiroa et al., 2021). According to Ormazabal et al. (2018) and Prieto-Sandoval et al. (2018), the CE is a paradigm change to the linear model, which minimizes resource consumption while preserving and recovering the value of waste and products. Adopting life cycle thinking in environmental management systems (Marrucci et al., 2022; Ormazabal & Sarriegi, 2014) and achieving sustainable consumption and production levels through cleaner production (Gupta et al., 2021; Lim et al., 2022) are essential goals of the CE at the corporate level to bring about harmony in the economy, environment, and society (Geissdoerfer et al., 2017; Prieto-Sandoval et al., 2018). Two of the primary goals of the CE are the adoption of cleaner production principles for waste and pollution prevention (Mora-Contreras et al., 2023). By continuously implementing preventive ecological techniques in manufacturing processes and products, efficiency can be improved, and risks, waste, and emissions can be reduced in a way that minimizes threats to people, the economy, and the environment (Afum et al., 2022; Bilitewski, 2012; da Silva et al., 2021). Redefining sustainability to satisfy human needs both now and in the future while honouring the presence of other species on Earth is necessary. The SDGs should ask if it is possible to advance global development while maintaining a balance between economic perspectives and the natural world (Horton & Horton, 2019).

## 5 Conceptual Framework for Sustainable Development Goals' (SDGs) Circular Harmony Ecosystem

A more sustainable development and a more peaceful society are anticipated outcomes of the CE as a new business model (Naustdalslid, 2014). As opposed to traditional business model implementation, long-term supply chain resilience is increased by switching to circular business models for net-zero goals as well as for SDGs (Di Vaio et al., 2024c; Massari et al., 2022). Resilience in supply chain literature can be defined in two ways: either dynamically, which focuses on the ability to evolve, moving towards the original but even new, more favourable equilibrium states, or statically, which focuses on the ability to absorb disturbance and bounce back to the original equilibrium state, maintaining core functions when shocked (Geissdoerfer et al., 2018). The concept of supply chain viability, which refers to the long-term adaptive capability toward interruptions guaranteeing the redesign of the supply chain structure and the replanning of economic performance, was proposed following the dynamic viewpoint of the supply chain (Ivanov & Dolgui, 2020). The literature supports the most suitable resilience approaches in the context of CE in the form of more sustainable resources (Rovanto & Bask, 2021), new dynamic capabilities ensuring reconfiguration towards waste-free ecosystem (Emanuelsson et al., 2021), minimizing risks in logistic networks (Flodén & Williamsson, 2016), fewer resources (Baars et al., 2021), better management of existing resources (Aurisicchio et al., 2021), reducing risks through collaborations (Ray & Mondal, 2016).

In order to improve the balance and harmony between the economy, environment, and society, CE seeks to promote resource efficiency by adopting acceptable decarbonization practices within an economic system. In order to assess SDGs, this chapter suggests a framework encompassing a wide variety of sustainability operations. This framework can help connect the firm's management, operations, and decision-making process regarding environmental resilience. This chapter covers underutilized topics in academia and practice that have gained more attention and exposure in a real-world setting concerning social, environmental, economic, and manufacturing ramifications. Corporate potential toward a more circular management framework is embraced by adopting decarbonization practices and 6Rs related framework balancing economic, social and environmental sustainability.

The conceptual framework of circular harmony for CE and environmental resilience to achieve SDGs is shown in Fig. 3. Hence, our chapter contributes to a thorough CE, circular harmony, and environment resilience—a practical approach to environmental challenges, a regenerative economy, and addressing society's social side. The unified framework highlights a shared view of resources, business operations, outputs, and results.

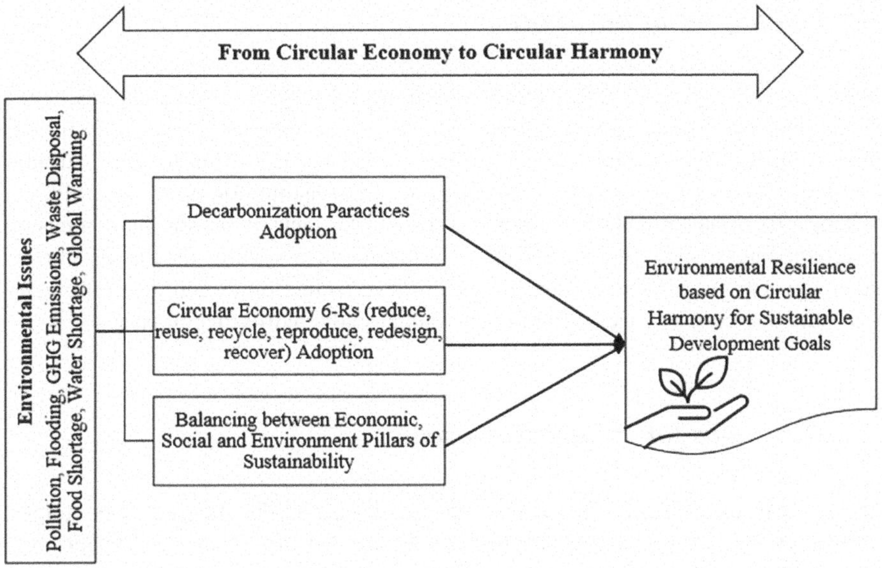


**Fig. 3** "Conceptual Framework of Circular Harmony for Circular Economy and Environmental Resilience to achieve Sustainable Development Goals". *Source* Authors' Compilation

## 6 Conclusion

The consequences of human-caused environmental devastation are now readily apparent, as demonstrated by the loss in the integrity of the biosphere and the record-breaking heat wave that struck Europe in the summer of 2022 (United in Science, 2022). As an alternative to the traditional linear take-make-waste technique, the CE has garnered much interest. Beyond just lessening the drawbacks of a linear economy, the CE is a fundamental change that creates opportunities for firms and the economy while offering social and environmental benefits.

Considering that the CE has supplanted the linear economy of resource usage due to several contemporary global issues, including resource depletion and scarcity, environmental contamination, and humanity's ever-increasing desires, it further stresses the relationship between green technologies and renewable energy resources in decarbonization processes as the worldwide use of technology gives a solution to maintain a balance between economic development with environmental and social well-being. Our research aims to establish guidelines for assessing the harmony of environmental development in the development of CE. When ecological growth aligns with the framework of the CE, there is an opportunity to establish a specific equilibrium of environmental conditions that further provides information on how the CE is developing to receive greater economic efficiency and ecological procedures. The necessary steps to counter the challenges and utilize the benefits of the resources without distorting the environment should be the new goal for everyone.

## 6.1 Limitations of the Study

This chapter aims to link CE with circular harmony to create environmental resilience by adopting decarbonization practices. However, the chapter has various limitations. First, traditional established practices hinder the transition towards circular principles. Second, the increasing natural resource consumption posits a significant challenge for a sustainable environment. Third, the lack of standards to measure the effectiveness of circular practices is a barrier to assessing the firms' performance towards sustainability. Finally, the costs of technologies and the complexity of circular systems may cause the failure to deal with environmental issues.

## 6.2 Directions for Future Research

Through institutional, legitimacy, resilience and stakeholder theories, this chapter outlines several creative avenues for future research that can concentrate on adopting various digital technologies, such as the Internet of Things and artificial intelligence, to optimize the use of resources better, enhancing the performance of firms. In addition, specific standards to assess environmental impacts against climate change will provide valuable intuitions towards sustainability. The researchers should concentrate more on regulatory frameworks and policy incentives to promote CE practices by contrasting the benefits and drawbacks of various existing resources. In order to better comprehend technical skills, researchers should also concentrate on environmental policies that institutions have regularized regarding adopting circular harmony to make the environment resilient. This approach will help firms fill the gap between theory and practice to fight against climate change and foster a more sustainable future.

**Acknowledgements** The authors would like to thank the book editor and reviewers for handling the chapter, providing helpful comments and suggestions, which led to improving the study.

This work is an outcome of the "BlueShipping&Cruise Lab" (BSCLab), Department of Law, University of Naples Parthenope, Italy.

# References

Afum, E., Agyabeng-Mensah, Y., Baah, C., Agyapong, G. K., Armas, J. A. L., & Al Farooque, O. (2022). Prioritizing zero-waste performance and green differentiation advantage through the prism of circular principles adoption: A mediated approach. *Journal of Cleaner Production, 361*, Article 132182.

Antikainen, M., Uusitalo, T., & Kivikytö-Reponen, P. (2018). Digitalization as an enabler of circular economy. *Procedia CIRP, 73*, 45–49.

Aurisicchio, M., Van Der Laan, A. Z., & Tennant, M. (2021). Material-service systems for sustainable resource management. *Sustainable Production, Life Cycle Engineering and Management,* (August), 89–101.

Awan, U., Sroufe, R., & Shahbaz, M. (2021). Industry 4.0 and the circular economy: A literature review and recommendations for future research. *Business Strategy and the Environment, 30*(4), 2038–2060.

Baars, J., Domenech, T., Bleischwitz, R., Melin, H. E., & Heidrich, O. (2021). Circular economy strategies for electric vehicle batteries reduce reliance on raw materials. *Nature Sustainability, 4*(1), 71–79.

Bansal, P., & Clelland, I. (2004). Talking trash: Legitimacy, impression management, and unsystematic risk in the context of the natural environment. *Academy of Management Journal, 47*(1), 93–103.

Barnabè, F., & Nazir, S. (2021). Investigating the interplays between integrated reporting practices and circular economy disclosure. *International Journal of Productivity and Performance Management, 70*(8), 2001–2031.

Barnabè, F., & Nazir, S. (2022). Conceptualizing and enabling circular economy through integrated thinking. *Corporate Social Responsibility and Environmental Management, 29*(2), 448–468.

Berkes, F., & Ross, H. (2016). Panarchy and community resilience: Sustainability science and policy implications. *Environmental Science & Policy, 61*, 185–193.

Bilitewski, B. (2012). The circular economy and its risks. *Waste Management, 32*(1), 1–2.

Birat, J. P. (2015). Life-cycle assessment, resource efficiency and recycling. *Metallurgical Research & Technology, 112*(2), 206.

Bjørnbet, M. M., Skaar, C., Fet, A. M., & Schulte, K. Ø. (2021). Circular economy in manufacturing companies: A review of case study literature. *Journal of Cleaner Production, 294*, Article 126268.

Boulding, K. E. (1966). The economics of the coming spaceship earth. In *Environmental quality in a growing economy: Essays from the sixth RFF forum* (pp. 3–14). RFF Press.

Burritt, R. L., Schaltegger, S., & Christ, K. L. (2023). Environmental management accounting–developments over the last 20 years from a framework perspective. *Australian Accounting Review, 33*(4), 1–16.

Cao, S., Ma, Z., Liu, Z., Guo, J., & Yuan, W. (2023). Achieving harmony between the economy and the environment. *Habitat International, 131*, Article 102733.

Chang, K. F., Lin, C. T., & Bin, Y. Q. (2024). Harmony with nature: Disentanglement the influence of ecological perception and adaptation on sustainable development and circular economy goals in country. *Heliyon*, 10(4).

Charonis, G. K. (2012, September). Degrowth, steady state economics and the circular economy: Three distinct yet increasingly converging alternative discourses to economic growth for achieving environmental sustainability and social equity. In *World economic association sustainability conference* (Vol. 24, pp. 1–18).

Chen, T. L., Kim, H., Pan, S. Y., Tseng, P. C., Lin, Y. P., & Chiang, P. C. (2020). Implementation of green chemistry principles in circular economy system towards sustainable development goals: Challenges and perspectives. *Science of the Total Environment, 716*, Article 136998.

Cooper, T. (2005). Slower consumption reflections on product life spans and the "throwaway society." *Journal of Industrial Ecology, 9*(1–2), 51–67.

da Silva, P. C., de Oliveira Neto, G. C., Correia, J. M. F., & Tucci, H. N. P. (2021). Evaluation of economic, environmental and operational performance of the adoption of cleaner production: Survey in large textile industries. *Journal of Cleaner Production, 278*, Article 123855.

Damschroder, L. J., Reardon, C. M., Widerquist, M. A. O., & Lowery, J. (2022). The updated consolidated framework for implementation research based on user feedback. *Implementation Science, 17*(75), 1–16.

De Jesus, A., Antunes, P., Santos, R., & Mendonça, S. (2018). Eco-innovation in the transition to a circular economy: An analytical literature review. *Journal of Cleaner Production, 172*, 2999–3018.

de Sousa Jabbour, A. L., Jabbour, C. J. C., Godinho Filho, M., & Industry, D. R. (2018). 4.0 and the circular economy: A proposed research agenda and original roadmap for sustainable operations. *Annals of Operations Research, 270*, 273–286.

Di Vaio, A., Zaffar, A., & Chhabra, M. (2024a). Intellectual capital and human dynamic capabilities in decarbonization processes for net-zero business models: An in-depth examination through a systematic literature review. *Journal of Intellectual Capital, 25*(7), 23-53.

Di Vaio, A., Zaffar, A., Chhabra, M., & Balsalobre-Lorente, D. (2024b). Carbon accounting and integrated reporting for net-zero business models towards sustainable development: A systematic literature review. *Business Strategy and the Environment,* Forthcoming. https://doi.org/10.1002/bse.3863

Di Vaio, A., Dell'Amura, G., Chhabra, M., & Garofalo, A. (2024c). Circular economy and waste production models for sustainable development goals 12 and 14: Evidence from cruise sustainability reporting. Sustainable Development.

Di Vaio, A., Hasan, S., Palladino, R., & Hassan, R. (2023a). The transition towards circular economy and waste within accounting and accountability models: A systematic literature review and conceptual framework. *Environment, Development and Sustainability, 25*(1), 734–810.

Di Vaio, A., Latif, B., Gunarathne, N., Gupta, M., & D'Adamo, I. (2023b). Digitalization and artificial knowledge for accountability in SCM: A systematic literature review. *Journal of Enterprise Information Management, 37*(2), 606–672.

DiMaggio, P. J., & Powell, W. W. (1983). The iron cage revisited: Institutional isomorphism and collective rationality in organizational fields. *American Sociological Review, 48*(2), 147–160.

Do, Q., Mishra, N., Colicchia, C., Creazza, A., & Ramudhin, A. (2022). An extended institutional theory perspective on the adoption of circular economy practices: Insights from the seafood industry. *International Journal of Production Economics, 247*, 108400.

Du, K., & Li, J. (2019). Towards a green world: How do green technology innovations affect total-factor carbon productivity. *Energy Policy, 131*, 240–250.

Emanuelsson, E. A. C., Charles, A., & Shivaprasad, P. (2021). A regenerative business model with flexible, modular and scalable processes in a post-covid era: The case of the spinning mesh disc reactor (smdr). *Sustainability (Switzerland), 13*(12).

Esposito, M., Tse, T., & Soufani, K. (2018). Introducing a circular economy: New thinking with new managerial and policy implications. *California Management Review, 60*(3), 5–19.

European Commission. (2014). Towards a circular economy: A zero-waste programme for Europe, COM (2014)398. Retrieved October 20, 2024, from https://www.eumonitor.eu/9353000/1/j9vvik7m1c3gyxp/vjl4tamjogxb

European Commission. (2015a). EU resource efficiency scoreboard. Retrieved October 23, 2024, from https://ec.europa.eu/environment/resource_efficiency/targets_indicators/scoreboard/index_en.htm

European Commission. (2015b). Closing the loops—an EU action plan for the circular economy. COM/2015/0614 final. Retrieved 22 October, 2024, from https://eur-lex.europa.eu/legal-content/EN/TXT/?uri=CELEX:52015DC0614

European Commission. (2015c). Communication from the commission to the European Parliament, the council, the European Economic and Social Committee and the committee of the regions. COM (2015) 614/2.

European Commission. (2017). The role of waste-to-energy in the circular economy. Retrieved 20 October, 2024, from https://ec.europa.eu/environment/waste/waste-to-energy.pdf

European Commission. (2018a). A European strategy for plastics in a circular economy. Retrieved 21 October, 2024, from https://ec.europa.eu/environment/circular-economy/pdf/plastics-strategy.pdf

European Commission. (2018b). EU raw material week 2018—Closing the loop on critical raw materials. Retrieved 20 October, 2024, from https://ec.europa.eu/easme/en/news/eu-raw-materials-week-2018-closing-loop-critical-raw-materials

European Parliament. (2024). Legislative train schedule: New EU strategy on adaptation to climate change in "A European green deal". Retrieved 27 October, 2024, from https://www.europarl.europa.eu/legislative-train/theme-a-european-green-deal/file-new-eu-strategy-on-adaptation-to-climate-change

Flodén, J., & Williamsson, J. (2016). Business models for sustainable biofuel transport: The potential for intermodal transport. *Journal of Cleaner Production, 113*, 426–437.

Folke, C., Carpenter, S. R., Walker, B., Scheffer, M., Chapin, T., & Rockström, J. (2010). Resilience thinking: integrating resilience, adaptability and transformability. *Ecology and Society, 15*(4).

Frosch, R. A., & Gallopoulos, N. E. (1989). Strategies for manufacturing. *Scientific American, 261*(3), 144–152.

Geissdoerfer, M., Morioka, S. N., de Carvalho, M. M., & Evans, S. (2018). Business models and supply chains for the circular economy. *Journal of Cleaner Production, 190*, 712–721.

Geissdoerfer, M., Savaget, P., Bocken, N. M., & Hultink, E. J. (2017). The circular economy–A new sustainability paradigm? *Journal of Cleaner Production, 143*, 757–768.

Geng, Y., Fu, J., Sarkis, J., & Xue, B. (2012). Towards a national circular economy indicator system in China: An evaluation and critical analysis. *Journal of Cleaner Production, 23*(1), 216–224.

Geng, Y., Fujita, T., Park, H. S., Chiu, A., & Huisingh, D. (2014). Call for papers: Towards post-fossil carbon societies: Regenerative and preventative eco-industrial development. *Journal of Cleaner Production, 68*, 4–6.

Ghinoi, S., Silvestri, F., & Steiner, B. (2020). The role of local stakeholders in disseminating knowledge for supporting the circular economy: A network analysis approach. *Ecological Economics, 169*, Article 106446.

Ghisellini, P., Cialani, C., & Ulgiati, S. (2016). A review on circular economy: The expected transition to a balanced interplay of environmental and economic systems. *Journal of Cleaner Production, 114*, 11–32.

Gupta, H., Kumar, A., & Wasan, P. (2021). Industry 4.0, cleaner production and circular economy: An integrative framework for evaluating ethical and sustainable business performance of manufacturing organizations. *Journal of Cleaner Production, 295*, 126253.

Gupta, S., Chen, H., Hazen, B. T., Kaur, S., & Gonzalez, E. D. S. (2019). Circular economy and big data analytics: A stakeholder perspective. *Technological Forecasting and Social Change, 144*, 466–474.

Hardies, K., Ohlrogge, F., Mentens, J., & Vandennieuwenhuysen, J. (2024). A guide for accounting researchers to conduct and report systematic literature reviews. *Behavioral Research in Accounting, 36*, 1–23.

Haunschild, P. R., & Miner, A. S. (1997). Modes of interorganizational imitation: The effects of outcome salience and uncertainty. *Administrative Science Quarterly*, 472–500.

Holling, C. S., Gunderson, L. H., & Peterson, G. (2002). Sustainability and panarchies. In L. H. Gunderson & C. S. Holling (Eds.), *Panarchy: Understanding transformations in human and natural systems* (pp. 63–102). Island Press.

Horton, P., & Horton, B. P. (2019). Redefining sustainability: Living in harmony with life on Earth. *One Earth, 1*(1), 86–94.

Ivanov, D., & Dolgui, A. (2020). Viability of intertwined supply networks: Extending the supply chain resilience angles towards survivability. A position paper motivated by COVID-19 outbreak. *International Journal of Production Research, 58*(10), 2904–2915.

Jabbour, C. J. C., de Sousa Jabbour, A. B. L., Sarkis, J., & Godinho Filho, M. (2019). Unlocking the circular economy through new business models based on large-scale data: An integrative framework and research agenda. *Technological Forecasting and Social Change, 144*, 546–552.

Jabbour, C. J. C., Seuring, S., de Sousa Jabbour, A. B. L., Jugend, D., Fiorini, P. D. C., Latan, H., & Izeppi, W. C. (2020). Stakeholders, innovative business models for the circular economy and sustainable performance of firms in an emerging economy facing institutional voids. *Journal of Environmental Management, 264*, Article 110416.

Jaca, C., Prieto-Sandoval, V., Psomas, E. L., & Ormazabal, M. (2018). What should consumer organizations do to drive environmental sustainability? *Journal of Cleaner Production, 181*, 201–208.

Jakhar, S. K., Mangla, S. K., Luthra, S., & Kusi-Sarpong, S. (2019). When stakeholder pressure drives the circular economy: Measuring the mediating role of innovation capabilities. *Management Decision, 57*(4), 904–920.

Kennedy, G. W., Sunday, S. E., Pandey, D., Kabari, A. D., Pandey, B. K., George, A. S., & Dadheech, P. (2024). Building information modeling: Harnessing innovations for professional proficiency. In *Emerging engineering technologies and industrial applications* (pp. 305–331). IGI Global.

Kennedy, S., & Linnenluecke, M. K. (2022). Circular economy and resilience: A research agenda. *Business Strategy and the Environment, 31*(6), 2754–2765.

Korhonen, J., Honkasalo, A., & Seppälä, J. (2018). Circular economy: The concept and its limitations. *Ecological Economics, 143*, 37–46.

Kryvda, O., Tulchynska, S., Smerichevskyi, S., Lagodiienko, N., Marych, M., & Naghiyeva, A. (2022). Harmony of ecological development in the conditions of the circular economy formation. *Environment and Ecology Research, 10*(1), 11–20.

Lakshmi, K. V. N., Rajan, R., Pandey, D., & Pandey, B. K. (2024). Circular economy: A catalyst for economic growth-an empirical study. *Circular Economy and Sustainability*, 1–22.

Lett, L. A. (2014). Global threats, waste recycling and the concept of circular economy. *Argentine Journal of Microbiology, 46*(1), 1–2.

Leventis, S., Kang, H., & Watson, L. (2024). Literature reviews exploring international aspects of accounting, auditing, and taxation. *Journal of International Accounting, Auditing and Taxation, 56*(100637), 1–3.

Lim, M. K., Lai, M., Wang, C., & Lee, Y. (2022). Circular economy to ensure production operational sustainability: A green-lean approach. *Sustainable Production and Consumption, 30*, 130–144.

Linnenluecke, M. K. (2017). Resilience in business and management research: A review of influential publications and a research agenda. *International Journal of Management Reviews, 19*(1), 4–30.

Liu, J., Feng, Y., Zhu, Q., & Sarkis, J. (2018). Green supply chain management and the circular economy: Reviewing theory for advancement of both fields. *International Journal of Physical Distribution & Logistics Management, 48*(8), 794–817.

Loeser, F. (2013). "Green IT and green IS: Definition of constructs and overview of current practices" presented at the Nineteenth Americas Conference on Information Systems, 15–17 August, Chicago, Illinois, available at Green IT and Green IS: Definition of Constructs and Overview of Current Practices (core.ac.uk). Retrieved 19 October 2024.

Lyle, J. T. (1996). *Regenerative design for sustainable development*. John.

MacArthur, E. (2013). Towards the circular economy. *Journal of Industrial Ecology, 2*(1), 23–44.

Marrucci, L., Daddi, T., & Iraldo, F. (2022). Do dynamic capabilities matter? A study on environmental performance and the circular economy in European certified organizations. *Business Strategy and the Environment, 31*(6), 2641–2657.

Masnavi, M. R., Gharai, F., & Hajibandeh, M. (2019). Exploring urban resilience thinking for its application in urban planning: A review of literature. *International Journal of Environmental Science and Technology, 16*, 567–582.

Massari, G. F., Annarelli, A., Primario, S., & Puliga, G. (2022). On the synergetic relationship between circular economy and resilience: Findings from a systematic literature review. *IFAC-PapersOnLine, 55*(10), 2869–2874.

Meyer, J. W., & Rowan, B. (1977). Institutionalized organizations: Formal structure as myth and ceremony. *American Journal of Sociology, 83*(2), 340–363.

Millar, N., McLaughlin, E., & Börger, T. (2019). The circular economy: Swings and roundabouts? *Ecological Economics, 158*, 11–19.

Mora-Contreras, R., Ormazabal, M., Hernández-Salazar, G., Torres-Guevara, L. E., Mejia-Villa, A., Prieto-Sandoval, V., & Carrillo-Hermosilla, J. (2023). Do environmental and cleaner production practices lead to circular and sustainability performance? Evidence from Colombian manufacturing firms. *Sustainable Production and Consumption, 40*, 77–88.

Murray, A., Skene, K., & Haynes, K. (2017). The circular economy: An interdisciplinary exploration of the concept and application in a global context. *Journal of Business Ethics, 140*, 369–380.

Naustdalslid, J. (2014). Circular economy in China–the environmental dimension of the harmonious society. *International Journal of Sustainable Development & World Ecology, 21*(4), 303–313.
Nazir, S., & Capocchi, A. (2024). Circular economy 6Rs and reporting practices: The role of institutional pressures. In *Sustainability reporting practices and the circular economy: Analysis and integrated strategies* (pp. 185–224). Springer Nature Switzerland.
Ormazabal, M., & Sarriegi, J. M. (2014). Environmental management evolution: Empirical evidence from Spain and Italy. *Business Strategy and the Environment, 23*(2), 73–88.
Ormazabal, M., Prieto-Sandoval, V., Puga-Leal, R., & Jaca, C. (2018). Circular economy in Spanish SMEs: Challenges and opportunities. *Journal of Cleaner Production, 185*, 157–167.
Park, J. Y., & Chertow, M. R. (2014). Establishing and testing the "reuse potential" indicator for managing wastes as resources. *Journal of Environmental Management, 137*, 45–53.
Prieto-Sandoval, V., Jaca, C., & Ormazabal, M. (2018). Towards a consensus on the circular economy. *Journal of Cleaner Production, 179*, 605–615.
Qin, Y., Harrison, J., & Chen, L. (2019). A framework for the practice of corporate environmental responsibility in China. *Journal of Cleaner Production, 235*, 426–452.
Ray, A., & Mondal, S. (2016). Study of collaborative PRM business model for sustainability. *Benchmarking: An International Journal, 24*(7), 1891–1911.
Rovanto, I. K., & Bask, A. (2021). Systemic circular business model application at the company, supply chain and society levels—a view into circular economy native and adopter companies. *Business Strategy and the Environment, 30*(2), 1153–1173.
Schöggl, J. P., Stumpf, L., & Baumgartner, R. J. (2020). The narrative of sustainability and circular economy-A longitudinal review of two decades of research. *Resources, Conservation and Recycling, 163*, Article 105073.
Schroeder, P., Anggraeni, K., & Weber, U. (2019). The relevance of circular economy practices to the sustainable development goals. *Journal of Industrial Ecology, 23*(1), 77–95.
Shocker, A. D., & Sethi, S. P. (1973). An approach to incorporating societal preferences in developing corporate action strategies. *California Management Review, 15*(4), 97–105.
Silvestri, C., Silvestri, L., Forcina, A., Di Bona, G., & Falcone, D. (2021). Green chemistry contribution towards more equitable global sustainability and greater circular economy: A systematic literature review. *Journal of Cleaner Production, 294*, Article 126137.
Sousa-Zomer, T. T., Magalhães, L., Zancul, E., Campos, L. M., & Cauchick-Miguel, P. A. (2018). Cleaner production as an antecedent for circular economy paradigm shift at the micro-level: Evidence from a home appliance manufacturer. *Journal of Cleaner Production, 185*, 740–748.
Stahel, W. R. (2014). Reuse is the key to the circular economy. *Eco-Innovation Action Plan–European Commission.* Retrieved 25 October, 2024, from https://green-business.ec.europa.eu/eco-innovation_en
Stahel, W. R. (1982). The product life factor. An inquiry into the nature of sustainable societies: The role of the private sector. Series:1982 Mitchell Prize Papers, NARC. Retrieved 22 October, 2024, from https://p2infohouse.org/ref/33/32217.pdf
Steffen, W., Richardson, K., Rockström, J., Cornell, S. E., Fetzer, I., Bennett, E. M., Biggs, R., Carpenter, S., de Vries, W., de Wit, C., Folke, C., Gerten, D., Heinke, J., Mace, G., Persson, L., Ramanathan, V., Reyers, B., & Sörlin, S. (2015). Planetary boundaries: Guiding human development on a changing planet. *Science, 347*(6223).
Steffen, W., Rockström, J., Richardson, K., Lenton, T. M., Folke, C., Liverman, D., Summerhayes, C., Barnosky, A., Cornell, C., Crucifix, M., Donges, J., Fetzer, I., Lade, S., Scheffer, M., Winkelmann, R., & Schellnhuber, H. J. (2018). Trajectories of the earth system in the Anthropocene. *Proceedings of the National Academy of Sciences, 115*(33), 8252–8259.
Suárez-Eiroa, B., Fernández, E., & Méndez, G. (2021). Integration of the circular economy paradigm under the just and safe operating space narrative: Twelve operational principles based on circularity, sustainability and resilience. *Journal of Cleaner Production, 322*, Article 129071.
Turkulainen, V., Kauppi, K., & Nermes, E. (2017). Institutional explanations: Missing link in operations management? Insights on supplier integration. *International Journal of Operations & Production Management, 37*(8), 1117–1140.

United in Science. (2022). A multi-organization high-level compilation of the most recent science related to climate change, impacts and responses. Retrieved 23 October, 2024, from https://wmo.int/publication-series/united-science-2022

Walker, B., Holling, C. S., Carpenter, S. R., & Kinzig, A. (2004). Resilience, adaptability and transformability in social–ecological systems. *Ecology and Society, 9*(2).

Weigend Rodríguez, R., Pomponi, F., Webster, K., & D'Amico, B. (2020). The future of the circular economy and the circular economy of the future. *Built Environment Project and Asset Management, 10*(4), 529–546.

Zu, L. (2023). Being in harmony with nature: A systemic view of Taoism. In *Responsible management and taoism, volume 2: Transforming management education for sustainable development goals (SDGs)* (pp. 165–173). Emerald Publishing Limited.

# The Sustainability Shift: Addressing Environmental Degradation Through Circular Practices

**Chukwuemeka Kingsley John, Fidelis Odedishemi Ajibade, Temitope Fausat Ajibade, Pankaj Kumar, Bashir Adelodun, Olaolu George Fadugba, James Rotimi Adewumi, and Christopher Oluwakunmi Akinbile**

**Abstract** The depletion of resources, pollution, and waste accumulation have led to significant environmental degradation, presenting a critical challenge for the planet's future and humanity's survival. The need for sustainable solutions is becoming more and more apparent as the strain on natural ecosystems increases. As a take-make-dispose model, the linear economy has been shown to be unsustainable, exacerbating environmental damage and resource scarcity. A promising substitute, on the other hand, is the circular economy, which emphasises recycling, reuse, and regeneration. This chapter investigates the ways in which the growing environmental catastrophe is propelling the global transition to circular practices. Mitigating climate change, cutting waste, and promoting long-term sustainability all depend on the shift to a circular economy as governments, corporations, and society come to understand the limited nature of resources. Innovation, green jobs, and smaller environmental footprints are all made possible by circular processes, which also prolong product

C. K. John
Department of Construction and Civil Engineering, University Centre Jersey Highlands College, St Saviour Jersey, UK

F. O. Ajibade (✉) · T. F. Ajibade · O. G. Fadugba · J. R. Adewumi
Department of Civil and Environmental Engineering, Federal University of Technology, Akure, Nigeria
e-mail: foajibade@futa.edu.ng

P. Kumar
School of Environmental Studies, Maa Shakumbhari University, Saharanpur, India

Research and Development Division, Society for Agro Environmental Sustainability, Dehradun, India

B. Adelodun
Arusha Climate and Environmental Research Centre, Aga Khan University, Arusha, Tanzania

Department of Agricultural and Biosystems Engineering, University of Ilorin, Ilorin, Nigeria

C. O. Akinbile
Department of Agricultural and Environmental Engineering, Federal University of Technology, Akure, Nigeria



P. Singh et al. (eds.), *Circular Economy and Environmental Resilience*,
https://doi.org/10.1007/978-3-031-93091-1_7

lifecycles. Circular economy models open the door to a more sustainable future by tackling the underlying causes of environmental deterioration, protecting ecosystems and raising upcoming generations' standard of living.

**Keywords** Environmental degradation · Circular economy · Resource depletion · Climate change · Sustainability · Waste reduction

## 1 Introduction

The unsustainable practices of global consumption and industrialization have caused several environmental concerns that the world faces today to reach catastrophic heights. The depletion of natural resources is one of the most pressing challenges we face today, as demand for raw materials for manufacturing and energy continues to rise. We are consuming freshwater, minerals, forests, and fossil fuels at a much faster rate than they can naturally replenish. This overconsumption has led to resource scarcity, biodiversity loss, and disruptions to ecosystems—key components that maintain ecological balance (Padilla & Kumar et al., 2022; Padilla, 2021). Alongside resource depletion, pollution poses a growing threat to both the environment and human health. Air, water, and land have been extensively contaminated by harmful substances like industrial chemicals, greenhouse gases, and plastic waste. This pollution is responsible for health risks faced by millions, particularly in vulnerable regions with limited access to clean water and air. Moreover, pollutants further exacerbate climate change (Chowdhary et al., 2020). Another major challenge is the rapid pace of climate change, which is disrupting weather patterns and damaging ecosystems. Rising global temperatures from excessive carbon emissions are causing sea levels to rise, glaciers to melt, and an increase in extreme weather events like storms, droughts, and floods. These climate shifts are severely affecting infrastructure, agriculture, and water availability, while also threatening food security in many parts of the world and contributing to economic instability. Together, these environmental crises have brought the planet to a critical juncture, underscoring the urgency of taking immediate action. The circular economy has emerged as one promising solution to address these issues, aiming to reduce waste, conserve resources, and promote regenerative practices across industries (Kumar et al., 2021; Nhemachena et al., 2020; Paudel et al., 2023; Singh et al., 2022).

The fundamental causes of environmental deterioration have been economic and industrial practices since the quest for profit and growth has frequently come at the expense of ecological sustainability. Without considering long-term environmental effects, the old linear economy, which is defined by a "take, make, dispose" paradigm, has mostly relied on the extraction of natural resources and the production of trash. Non-renewable resources like fossil fuels, metals, and minerals are depleted due to the massive energy and material consumption of industrial operations. Significant environmental effects of this paradigm include habitat destruction, deforestation, and soil degradation, all of which compromise the planet's capacity to sustain biodiversity and deliver vital ecosystem services. Industries frequently overlook the environmental costs of their production systems because they place a higher priority on short-term profits than on sustainable practices (Kumar et al., 2022; Nathaniel et al., 2021; Paliwal, 2022). Furthermore, a sizable amount of the world's greenhouse gas emissions, which fuel climate change and air pollution, are caused by industrial operations. The atmosphere is exposed to significant amounts of carbon dioxide ($CO_2$), methane ($CH_4$), and nitrous oxide ($N_2O$) from fossil fuel-powered factories, power plants, and transportation infrastructure. The greenhouse effect is made worse by these emissions, which raises global temperatures and modifies weather patterns (Reijnders, 2022; Shen et al., 2020). Waste and industrial waste disposal also accelerates environmental degradation. Waste management done poorly contaminates water and land, threatening human health and damaging ecosystems. The disposal of hazardous chemicals, plastics, and electronic waste falls under this category. We must quickly transition to circular economic models, which prioritise resource efficiency, waste reduction, and the use of renewable energy sources at the core of industrial operations, as demonstrated by the cumulative effects of these unsustainable practices. By implementing circular concepts, we may lessen the environmental damage caused by old economic systems and create a more resilient and sustainable global economy (Kolawole et al., 2021; Iyiola, 2023).

The urgency of global environmental deterioration has never been higher, pushing ecosystems and climatic systems to the verge of irreversible harm. Vital ecosystems like coral reefs, wetlands, and rainforests are being destroyed, endangering biodiversity and upsetting the natural mechanisms that sustain life on Earth. The pace of extinction of species is increasing; because of habitat degradation, pollution, and climate change, many plant and animal populations might become extinct in a matter of decades. In addition to upsetting ecological equilibrium, biodiversity loss jeopardizes vital ecosystem functions like carbon sequestration, pollination, and water purification. Human societies are more vulnerable to natural disasters, food shortages, and water scarcity when ecosystems deteriorate. The swift rate of environmental degradation necessitates prompt, coordinated worldwide action to stop additional harm and repair damaged ecosystems before crucial thresholds are reached (Nair, 2023; Rani et al., 2020; Upreti, 2023). Given that the effects of climate change are already being seen, such as increased sea levels, catastrophic weather events, and changes in global agriculture patterns, it is especially critical to address this environmental disaster. The poorest and most vulnerable people on the planet are most severely affected by the effects of global warming, which worsen social inequality and pose a threat to millions of people by making entire regions uninhabitable and raising sea levels. Furthermore, as the harm to public health, agriculture, and infrastructure keeps getting worse, delaying action to address environmental degradation will raise the costs of adaptation and mitigation. Significant risks to economic stability are posed by the increasing frequency of climate-related disasters, which also put a strain on public resources and disrupt global supply systems. This increased urgency highlights how important it is to move away from a linear economic paradigm and towards a circular economy that puts an emphasis on waste reduction, resource efficiency, and the restoration of natural systems. Societies can promote long-term resilience and sustainability while mitigating environmental deterioration by adopting circular practices (Hussain et al., 2020; D'amato & Korhonen, 2021; Anukwonke et al., 2022; Abdullahi et al., 2024).

This chapter explores how the ongoing environmental degradation can serve as a powerful motivator for adopting circular economy models. Due to the growing problems of pollution, resource depletion, climate change, and biodiversity loss, the world's current industrial and economic systems are obviously unsustainable. As this chapter will show, the circular economy encourages a move towards more sustainable practices by emphasising resource efficiency, waste reduction, and ecosystem regeneration. By exploring the relationship between environmental problems and circular solutions, this chapter hopes to add to the larger discussion on sustainable development and resource management. The following are the chapter's main goals: should first determine and examine the main factors contributing to environmental deterioration, such as pollution, resource exploitation, and unsustainable industrial practices. The comprehension of these elements is crucial to appreciating the scope and gravity of the environmental catastrophe. Along with evaluating the ecological and economical effects of environmental deterioration, the chapter will also emphasise how it affects ecosystems, human health, and international economies. It will highlight how important it is to implement circular practices to prevent additional environmental damage by examining these effects. Furthermore, the chapter will give a thorough introduction to the circular economy, outlining its fundamental ideas and contrasting them with the linear economic model that now governs industrial activities. Lastly, the chapter will include case studies of global circular economy projects that have been successful, demonstrating how the environmental crisis can spur the broad adoption of circular practices. The chapter hopes to provide readers a better grasp of how environmental issues around the world not only endanger life as we know it but also offer a chance for revolutionary change in business and the economy.

## 2 Key Drivers of Global Environmental Degeneration

### *2.1 Industrialization and Resource Depletion*

Global economic growth has been largely fuelled by industrialization, but this progress has come at a high environmental cost, mostly in the form of natural resource depletion. Throughout the previous 200 years, as industries grew, so did the need for raw materials such as minerals, metals, wood, and fossil fuels (Okkerse & Van Bekkum, 1999; Rankin, 2011; Smil, 2013). The depletion of vital reserves, like coal, oil, and natural gas, which power most of the industrial activity, is the result of this over-reliance on limited resources. One factor contributing to the depletion of these resources is the rate at which fossil fuels are being utilized, for instance, considerably outpacing their rate of natural regeneration. Long-term sustainability issues arise from the depletion of non-renewable resources caused by the exploitation of metals and minerals utilized in manufacturing. Moreover, deforestation and soil degradation brought on by industrial-scale forestry and agriculture practices have further strained ecosystems and decreased the planet's capacity for self-repair (Jovine, 2021; Nathaniel et al., 2021; Okoh & Okoh, 2021; Manić & Lutovac, 2022). Unsustainable resource extraction has serious ecological repercussions as well. For example, mining operations have the potential to destroy local ecosystems and landscapes, resulting in the loss of biodiversity and habitats. Water availability for communities and ecosystems is reduced by the ongoing industrial exploitation of water resources, particularly in water-scarce regions, endangering the security of food and water. Furthermore, non-renewable resources are used extensively in the manufacturing and use of things like metals and plastics, which degrades the environment when these items are disposed of as waste. The planet's long-term health is at risk due to the cumulative effects of resource depletion brought on by industrialization. If circular practices which prioritize resource conservation and sustainable production are not embraced, the world will eventually face resource shortages and environmental collapse (Bano et al., 2020; Salem et al., 2022; Tiamgne et al., 2022). The circular economy presents a strong substitute for resource-intensive industrialization in the face of this issue. Reusing, recycling, and remanufacturing materials is prioritized by the circular model, which minimizes waste and lowers the need for virgin resources. When items are made with lifetime and repurposing in mind, closed-loop systems are encouraged, and this can greatly slow down the loss of natural resources. This strategy ensures that industries can prosper without overtaxing the planet's limited resources, in line with the requirement to strike a balance between environmental sustainability and economic growth. Industrialization can stimulate economic growth and innovation while preventing more environmental damage if it is reorganized according to circular principles (Braun et al., 2021; Camilleri, 2021; Godfrey et al., 2021).

## 2.2 *Pollution and Contamination: Land, Air, and Water*

One of the biggest environmental problems of the modern era is pollution, which has dangerous effects on the land, the air, and the water. Urbanization and industrialization have produced enormous volumes of waste and pollution that have contaminated ecosystems all over the world. The main source of land pollution is the inappropriate disposal of domestic and industrial waste, which contaminates soil and leads to the buildup of non-biodegradable items like plastics. Wildlife is put in danger, agricultural output is impacted, and soil quality is deteriorated by this trash. Hazardous trash poses serious health dangers to both humans and animals. This includes hazardous chemicals and electronic debris, or "e-waste," which exacerbates land pollution. Because landfills are frequently overflowing, they discharge dangerous pollutants into the environment, such as methane, a strong greenhouse gas that accelerates climate change (Ajibade et al., 2021; Sharma et al., 2021). Air pollution is another significant effect of industrialization, primarily caused by emissions from automobiles, power plants, industry, and agricultural operations. Hazardous gases that are released into the atmosphere, such as carbon dioxide ($CO_2$), sulfur dioxide ($SO_2$), and nitrogen oxides ($NO_x$), are responsible for acid rain and global warming, both of which have catastrophic effects on ecosystems. Particulate matter (PM) and volatile organic compounds (VOCs) are two examples of airborne pollutants that seriously endanger human health and cause cardiac problems, respiratory disorders, and early mortality. Cities and industrial areas are major sources of air pollution, especially in poorer nations where there are sometimes insufficient legal frameworks to limit emissions. Urgent adjustments in industrial and transportation practices are needed in response to the worldwide air quality issue. These changes should include a move towards cleaner, renewable energy sources and tighter emission restrictions (Abdullah, 2021; Anwar et al., 2021; John et al., 2023; Zhang et al., 2021). The environmental contamination trio is completed by water pollution. Freshwater and marine environments are contaminated by untreated sewage, agricultural runoff, and industrial discharges. Heavy metals, fertilizers, and pesticides from agricultural fields find their way into rivers, lakes, and the ocean where they cause eutrophication, a condition that results in dead zones devoid of aquatic life. Furthermore, plastic debris has spread throughout the world's oceans, creating enormous rubbish patches and posing a hazard to marine life through ingestion and entanglement. Oil leaks and industrial spills further destroy aquatic habitats, causing long-term environmental harm. Industries need to embrace more sustainable methods, such as wastewater treatment, pollution controls, and a decrease in the usage of hazardous materials, to fight pollution. Through the design of waste out of industrial systems and the creation of cleaner production cycles that minimize pollution, circular economy principles can help ameliorate these concerns (Aa et al., 2022; Eze et al., 2024; Ogidi & Akpan, 2022).

## 2.3 *Deforestation and Habitat Destruction*

Since human activities continue to encroach onto and destroy important ecosystems, deforestation, and habitat destruction are major contributions to the degradation of the global environment. Large-scale forest clearing has been caused by the growth of agriculture, logging, and infrastructural development, especially in tropical areas like the Amazon, Southeast Asia, and Central Africa. The need for lumber, agricultural land, and urbanization is causing an alarming rate of forest loss. Forests are hotspots for biodiversity and act as carbon sinks. Numerous species are driven to the verge of extinction due to this deforestation, which destroys habitats supporting a variety of species. The carbon cycle is also upset by the loss of forests because the carbon stored in trees is released into the atmosphere as carbon dioxide ($CO_2$), which fuels global warming (Jayathilake et al., 2021; Kabir et al., 2023; Srivastav, 2024; Wassie, 2020). Not only do forests suffer from habitat degradation, but wetlands, grasslands, and marine habitats are also impacted. Dragging or conversion of wetlands for agricultural and industrial use is occurring. These wetlands offer vital ecosystem services like flood control, water filtering, and biodiversity support. Climate change, pollution, and coastal development are destroying coastal habitats that support marine life and safeguard shorelines, such as mangroves and coral reefs. The resilience of ecosystems is weakened by the widespread destruction of habitats, leaving them more susceptible to external stresses like invading species and climate change. Reduced space and resource availability for species results from habitat shrinkage and fragmentation, which also affects ecosystem services that are essential to human existence and biodiversity (Choudhury et al., 2021; Kattel, 2022; Ogidi & Akpan, 2022). Deforestation and habitat degradation have far-reaching effects. Deforestation causes soil erosion, desertification, and disturbance of water cycles in addition to reducing biodiversity and ecosystem services. Land productivity gradually decreases due to nutrient depletion and soil degradation, which are frequently caused by deforestation and the conversion of natural ecosystems into agricultural land. In addition to harming the soil, the loss of these ecosystems pushes wildlife into human areas in quest of food and shelter, increasing tensions between people and the natural world. Global collaboration is needed to address this escalating issue to preserve the few surviving natural habitats, restore damaged ecosystems, and advance sustainable land-use practices. Regenerative agriculture, sustainable forestry, and ecosystem restoration are examples of circular economy techniques that can be implemented to mitigate the negative consequences of habitat loss and preserve biodiversity. Instead of focussing on depleting natural resources, these methods aim to refill them, establishing a regenerative and balanced system that promotes long-term agricultural productivity and environmental health (Wassie, 2020; D'amato & Korhonen, 2021; Ekka et al., 2023; van Genuchten, 2024).

## *2.4 Climate Change and Its Consequences*

Global environmental deterioration is largely caused by climate change, which has significant effects on ecosystems, the economy, and human society. Increased greenhouse gas emissions from industrial processes, deforestation, and agriculture are causing the globe to warm, which is affecting weather patterns, escalating natural disasters, and endangering biodiversity worldwide. Sea levels are rising due to the melting of polar ice caps and glaciers brought on by rising global temperatures, endangering low-lying areas and coastal communities. Millions of people are at risk of flooding and displacement due to this occurrence, especially in areas like Southeast Asia, the Pacific Islands, and some sections of the United States. Higher temperatures are also causing heatwaves, droughts, and wildfires to occur more frequently and with greater severity. These events have a devastating impact on human populations as well as natural ecosystems (Clarke et al., 2022; Palinkas & Palinkas, 2020; Wassie, 2020). The effect of climate change on global food security is among its most alarming side effects. Weather patterns are changing, and this is causing disruptions to agricultural output, particularly in areas where farming is dependent on stable climates. These changes include extended droughts, erratic rainfall, and shifting growing seasons. Millions of people are already impacted by crop failures, water scarcity, and lower yields, especially in developing nations where agriculture is the main source of income. Freshwater resources are under stress due to a combination of rising temperatures and water scarcity, making it more challenging for communities to obtain clean water for agriculture, drinking, and sanitary purposes. This puts more pressure on international organizations and countries to deal with the humanitarian issues brought on by climate change, such as food shortages, water shortages, and migration brought on by the phenomenon (Paudel et al., 2023; Shemer et al., 2023; Srivastav et al., 2021). Furthermore, the current environmental issues of habitat degradation and biodiversity loss are made worse by climate change. Many species are unable to adjust to the changing conditions quickly enough when temperatures increase and ecosystems are disrupted, which results in a loss of biodiversity and the collapse of ecosystems. For example, coral bleaching and subsequent die-off are brought on by rising sea temperatures, which pose a special threat to coral reefs. The millions of marine animals that rely on these reefs for survival are therefore impacted. The melting of ice caps in polar regions poses a hazard to species that are fast losing their habitat, such as seals and polar bears. The wider effects of climate change on the environment and society emphasize how vital it is for the world to take action to cut greenhouse gas emissions and move towards a low-carbon economy. With its emphasis on waste reduction, renewable energy, and resource efficiency, the circular economy provides businesses and governments with a strategy to lessen the consequences of climate change while promoting sustainable growth (Johnson et al., 2020; Lamb et al., 2022; Tan et al., 2022).

## 3 The Socio-Economic Impacts of Environmental Degradation

### *3.1 Impacts on Human Health and Livelihoods*

Human health is directly and severely impacted by environmental degradation, mostly due to increased pollution exposure and the breakdown of ecosystems that produce clean water, food, and air. One of the biggest threats to public health is air pollution, which is brought on by deforestation, automobile exhaust, and industrial pollutants. Particularly in cities with significant industrial activity, respiratory illnesses, cardiovascular disorders, and early mortality have been related to fine particulate matter ($PM_{2.5}$) and other airborne pollutants. Vulnerable groups, such as children, the elderly, and the poor, are disproportionately affected by poor air quality and frequently have limited access to healthcare. Furthermore, in areas with insufficient sanitary infrastructure, the depletion of clean water resources brought on by pollution from industrial runoff, agricultural pesticides, and untreated sewage promotes the development of waterborne illnesses including cholera, diarrhea, and typhoid. Human health is still seriously threatened by unclean drinking water, especially in underdeveloped nations without adequate water treatment infrastructure (Ajibade et al., 2021; Anwar et al., 2021; Zhang et al., 2022). Environmental degradation also affects livelihoods by upsetting the ecosystems that support local economies and means of subsistence. Food insecurity and the uprooting of farming communities are caused by agricultural production being compromised in rural areas by deforestation, soil erosion, and desertification. Particularly at risk are small-scale farmers in developing nations who depend on the land for both their subsistence and means of subsistence. Fishing communities suffer similar difficulties as natural resources are diminished because overfishing and pollution lower fish supplies, endangering food security and the livelihoods of millions of people who depend on the oceans for their livelihoods. Environmental degradation in metropolitan settings can exacerbate poverty by increasing the cost and decreasing the accessibility of essential commodities like energy, clean water, and air. Social injustices are exacerbated by this deterioration since people with lower incomes are disproportionately affected and frequently reside in more environmentally hazardous places (Nkonya et al., 2020; Ajibade et al., 2021; Herrera Arango et al., 2022; Parsonset al., 2024). Environmental deterioration, which is made worse by climate change, increases the risks to human health and way of life. Because rising temperatures extend the habitats of mosquitoes and other disease-carrying organisms, they accelerate the development of vector-borne diseases like dengue fever and malaria. Hurricanes, floods, and droughts are examples of extreme weather occurrences that cause millions of people to be displaced, demolish homes and infrastructure, and interfere with economic activity. This has resulted in the rise of "climate refugees," people who have been compelled to leave their homes because of natural calamities. Low-income communities and developing countries are disproportionately affected by the worsening effects of environmental degradation, hence it is crucial to address these challenges through the adoption of circular economic models and sustainable practices (Marazziti et al., 2021; Menton et al., 2020; Yadav & Upadhyay, 2023).

## *3.2 Economic Costs of Environmental Degradation*

Long-term economic stability is threatened by environmental degradation, which has significant financial costs for communities, businesses, and countries alike. The higher expense of healthcare resulting from illnesses linked to pollution is one of the most noticeable effects right away. Healthcare costs are on the rise due to the increased prevalence of chronic respiratory diseases, malignancies, and other pollution-related ailments brought on by air and water pollution. The cost of treating these illnesses falls on governments and health systems and lost productivity from illness further lowers economic output. For instance, the expenses of treating respiratory conditions and missed workdays owing to poor air quality are enormous in highly polluted cities, and this has a direct impact on the economic development of a nation. These expenses, which are frequently overlooked in economic analyses, draw attention to the unstated cost of environmental degradation to national economies and public health systems (Chen & Chen, 2021; Sun et al., 2022; Wang et al., 2022). There are large financial losses because of the depletion of natural resources, especially in industries like forestry, fishery, and agriculture. Industries that depend on arable land and forest products are impacted by land degradation, desertification, and deforestation, which decrease their availability. In addition to governments losing money on agricultural exports, farmers experience decreased incomes and food insecurity due to soil erosion, lower crop yields, and unpredictable weather patterns. Similarly, fisheries are disrupted by overfishing and marine pollution, which results in falling fish stocks and the collapse of fishing-dependent local economies. The loss of ecosystems results in a reduction in the ecosystems' capacity to deliver economically valuable services like carbon sequestration, flood control, and pollination, which raises the costs of biodiversity loss significantly (Elisha & Felix, 2021; Kattel, 2022; Paudel et al., 2023). Along with being a major cause of environmental degradation, climate change has a huge financial impact due to increasing expenses associated with disaster recovery, damage to infrastructure, and disruptions to the economy. Every year, homes, companies, and vital infrastructure are destroyed by extreme weather events like hurricanes, floods, and wildfires, resulting in billions of dollars in losses. Rebuilding and recovery come at a huge financial cost to communities, governments, and insurance companies. In addition, long-term economic difficulties brought on by population displacement due to natural catastrophes include lost livelihoods, a rise in the need for social services, and a decline in economic productivity. These financial costs will only increase as environmental degradation worsens, so to lessen the long-term financial effects of environmental decline, businesses and governments must embrace circular economy principles that encourage resource efficiency, waste reduction, and environmental resilience (Chaudhary & Piracha, 2021; Short & Short, 2021; Wassie, 2020).

### *3.3 Environmental Justice: Inequalities in the Global South and North*

Deforestation, resource exploitation, and industrial activities generate an environmental impact that disproportionately affects disadvantaged populations in the Global South. Developing nations are more vulnerable to the damaging effects of unsustainable economic activities because they frequently have laxer environmental laws and enforcement. Global South natural resources are routinely exploited by multinational firms, who take use of them by mining, burning fossil fuels, and harvesting wood without thinking about the effects on the environment and society. Deforestation, soil erosion, and water pollution result from this, severely harming regional ecosystems and the means of subsistence for the communities that depend on them. Moreover, a lot of developing nations act as disposal sites for hazardous waste, such as toxic chemicals and electronic trash, from wealthy countries, which exacerbates environmental and health disparities (Menton et al., 2020; Wassie, 2020; Ajibade et al., 2021; Büscher et al., 2024). The idea of environmental justice draws attention to the glaring disparity between the advantages enjoyed by the Global North and the environmental responsibilities imposed on the Global South. The Global North's industrialized countries are mostly at blame for past and present greenhouse gas emissions, but developing countries bear the brunt of climate change's effects. Extreme weather, rising sea levels, and shifts in agricultural productivity disproportionately affect low-income nations, whose economies are more reliant on climate-sensitive industries like fisheries and agriculture and have less capacity for adaptation. Moreover, many developing nations are more susceptible to the negative consequences of environmental deterioration because they lack the technological infrastructure and financial means to implement mitigation and adaptation techniques (Kumar & Donnelly, 2024; Maina et al., 2023; Nguyen et al., 2023). Environmental justice concerns also develop inside nations, in addition to the North–South split, since low-income and marginalized groups, both in the Global North and South, are frequently found in places most impacted by pollution, industrial activity, and environmental risks. Communities of color and low-income people are more likely to reside close to industrial zones, landfills, and polluted metropolitan areas in the Global North, where they are more likely to suffer from cancer, respiratory illnesses, and other pollution-related health problems. Additionally, these communities lack the means and political clout to promote healthier environments or to move away from dangerous locations. To address environmental justice, international cooperation is needed to make sure that those who are most at risk from environmental degradation do not suffer disproportionately from it and that everyone benefits equally from sustainable development. Through the promotion of sustainability, resource efficiency, and cleaner manufacturing techniques, the implementation of circular economy principles, particularly in industries operating in the Global South, can aid in the reduction of environmental inequities (Kekana et al., 2023; Njoku & Sampson, 2023; Pastor et al., 2024).

## 4 The Rise of the Circular Economy Concept

### *4.1 Definition and Principles of the Circular Economy*

The circular economy creates closed-loop systems where products, materials, and energy are continuously recycled, reused, or repurposed to reduce waste and maximise resource utilisation. The circular economy aims to recover and renew materials and products at the end of their life cycle, maximize value extraction from resources, and prolong the useful life of resources as opposed to the conventional linear economy, which uses a "take-make-dispose" approach. This strategy promotes sustainable development and lessens the impact on the environment by assisting in the decoupling of economic growth from resource use. The main concept is to eliminate waste from the economy by planning for products' end of life early on and allowing materials to be recycled back into the production cycle as opposed to ending up in landfills or incinerators (Hussain et al., 2023; Kirchherr et al., 2023). The application of the circular economy is guided by several fundamental ideas (Fig. 1). Firstly, it places a strong emphasis on designing for modularity and longevity, which means that goods are made to be easily repaired, long-lasting, and versatile. This idea lessens the requirement for ongoing material manufacture and increases the lifespan of products. Secondly, the circular economy encourages a regenerative approach by building systems that replenish and restore natural capital and support the use of renewable energy and materials. Thirdly, it emphasizes using waste as a resource, which entails recycling, composting, or other methods to reintegrate materials and byproducts that were previously regarded as garbage into industrial systems. To provide a comprehensive transition to a more sustainable economy, it also encourages collaboration across value chains to optimize resource flows, from manufacturers and consumers to recyclers and policymakers (Liow Li Kandpal et al., 2024; Sa & Choon-Yin, 2023). This concept aligns with broader sustainability goals, including reducing carbon emissions, conserving biodiversity, and minimizing the depletion of natural resources. Industries can shift from conventional resource-intensive techniques to more effective and waste-free manufacturing methods by embracing the concepts of the circular economy. The circular economy encourages a regenerative model that places an emphasis on durability, reuse, and recycling in place of the traditional linear method, which results in excessive waste and environmental deterioration. This approach provides a more resilient and sustainable path to economic growth by reconsidering the production, use, and recycling of products. By lowering reliance on limited resources and promoting innovation in sustainable methods, it not only lessens the environmental problems caused by industrial activity but also promotes long-term economic stability.

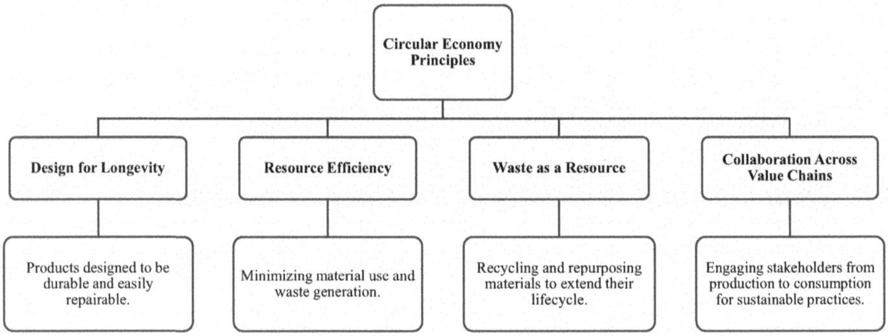


**Fig. 1** Core principles of the circular economy

### *4.2 Circular Economy Verus Linear Economy*

The production, use, and management of resources in society are radically different in the circular and linear economies. Since the Industrial Revolution, the linear economy has dominated the world economy. It's simple, the extractive paradigm involves taking raw materials from the Earth, converting them into products, and then discarding the finished goods. This "take-make-dispose" mentality has contributed to economic expansion but at the cost of serious environmental damage. Since items in the linear economy typically have short life cycles, which cause natural resources to be depleted quickly and waste to accumulate, the system is fundamentally inefficient. Additionally, this paradigm makes pollution worse by using fossil fuels and producing materials like hazardous chemicals and some plastics that are hard or impossible to recycle (Kandpal et al., 2024; Upreti, 2023). The circular economy, on the other hand, aims to end this wasteful loop by changing the way resources are handled over their whole life cycle. Products in a circular economy are made to be recycled, mended, remanufactured, and reused to prolong the useful life of resources. This minimizes the environmental impact of production and disposal while drastically reducing the demand for fresh resource extraction. At the conclusion of their life cycle, materials are recovered and reintegrated into new production processes rather than being destroyed, producing a continuous loop of usage. For instance, the circular economy would promote repairing or recycling electronic device components rather than discarding them when their useful lives are up. This would cut down on e-waste and preserve important resources like metals and rare earth elements (Bandh et al., 2024; Kandpal et al., 2024). The circular economy is founded on the principles of resource efficiency and sustainability, while the linear economy is based on intentional obsolescence and excessive consumption. There are several advantages to switching from a linear to a circular paradigm, both economically and environmentally. Financially speaking, it lowers expenses associated with energy use, the acquisition of raw materials, and waste management while promoting innovation in logistics, product design, and recycling software. The circular economy also contributes significantly to the reduction of greenhouse gas emissions by promoting energy efficiency, minimising reliance on virgin materials, and giving priority to renewable resources. In contrast to the wasteful and damaging behaviours of the conventional linear model, the circular economy provides a more sustainable and progressive method. The transformation of industrial practices to prioritise long-term value preservation over short-term resource depletion, the circular economy fosters a more resilient and regenerative economic model—one that harmonizes economic growth with environmental stewardship (Kirchherr et al., 2023; Kumar et al., 2024).

## *4.3 Historical Development of Circular Practices*

The concept of circular practices has a long history, even if the formal term "circular economy" has only gained traction in recent decades. Resource efficiency and reuse have historically been fundamental to many traditional civilizations, particularly in pre-industrial and agricultural contexts where groups relied on the prudent management of few resources (Blomsma & Brennan, 2017; Jackson et al., 2014). Waste was frequently reduced by methods like recycling things like metal and glass, mending equipment and clothing, and composting organic matter prior to the industrialization of economies. The concepts of the present circular economy are largely derived from the resourcefulness and frugal living of past societies. The rise of mass production and consumerism in the twentieth century, which put convenience and efficiency ahead of sustainability, was largely responsible for the trend towards wastefulness and disposability (Berry et al., 2023; Roy et al., 2024). As awareness of Earth's limited resources and growing concerns about the effects of industrialization on the environment grew in the 1970s, the contemporary concept of the circular economy started to take shape. The importance of managing the planet's resources in a manner that emulates natural ecosystems, where nothing is wasted and everything is recycled, was highlighted by seminal books like Kenneth Boulding's "The Economics of the Coming Spaceship Earth" (1966). The "cradle to cradle" design theory, put forth by American architect William McDonough and German scientist Michael Braungart, gained popularity at about the same period. To reduce waste and protect the environment, this strategy promoted the development of items that could be completely recycled or composted at the end of their life cycle. Since then, the circular economy has been built on the cradle-to-cradle design principles, which have shaped how companies and decision-makers view sustainability (Kabir et al., 2023; Weigend Rodríguez, 2023; Dobrauz-Saldapenna & Huras, 2023). Due to both economic and environmental concerns, the circular economy notion began to take shape in the late 20th and early twenty-first centuries. The linear economy is not sustainable, as evidenced by the growing evidence of climate change, resource shortages, and global financial crises. Groups like the Ellen MacArthur Foundation—which was established in 2010—were instrumental in advancing the circular economy as a workable answer to these problems. The circular economy movement has developed into a global paradigm for reconsidering production, consumption, and waste through research, lobbying, and collaborations with companies, governments, and academic institutions. The circular economy is now viewed as a chance for long-term growth, resilience, and economic innovation in addition to being a tool for environmental sustainability (Kirchherr et al., 2023; Kumar et al., 2024).

# 5 Environmental Degeneration as a Catalyst for Circular Economy Practices

## *5.1 Waste Reduction and Resource Efficiency*

The circular economy concept, which represents a paradigm change from the conventional linear economy, where waste is considered an inevitable result of production and consumption, is centered on waste reduction and resource efficiency. Waste is reduced in a circular economy by creating systems and products that consume less materials and put efficiency first throughout the whole manufacturing process. This strategy pushes industries to reconsider how products are produced, used, and disposed of. The circular economy places more emphasis on upstream modifications, such as designing items to last longer, be easily repaired, and use less raw materials, than it does on end-of-life solutions, such as recycling. This shift lessens the amount of garbage that is dumped in landfills and the need for additional resources, both of which contribute to the preservation of natural ecosystems and the slowing down of environmental degradation (Kirchherr et al., 2023; Wiebe et al., 2023). Additionally, resource efficiency encourages industrial processes to use water and energy as efficiently as possible. Industries can minimize their environmental impact and cut expenses by optimizing resource utilization. Lean manufacturing, for instance, aims to eliminate waste throughout the production process by streamlining operations, lowering material input requirements, and recycling byproducts. As a result, a lot of businesses are implementing these principles. Similarly, resource efficiency is being attained in industries like construction and agriculture through methods like precision farming, which uses less water, fertilizer, and pesticides, and green building strategies that make use of energy-efficient technologies and sustainable building materials. These strategies assist companies in remaining profitable while reducing their environmental effect (Kirikkaleli & Ali, 2024; Sharma et al., 2023). Not only are waste reduction and resource efficiency good for the environment, but they also present substantial business potential. Companies can cut waste management costs, save the cost of raw materials, and generate new revenue streams by selling recycled or refurbished goods when they transition to a circular business model. Furthermore, as companies engage in innovative methods and technologies that optimize their use of materials and energy, resource efficiency can stimulate innovation. Waste reduction and resource efficiency are becoming essential elements of the worldwide shift to a circular economy as firms grapple with mounting demand from consumers, governments, and international organizations to embrace more sustainable practices (Kandpal et al., 2024; Majid et al., 2023).

## 5.2 Renewable Energy and Sustainable Production

Renewable energy plays a pivotal role in enabling the transition to a circular economy by reducing the reliance on fossil fuels and minimizing the environmental impact of production processes. Renewable energy sources, such as solar, wind, hydro, and geothermal power, take the place of non-renewable energy inputs in a circular economy, reducing greenhouse gas emissions and battling climate change. Industries may function more sustainably and drastically lower their carbon footprints by incorporating renewable energy into their production systems. This is especially crucial for companies that use a lot of energy, such as manufacturing, transportation, and energy-intensive sectors, as their use of fossil fuels contributes significantly to environmental deterioration. Because renewable energy sources are plentiful and less susceptible to price swings and geopolitical threats, their use not only supports the concepts of the circular economy but also enhances energy security and resilience (Jahanger et al., 2023; Kandpal et al., 2024). Renewable energy and sustainable production go hand in hand because they both prioritize reducing environmental damage during the production process. This involves reducing waste and pollution, using resources wisely, and procuring materials responsibly. By implementing circular design concepts, employing eco-friendly materials, and implementing cleaner technology, many businesses are embracing sustainable production practices. Examples of sustainable production projects include the use of closed-loop water systems in manufacturing, the use of biodegradable materials in packaging, and the sustainable sourcing of raw materials in the fashion sector. These methods assist in lessening the environmental impact of industries while producing goods that better satisfy the demands of sustainable consumers (Batra, 2023; Mondal et al., 2023; Roy et al., 2024). Furthermore, by lowering operating costs and raising efficiency, sustainable production and renewable energy provide long-term financial advantages. Industries can reduce their energy costs and enhance their environmental performance by utilizing renewable energy technologies as they become more accessible and economical. Aside from helping businesses comply with regulations, adopting sustainable production processes can lower their risk of environmental fines and improve their brand's reputation among consumers who care about the environment. As they realize how crucial renewable energy and sustainable production methods are to meeting climate targets and advancing a more sustainable global economy, governments and international organizations are pushing companies to use them (Algarni et al., 2023; Santos et al., 2024).

## 5.3 Innovations in Recycling, Reuse, and Remanufacturing

The circular economy depends on innovations in recycling, reuse, and remanufacturing because they offer the means of extending the life of products and materials, lowering the demand for virgin resources, and cutting waste. Recent substantial advancements in recycling technologies have made it possible to break down and reprocess a wider range of materials, including complex and composite goods, into new materials. The drawbacks of conventional recycling methods are being addressed by innovations like chemical recycling, which disassembles polymers into their constituent chemicals for subsequent use, and sophisticated metal recovery methods. These developments allow enterprises to construct more sustainable supply chains, recover valuable materials from waste, and lessen their impact on the environment (Kandpal et al., 2024; Santos et al., 2024). Another important component of circular economy innovation is reuse, which increases product life cycles by promoting maintenance, refurbishing, and repurposing. Reuse over disposal is becoming the norm in many businesses, as evidenced by the rising popularity of leasing goods rather than buying them. Under this approach, businesses own the product outright and oversee maintaining it so that it can be reused several times before its life cycle ends. Furthermore, platforms that make it easier to share or resell commodities like online markets for used goods are growing in popularity and allowing both consumers and businesses to offer goods a second chance at life. These programs decrease waste while simultaneously opening new commercial prospects for the circular economy (Braun et al., 2021; Cerio et al., 2024; Kandpal et al., 2024). Remanufacturing is an additional invention that promotes the circular economy. It is the process of returning discarded things to their original state. Within industries like electronics, heavy machinery, and automotive, remanufacturing enables businesses to reclaim outdated products, upgrade their components, and resale them for less money than they would charge for new ones. This lowers the energy and emissions involved in creating new items from scratch in addition to conserving resources. Businesses may meet the growing demand from consumers for sustainable products, save production costs, and lessen their environmental effects by investing in remanufacturing. Innovations in remanufacturing, recycling, and reuse work together to close the circle in the circular economy and move away from a wasteful, linear paradigm of production and consumption (Chaudhari et al., 2024; Reddy et al., 2023).

## *5.4 The Role of Governments and International Agreements in Promoting Circular Models*

Governments and international agreements are essential in encouraging the implementation of circular economy principles because they establish the laws, rewards, and policies required to propel the change. As a result of growing awareness of the financial and environmental advantages of circular models, national and local governments are enacting laws to promote resource efficiency, waste minimization, and sustainable production. For example, the Circular Economy Action Plan of the European Union describes a thorough plan to lower waste, increase recycling rates, and support circular business models in all industries. Governments can incentivize industry to design products with the end-of-life phase in mind by establishing explicit standards and targets, such as extended producer responsibility (EPR) schemes, which guarantee that materials can be readily recycled or reused (Kandpal et al., 2024; Möslinger et al., 2023). Governments are offering financial incentives to companies that use circular economy principles in addition to legal frameworks. Companies wishing to innovate in sectors like recycling, remanufacturing, and sustainable production find it easier to get started when there are subsidies, tax breaks, and grants available for research and development in circular technologies. A market for sustainable commodities is created by public procurement regulations that give priority to circular products and services, which incentivizes companies to participate in circular processes. In addition, governments are attempting to increase public knowledge of the advantages of the circular economy to influence consumer behavior in the direction of more environmentally friendly purchase habits (Kandpal et al., 2024; Möslinger et al., 2023; Siderius & Zink, 2023). The global promotion of circular models necessitates international agreements and alliances since resource depletion and environmental degradation are transnational challenges requiring coordinated responses. International cooperation on environmental protection and resource management has its frameworks established by accords like the Basel Convention on hazardous waste and the Paris Agreement on climate change. In addition, groups like the World Economic Forum and the United Nations are actively pushing the concepts of the circular economy as part of larger objectives for sustainable development. The adoption of circular practices is accelerated globally by several global efforts that promote cooperation between enterprises, governments, and civil society. Governments can facilitate a more harmonious and efficient global shift towards a circular economy by harmonizing domestic policies with international accords (Kandpal et al., 2024; Naqash & Bhat, 2023).

## 6 Case Studies of Circular Practices in Response to Environmental Crisis

### *6.1 Circular Economy in the European Union*

The European Union (EU) has spearheaded the worldwide transition to a circular economy by enacting a comprehensive policy framework aimed at curbing waste, enhancing resource efficiency, and cultivating sustainable production and consumption practices (Fig. 2). Launched in 2015 and revised in 2020 as part of the European Green Deal, the EU Circular Economy Action Plan serves as the cornerstone of this endeavor. With an emphasis on industries with significant resource consumption and waste creation, such as electronics, textiles, plastics, and construction, this plan lays out important steps and goals to move from a linear to a circular economy. Ensuring that materials are used for as long as feasible is one of the fundamental goals of product design, which involves making them more durable, repairable, and recyclable. Producers are encouraged to embrace circular design ideas by the EU, which

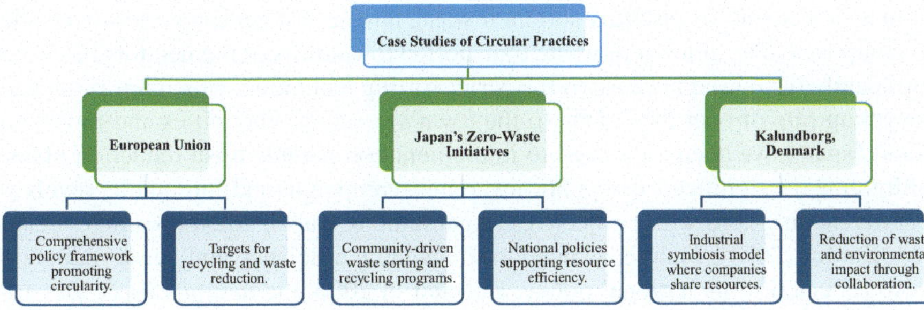


**Fig. 2** Successful implementations of circular economy practices in EU, Japan, and Denmark

makes them responsible for the full lifespan of their products through the implementation of rules such as the Extended Producer Responsibility (EPR) scheme (Doumkou, 2024; Gorbatcheva et al., 2024). Apart from enacting laws, the European Union has also provided financial incentives to encourage circular behaviors. One such initiative is the Horizon Europe research program, which provides funding for innovative projects in sustainable production, recycling, and remanufacturing. A market for sustainable commodities is also created by public procurement regulations that give priority to circular goods and services, which incentivizes companies to invest in circular business models. Since the shift to a circular economy is anticipated to drastically cut greenhouse gas emissions by lowering the demand for energy and raw materials, the EU's campaign for circularity is also connected to its climate goals. The European Commission, for instance, projects that by 2030, the circular economy will have reduced carbon emissions in the EU by 450 million tonnes yearly (Hartley et al., 2023; Möslinger et al., 2023; Siderius & Zink, 2023). Waste management is included in the EU's circular economy policy, which places a major emphasis on increasing recycling rates and decreasing landfilling. Aggressive recycling goals, like recycling 55% of municipal garbage by 2025 and 65% by 2035, must be met by member states. To make this easier, the EU has implemented laws that are meant to enhance garbage collection and sorting, outlaw single-use plastics, and improve recycling facilities. The EU is a global pioneer in circular economy practices thanks to its all-encompassing strategy, and other regions looking to strike a balance between environmental sustainability and economic growth can look to it as an example (Möslinger et al., 2023; Novarlić & Đurić, 2024).

## 6.2 *Japan's Zero-Waste Initiatives*

Japan has been a global pioneer in waste management and circular economy practices, particularly through its innovative zero-waste initiatives (Fig. 2). The town of Kamikatsu is one of the most well-known instances; it has enacted a comprehensive zero-waste policy with the goal of eliminating garbage by 2030. It is mandatory for

Kamikatsu residents to carefully sort their waste into 45 distinct categories to recycle and compost it. The amount of waste transported to landfills or incinerators has been significantly decreased because to this strict sorting technique, which has produced a recycling rate of over 80%. Due to the town's success, other cities and towns all around Japan have been motivated to implement comparable trash reduction plans, highlighting the significance of public awareness campaigns and community involvement in reaching zero-waste objectives (Agamuthu & Babel, 2023; Möslinger et al., 2023; Zorpas, 2024). Japan's dedication to resource efficiency and waste reduction is also reflected in its national policy. Enacted in 2000, the nation's Basic Act for Establishing a Sound Material-Cycle Society offers a thorough legal framework for advancing resource sustainability, waste reduction, and recycling. The "3R" (Reduce, Reuse, Recycle) philosophy, which has been incorporated into a variety of industries, including consumer goods and manufacturing, was established in Japan thanks to this rule. Extended producer responsibility (EPR) programs, which force producers to assume accountability for the recycling and disposal of their products at the end of their life cycles, have also been introduced by the Japanese government. As a result, there have been notable developments in sectors like electronics and autos, where goods are now developed with a stronger focus on resource efficiency and recyclability (Agamuthu & Babel, 2023; Jimura, 2023). Japan has welcomed technology innovation in addition to national measures to assist its zero-waste goals. The nation is a pioneer in the application of cutting-edge recycling techniques, such as chemical recycling, which disassembles plastics and other materials into their constituent chemicals for subsequent use. Additionally, Japan has developed advanced waste-to-energy (WtE) systems that use incineration to turn non-recyclable garbage into energy. These programs meet the nation's energy requirements while simultaneously assisting in lessening the negative environmental effects of garbage disposal. Japan's zero-waste programs show how effective circular economy practices and environmental degradation can be promoted by combining government policy, community involvement, and technical innovation (Jimura, 2023; Kumar et al., 2023; Harjam et al., 2024).

## *6.3 Industrial Symbiosis in Kalundborg, Denmark*

One of the most well-known instances of industrial symbiosis—a circular economy strategy in which many companies collaborate to share resources, energy, and byproducts occurs in Kalundborg, Denmark (Fig. 2). When local industries realized, they could cut costs and environmental impact by exchanging waste materials and energy, the Kalundborg Symbiosis initiative got underway in the 1970s. Currently, a power plant, an oil refinery, a pharmaceutical firm, and a maker of gypsum board are among the many industries that are part of the network. By working together, waste products from one business are turned into raw materials for another, resulting in a closed-loop system that uses fewer natural resources and produces less trash (Khan et al., 2023;

Ventura et al., 2024). The Asnaes Power Station, which provides steam to industrial partners and excess heat to the neighboring community for district heating, is one of the main exchanges in Kalundborg. The adjacent refinery provides treated wastewater to the power plant in exchange, which it utilizes for cooling purposes. By decreasing the need for energy and virgin materials, this kind of resource sharing greatly lessens the environmental impact of the participating industries. Furthermore, fly ash, sulfur, and gypsum are recycled by the symbiotic network and turned into useful products like plasterboard and cement. In addition to lowering waste and pollution, this industrial cooperation raises the participating companies' overall productivity (Khan et al., 2023; Nighot & Kumar, 2023). The Kalundborg Symbiosis has been a model for industrial sustainability and has sparked copycat projects worldwide due to its success. As part of its larger circular economy policy, the Danish government has actively pushed industrial symbiosis, realizing its potential to both create economic benefit and lessen environmental damage. Kalundborg has shown that circular economy methods can result in both environmental and economic benefits by encouraging collaboration amongst industries, demonstrating that sustainable industrial development is feasible. As new businesses join the network and investigate new avenues for resource sharing and the development of a more sustainable industrial ecosystem, the project keeps evolving (Al-Kuwari et al., 2024; Vitale & Ziegler, 2024).

### *6.4 Waste-To-Energy Practices in Developing Countries*

The utilization of waste-to-energy (WtE) techniques is becoming more popular in developing nations as a means of addressing issues related to waste management and energy scarcity. Large volumes of waste are dumped in open landfills in many poor countries due to inadequate infrastructure for waste disposal, which increases environmental pollution and poses health hazards. By turning non-recyclable garbage into energy, usually by anaerobic digestion or burning, waste-to-energy technologies provide an alternative. This method decreases the amount of waste that is dumped in landfills while simultaneously producing heat or electricity, making it an important renewable energy source in areas with poor access to consistent power (Ameen et al., 2023; Rezania et al., 2023). The capacity of WtE systems in developing nations to concurrently handle several environmental and socioeconomic concerns is one of their main advantages. For instance, WtE facilities support both the national energy grid and the rapidly urbanizing nations of India and Indonesia, where waste output has surged due to overcrowding in landfills. To enhance waste management and boost the use of renewable energy, the Indian government has been pushing WtE technologies through its Swachh Bharat (Clean India) initiative. Numerous waste-to-energy (WtE) plants have been set up in places like Delhi and Pune, generating power from municipal solid waste and lowering the need for fossil fuels (Rezania et al., 2023; Singh, 2024). Despite its potential, WtE methods in developing nations confront several obstacles, such as the high upfront costs associated with constructing WtE plants, a lack of technical know-how, and worries about air pollution from incineration procedures. Governments and international organizations are offering financial and technical support to help scale up WtE technology to address these problems. For example, the World Bank and the United Nations Environment Programme (UNEP) have been collaborating with governments in Asia and Africa to create sustainable WtE solutions that meet public health and environmental criteria. Trash-to-energy techniques have the potential to be a major component of developing nations' circular economies with the right funding and regulation, providing a long-term answer to the problems associated with energy and trash management (Agamuthu & Babel, 2023; Kibria et al., 2023).

## 7 Challenges and Barriers to Implementing Circular Practices

### *7.1 Technological Limitations*

The impediments to efficient material recovery, recycling, and reuse posed by technological constraints represent a major obstacle to the adoption of circular processes. Electronics and some plastics are examples of sophisticated, multi-material items that are too complex for many of the recycling processes now in use. For example, separating elements like glass, metals, and plastics from electronic waste frequently calls for expensive, complex procedures that aren't available in many places. In a similar vein, scalability, energy efficiency, and cost-effectiveness issues are still being worked out in the early stages of chemical recycling technologies, which break down materials like plastics to their molecular level for reuse. It is challenging to attain the high recovery rates required to make the transition to a completely circular economy due to the lack of innovation in recycling technology (Agrawal et al., 2023; Aquatar et al., 2023; Tavares et al., 2023). Designing products with circularity in mind is another technological constraint. Easy disposal and recycling are not given priority while designing many products, which is still done using a linear approach. It is challenging to disassemble or recycle products with intricate packaging or those composed of a mix of incompatible materials, for instance. Circular design is still a relatively new concept, but some businesses are starting to embrace eco-design concepts that prioritize longevity, reparability, and recyclability. The inability to shut the loop on resource consumption is made worse by the lack of standardized criteria or incentives for designing goods that work well in circular systems (Rashid & Malik, 2023; Tamizhdurai et al., 2024). In addition, sectors like building and textiles that mostly depend on natural resources have a difficult time implementing circular business models due to technology. Technologies that enable the recovery and repurposing of materials such as textiles, concrete, or specific metals are either in their early stages of development or not commercially feasible when used on a wide scale. As a result, many businesses are taking a long time to adopt circular models since the required technologies are either too expensive or non-existent. To ensure that the concepts of the circular economy are successfully adopted worldwide, closing these technology gaps will be essential (Wójcik-Karpacz et al., 2023; Zante et al., 2024).

## *7.2 Economic and Market Barriers*

Another major obstacle to the widespread adoption of circular economy methods is the existence of market and economic obstacles. The high initial cost of switching to circular business models is one of the main economic obstacles. It is frequently necessary to make large investments in new infrastructure, training, and technology to transition from linear to circular processes. For the collecting and redistribution of used items, for instance, businesses could need to make investments in cutting-edge recycling facilities, rethink their products, or put in place new logistics systems. These expenditures can be unaffordable for many organizations, especially small and medium-sized enterprises (SMEs), especially in the lack of government support or financial incentives. Despite the long-term advantages, many businesses are hesitant to adopt circular processes (Kandasamy et al., 2023; Tavares et al., 2023; Wójcik-Karpacz et al., 2023). The adoption of circular economy methods is hampered by market restrictions in addition to high prices. A significant obstacle is the absence of consumer interest in recycled or remanufactured goods, which are frequently thought to be of inferior quality to brand-new goods. This is especially true in industries like electronics and textiles, where customers prefer new products over ones that have been recycled or refurbished. Finding viable markets for their circular products can be challenging for businesses as many countries have minimal markets for recycled materials and second-hand goods. Furthermore, as virgin materials frequently become less expensive than their recycled counterparts when commodity prices decline, shifting raw material prices might reduce the competitiveness of recycled materials (Kandasamy et al., 2023; Koide et al., 2023). The structure of the current economic system also supports linear production models, which frequently externalize the costs of resource extraction and waste disposal and do not show up in product prices. Because the financial benefits of resource efficiency or waste reduction are not always evident, firms are not motivated to implement circular processes. This problem is made worse by the absence of financial tools, such as resource taxes or carbon pricing, that take into consideration the environmental costs of resource extraction and waste production. Stronger financial incentives, increased customer knowledge, and regulations that level the playing field between circular and linear business models are required to get over these commercial and economic obstacles (Hossain et al., 2024; Majid et al., 2023; Rashid & Malik, 2023).

### *7.3 Cultural and Behavioural Resistance*

Since changing society norms and consumer behaviours can be challenging and time-consuming, cultural and behavioural resistance poses a substantial hurdle to the adoption of circular economy techniques. The concept of ownership and the predilection for new overused goods are strongly embedded in many cultures. The demand for circular goods is constrained because consumers frequently consider reconditioned, recycled, or second-hand goods to be inferior to brand-new ones. This is especially true in sectors where trends and the newest models are highly prized, such as fashion and electronics, creating a "throwaway culture." The demand for novelty combined with the fast pace of consumption creates a significant obstacle to circular behaviours like product reuse, repair, and remanufacturing (Kandasamy et al., 2023; Prasad et al., 2023). Businesses themselves could be hesitant to implement circular models due to industry-specific cultural beliefs, in addition to customer reluctance. Many businesses still place a high priority on quick profits and linear growth plans, which concentrate on creating and marketing as many new goods as possible. Adopting circular methods frequently necessitates a mental shift on the part of firms, who must go beyond the sale of their products and instead examine the whole lifecycle of their products, including end-of-life management. Companies used to traditional business methods may find this transition challenging, and internal opposition to change may hinder the adoption of circular ideas (Behera et al., 2023; Kanda et al., 2024). In addition, public education and awareness campaigns are essential for overcoming cultural barriers. The advantages of circular economy techniques and their ability to mitigate environmental deterioration are not well understood in many areas. Customers and businesses are less inclined to demand or support circular solutions if they are not aware of the environmental and financial benefits of circular models. Raising awareness via media campaigns, community efforts, and education is crucial to influencing consumer behaviour and getting beyond cultural obstacles that stand in the way of the shift to a circular economy (Aslam et al., 2023; Siderius & Zink, 2023; Wójcik-Karpacz et al., 2023).

## 8 Regulatory and Policy Constraints

One of the biggest obstacles to adopting circular practices is the lack of appropriate laws and regulations, which are usually created to promote linear economies rather than circular ones. Waste management laws in many nations are antiquated and do not provide incentives for recycling, reusing, or remanufacturing. For instance, there is no governmental impetus to embrace more environmentally friendly alternatives; landfilling and incineration remain the most popular garbage disposal techniques in many locations. Companies can continue operating linearly without addressing the environmental effects of their products at the end of their lifecycle because there are no strict regulations, such as extended producer responsibility (EPR) laws, requiring businesses to take responsibility for the lifecycle of their products (Kandasamy et al., 2023; Taveres et al., 2023; Wójcik-Karpacz et al., 2023). Furthermore, a lot of regulatory regimes, especially those pertaining to permits and requirements for recycled materials, do not sufficiently encourage innovation in circular processes. Even in cases where using recycled materials is both environmentally and financially feasible, businesses may find it challenging to comply with stringent laws in certain areas on the safety and quality of items manufactured from these resources. For instance, the market for recycled plastics is typically constrained by severe safety regulations for food packaging, which prohibit the use of such materials. Regulating barriers can also hinder innovation in industries that stand to gain from circular practices by prohibiting the use of construction materials generated from recycled or reused components (Nguyen et al., 2024; Turunen et al., 2023). The implementation of circular economy methods is made more difficult by regional policy disparities, especially for multinational corporations. Businesses find it difficult to apply circular models globally due to a lack of standardized norms and laws across national borders. Different interpretations of what defines "waste" or "recycled materials," for instance, can result in legal obstacles that obstruct the cross-border movement of raw materials and recycled commodities. Governments must update waste management regulations, create stronger incentives for circular practices, and strive towards international policy harmonization to overcome these legal and legislative barriers and support the worldwide shift to a circular economy (Kandpal et al., 2024; Wójcik-Karpacz et al., 2023; Yang et al., 2023).

## 9 The Future of Circular Practices in Combating Global Environmental Degeneration

### *9.1 Advancing Circular Innovations and Technologies*

The progression of circular inventions and technologies is essential for the future of the fight against environmental degradation on a worldwide scale. Overcoming the existing constraints of circular practices will require innovations in material science, recycling technology, and sustainable product design. For example, the development of more effective chemical recycling techniques may make it possible to recover materials like complicated electronics and multi-layered polymers that are currently hard to recycle. By preventing materials from being thrown away as trash and instead being continuously cycled through the economy, these technological innovations can help shut the loop on material use. Furthermore, prolonging product lifecycles and lowering the overall demand for raw materials depends on design advances that prioritize durability, modularity, and reparability (Kandpal et al., 2024; Tavares et al., 2023). The application of digital technologies, including artificial intelligence (AI), blockchain, and the Internet of Things (IoT), presents a significant opportunity for circular innovation. By improving supply chains' efficiency and transparency, these technologies can facilitate the tracking of materials and goods throughout their entire lifecycle. IoT sensors, for instance, can be used to track a product's status in real-time, allowing for predictive maintenance and decreasing the need for premature replacements. To promote trust in circular supply chains, blockchain can also help make sure that recycled materials are traceable and up to quality standards. Artificial intelligence (AI) can improve recovery rates and lower contamination in recycling streams by streamlining waste sorting and recycling procedures (Pathan et al., 2023; Zhang et al., 2023). The development of circular business models is essential for the future of circular practices, in addition to improvements in products and processes. Product-as-a-service (PaaS) business models, in which customers rent or lease goods rather than buy them, are becoming more and more popular in a variety of sectors, including fashion and electronics. Since these models provide corporations ownership and maintenance responsibility for their products, they are incentivized to develop products with longevity and reparability in mind. Through the promotion of technological and business model innovation, the circular economy has the potential to evolve into a more feasible and expansive approach to tackling worldwide environmental issues (Ding et al., 2023; Godfrey et al., 2021; Gorbatcheva et al., 2024).

## *9.2 Policy Recommendations and Global Cooperation*

For circular practices to be widely adopted, robust legislative frameworks and international collaboration are necessary. Governments are essential in establishing rules and rewards that push companies and customers to adopt circular business models. The adoption of extended producer responsibility (EPR) regulations, which hold manufacturers responsible for the whole lifecycle of their products, from design to disposal, is a crucial policy proposal. By encouraging businesses to create goods that are simpler to reuse, repair, and recycle, EPR can relieve customers and local governments of some of the responsibility associated with waste management. Furthermore, policies that level the playing field by raising taxes on virgin materials and offering subsidies for the use of recycled materials might increase the economic appeal of circular processes (Brown et al., 2023; Ezeudu, 2024). Due to the transnational nature of resource depletion and environmental deterioration, global cooperation is equally crucial. International accords, like the Paris Climate Agreement, offer a structure for nations to collaborate in decreasing their ecological impact. Similarly, criteria for the circular economy must be harmonized internationally, especially regarding how trash and recovered products are classified. Multinational corporations can find it easier to use circular practices globally if they adopt a uniform global approach that facilitates the cross-border flow of circular products and materials. International organizations such as the World Trade Organisation and the United Nations can be vital in promoting international cooperation by pushing countries to harmonize their policies and exchange best practices (Aisbett et al., 2023; Lu & Wang, 2023). Furthermore, international cooperation is essential to resolving the unique obstacles that developing nations encounter while implementing circular practices. Developing countries frequently lack the financial, technological, and infrastructure resources required to successfully deploy circular models. Thus, to facilitate the transformation in these places, finance, technical aid, and knowledge-sharing should come from wealthier states and international development organizations. Countries can work together to lessen climate change, lessen their own environmental effect, and protect natural resources for future generations by promoting a global circular economy (Hussain et al., 2023; Kandasamy et al., 2023; Kandpal et al., 2024).

## *9.3 The Role of Businesses and Consumers in Driving Change*

Businesses and consumers are central to the success of circular practices, as they are the primary actors in both production and consumption cycles. Businesses must fundamentally change the way they develop, manufacture, and sell their products to implement circular models. Businesses need to adopt more sustainable methods including closed-loop supply chains, resource efficiency, and product redesign in place of the conventional linear "take, make, dispose" paradigm. Manufacturers may both capture new market opportunities and lessen their environmental impact by working with suppliers and investing in circular ideas. Businesses that adopt circularity can also reap the benefits of improved consumer loyalty and brand reputation, as more and more consumers choose sustainable products and are environmentally conscientious (Kirchherr et al., 2023; Rashid & Malik, 2023). Additionally, consumers have a significant influence on the transition to a circular economy. Customers have the power to shape market demand for circular goods and services by making informed purchasing decisions. Consumers may tell businesses that sustainability matters by choosing products that are manufactured from recycled materials, long-lasting designs, or both. Furthermore, consumer actions like choosing used goods, taking part in recycling programs, or adopting product-as-a-service models directly contribute to waste reduction and product longevity. Education and awareness campaigns can further empower consumers to make more informed decisions and adopt sustainable habits that support circular practices (Kandpal et al., 2024; Vidal-Ayuso et al., 2023). However, companies and consumers will need to work together to make the shift to a circular economy. Industry, supply chain, and stakeholder collaboration are essential to building the infrastructure required for the success of circular processes. For instance, organizations might create reverse logistics systems that facilitate the collection and reprocessing of discarded goods by collaborating with recycling companies, shipping providers, and other enterprises. In a similar vein, companies can interact with customers to promote the advantages of circular products by providing rewards for returning worn items or taking part in repair initiatives. Businesses and consumers working together have the power to accelerate the adoption of circular practices, paving the way for a more sustainable future (McKie et al., 2023; Butt et al., 2024; Kandpal et al., 2024).

## 9.4 Pathways Towards a Sustainable Future

A more sustainable future can be achieved with ease by making the shift to a circular economy. Systems that prioritize sustainability at every stage of the product lifecycle must be developed in collaboration by governments, corporations, and individuals to accomplish this shift. Enabling this transformation requires policies that promote recycling, resource efficiency, and renewable energy. It is imperative for governments to provide backing for the advancement of circular technology and provide regulatory structures that encourage corporations to use sustainable practices. Since the circular economy is a worldwide idea that needs standardized guidelines and coordinated activity to be successful, international cooperation will be crucial. Through international policy and regulatory alignment, nations may facilitate the growth of cross-border circular supply chains and advance sustainable practices worldwide (Alkaraan et al., 2023; Siderius & Zink, 2023). Businesses' and consumers' active involvement will be crucial to the shift to a circular economy. Businesses need to reconsider their business models, moving away from the conventional linear approach and towards strategies that prioritise reparability, durable product design, and closed-loop supply networks. Businesses can minimise waste, maximise resource utilisation, and establish more sustainable production cycles by incorporating circular ideas into their operations. To increase the effectiveness and widespread adoption of circular practices, innovations in waste management, recycling technology, and environmentally friendly product design will be crucial. The demand for sustainable products and services is also greatly influenced by consumers. People can influence industries to adopt more environmentally friendly practices by promoting responsible consumption, supporting businesses that prioritise sustainability, and making thoughtful purchasing decisions. Increasing knowledge and educating people about the advantages of circular economy models will contribute to the development of a society in which sustainability is the rule rather than the exception. Ultimately, integrating circular practices into daily life will depend on encouraging cooperation between companies, customers, legislators, and other stakeholders. The adoption of these ideas can build a more robust and sustainable economy that protects the environment for coming generations while also promoting social advancement and economic prosperity (Kandpal et al., 2024; Sharma et al., 2023). The path to a sustainable future ultimately lies in the adoption of circular models that balance environmental health with social and economic growth. By creating innovative technologies, passing legislation that supports them, and collaborating with companies and consumers, we can reduce the environmental impact of our actions while preserving the long-term availability of resources. This approach addresses the urgent problems caused by environmental deterioration globally while also laying the foundation for a more equitable and sustainable world. By incorporating circular concepts into the framework of our economic systems, we may ensure that the planet that the following generation inherits is healthier, more affluent, and capable of supporting human life (Bibri et al., 2024; Kandpalet al., 2024).

## 10 Conclusion

The urgent necessity to switch to circular practices has been brought to light by the increasing deterioration of the environment worldwide. Resource depletion, pollution, deforestation, and climate change are just a few of the serious environmental problems that have been exacerbated by linear economic models, which are based on the extraction of raw materials, the production of goods, and the disposal of trash. These issues jeopardize human health, way of life, and economic stability in addition to ecosystems. Reducing trash, recycling, and using sustainable production methods are examples of circular practices that provide a solution by establishing systems that minimize the environmental effect and constantly utilize resources. Making the shift to a circular economy is crucial for lowering the industry's environmental impact, protecting natural resources, and lessening the effects of climate change in a world with finite resources and growing ecological strain. The implementation of circular practices has the potential to revolutionize economic processes, promoting robust and enduring growth.

One of the main drivers for the widespread adoption of circular economy methods is the continuous environmental degradation we currently confront. Circular models are viewed as an essential response as people, organisations, and governments grow more conscious of the planet's limited resources and the negative effects of excessive consumption. From encouraging more sustainable resource management to decreasing trash output, the circular economy tackles several facets of environmental degradation. Circular practices are a practical way to lessen the harm brought about by decades of linear production and consumption by reevaluating the way items are made, utilised, and disposed away. To restore natural equilibrium and guarantee a healthier planet for future generations, adopting circular models is not merely a choice, but it is an essential step.

In conclusion, it is evident how urgent it is that we change the way we manage the resources of the planet. A major factor in the growing environmental catastrophe, which is characterised by pollution, deforestation, resource depletion, and climate change, is the old linear economy, which is founded on the "take, make, dispose" philosophy. Without significant reform, the repercussions of this unsustainable system, which has been largely responsible for the ecological harm done to the world, will only get worse. The circular economy, which prioritises resource efficiency, material recycling, and waste reduction, provides a clear route forward. The circular economy fosters sustainable production and consumption, which not only lessens environmental damage but also creates opportunities for economic expansion and innovation. Circular methods reduce the strain on ecosystems and promote resilient economies that can coexist peacefully with a healthy environment by turning waste into a resource.

## References

Aa, I., Op, A., Ujj, I., & Mt, B. (2022). A critical review of oil spills in the Niger Delta aquatic environment: Causes, impacts, and bioremediation assessment. *Environmental Monitoring and Assessment, 194*(11), 816.

Abdullah, A. A. (2021). Atmospheric emissions effects and mechanism. *NeuroQuantology, 19*(5), 25.

Abdullahi, A. M., Kalengyo, R. B., & Warsame, A. A. (2024). The unmet demand of food security in East Africa: Review of the triple challenges of climate change, economic crises, and conflicts. *Discover Sustainability, 5*(1), 1–15.

Agamuthu, P., & Babel, S. (2023). Waste management developments in the last five decades: Asian perspective. *Waste Management & Research, 41*(12), 1699–1716.

Agrawal, D., Dwivedi, A., Patil, A., & Paul, S. K. (2023). Impediments of product recovery in circular supply chains: Implications for sustainable development. *Sustainable Development, 31*(3), 1618–1637.

Aisbett, E., Raynal, W., Steinhauser, R., & Jones, B. (2023). International green economy collaborations: Chasing mutual gains in the energy transition. *Energy Research & Social Science, 104*, Article 103249.

Ajibade, F. O., Adelodun, B., Lasisi, K. H., Fadare, O. O., Ajibade, T. F., Nwogwu, N. A., & Wang, A. (2021). Environmental pollution and their socioeconomic impacts. In *Microbe Mediated Remediation of Environmental Contaminants* (pp. 321–354). Woodhead Publishing.

Algarni, S., Tirth, V., Alqahtani, T., Alshehery, S., & Kshirsagar, P. (2023). Contribution of renewable energy sources to the environmental impacts and economic benefits for sustainable development. *Sustainable Energy Technologies and Assessments, 56*, Article 103098.

Alkaraan, F., Floyd, D., Rahman, M., & Nuery, N. (2023). A new era of strategic investment decision-making practices in UK companies: Towards sustainable supply chains and circular economy. *Theoretical Economics Letters, 13*(3), 666–682.

Al-Kuwari, A., Kucukvar, M., & Onat, N. C. (2024). Uncovering the role of sustainable value chain and life cycle management toward sustainable operations in electricity production technologies. *Operations Management Research*, 1–20.

Ameen, M., Anwar-Ul-Haq, M., Sohail, M. I., Akmal, F., & Siddiqui, A. (2023). Mismanagement of waste in developing countries. In *Waste Problems and Management in Developing Countries* (pp. 31–72). Apple Academic Press.

Anukwonke, C. C., Tambe, E. B., Nwafor, D. C., & Malik, K. T. (2022). Climate change and interconnected risks to sustainable development. *Climate change: The Social and Scientific Construct* (pp. 71–86). Springer International Publishing.

Anwar, M. N., Shabbir, M., Tahir, E., Iftikhar, M., Saif, H., Tahir, A., & Nizami, A. S. (2021). Emerging challenges of air pollution and particulate matter in China, India, and Pakistan and mitigating solutions. *Journal of Hazardous Materials, 416*, Article 125851.

Aquatar, M. O., Meena, M., Meena, P. L., & Banerjee, M. (2023). *Plastic recycling & waste management*. Academic Guru Publishing House.

Aslam, F., Yue, Y., Jafri, N., & Babar, Z. U. D. (2023). Research evidence in improving vaccine practices in low-and middle-income countries: Examples of community engagement, barriers, and implementation strategies. *Encyclopedia of Evidence in Pharmaceutical Public Health and Health Services Research in Pharmacy* (pp. 1241–1254). Springer International Publishing.

Bandh, S. A., Malla, F. A., Wani, S. A., & Hoang, A. T. (2024). Waste management and circular economy. *Waste Management in the Circular Economy* (pp. 1–17). Springer International Publishing.

Bano, N., Younas, T., Shoaib, F., Rashid, D., & Jaffri, N. (2020). Plastic: Reduce, recycle, and environment. In *Environmentally Benign Energy Solutions* (pp. 191–208). Springer International Publishing.

Batra, G. (2023). Renewable energy economics: Achieving harmony between environmental protection and economic goals. *Social Science Chronicle, 2*(2), 1–32.

Behera, R. K., Bala, P. K., & Rana, N. P. (2023). Creation of sustainable growth with explainable artificial intelligence: An empirical insight from consumer-packaged goods retailers. *Journal of Cleaner Production, 399*, Article 136605.

Berry, B., Haverkamp, J., Isenhour, C., Bilec, M. M., & Lowden, S. S. (2023). Is convergence around the circular economy necessary? Exploring the productivity of divergence in US circular economy discourse and practice. *Circular Economy and Sustainability, 3*(3), 1597–1622.

Bibri, S. E., Krogstie, J., Kaboli, A., & Alahi, A. (2024). Smarter eco-cities and their leading-edge artificial intelligence of things solutions for environmental sustainability: A comprehensive systematic review. *Environmental Science and Ecotechnology, 19*, Article 100330.

Blomsma, F., & Brennan, G. (2017). The emergence of circular economy: A new framing around prolonging resource productivity. *Journal of Industrial Ecology, 21*(3), 603–614.

Braun, A. B., Trentin, A. W. D. S., & Thomé, A. (2021). Sustainable remediation: A new environmentally sustainable paradigm in urbanization and industrialization. *Industry, Innovation and Infrastructure* (pp. 1102–1113). Springer International Publishing.

Brown, A., Laubinger, F., & Börkey, P. (2023). New Aspects of EPR: Extending producer responsibility to additional product groups and challenges throughout the product lifecycle.

Büscher, B., Koot, S., & Thakholi, L. (2024). Fossilized conservation, or the unsustainability of saving nature in South Africa. *Environment and Planning E: Nature and Space, 7*(1), 50–68.

Butt, A. S., Ali, I., & Govindan, K. (2024). The role of reverse logistics in a circular economy for achieving sustainable development goals: A multiple case study of retail firms. *Production Planning & Control, 35*(12), 1490–1502.

Camilleri, M. A. (2021). A circular economy strategy for sustainable value chains: A European perspective. *Global Challenges to CSR and Sustainable Development: Root Causes and Evidence from Case Studies*, 141–161.

Cerio, E., Debenedetti, A., & Sophie, R. (2024). When the secondhand economy is not as good as it seems: understanding conflicts and their (ir) resolutions between users on secondhand resale platforms. *Qualitative Market Research: An International Journal.*

Chaudhari, S., Nimbalkar, S., Lung, B., Gonzalez, M., Hill, B., & Esch, B. (2024). Key strategies in industry for circular economy: analysis of remanufacturing and beneficial reuse. *Technology Innovation for the Circular Economy: Recycling, Remanufacturing, Design, Systems Analysis and Logistics*, 279–296.

Chaudhary, M. T., & Piracha, A. (2021). Natural disasters—origins, impacts, management. *Encyclopedia, 1*(4), 1101–1131.

Chen, F., & Chen, Z. (2021). Cost of economic growth: Air pollution and health expenditure. *Science of the Total Environment, 755*, Article 142543.

Choudhury, M., Sharma, A., Singh, P., & Kumar, D. (2021). Impact of climate change on wetlands, concerning Son Beel, the largest wetland of Northeast, India. In *Global Climate Change* (pp. 393–414). Elsevier.

Chowdhary, P., Bharagava, R. N., Mishra, S., & Khan, N. (2020). Role of industries in water scarcity and its adverse effects on environment and human health. *Environmental Concerns and Sustainable Development: Volume 1: Air, Water and Energy Resources*, 235–256.

Clarke, B., Otto, F., Stuart-Smith, R., & Harrington, L. (2022). Extreme weather impacts of climate change: An attribution perspective. *Environmental Research: Climate, 1*(1), Article 012001.

D'amato, D., & Korhonen, J. (2021). Integrating the green economy, circular economy and bioeconomy in a strategic sustainability framework. *Ecological Economics, 188*, Article 107143.

Ding, S., Tukker, A., & Ward, H. (2023). Opportunities and risks of internet of things (IoT) technologies for circular business models: A literature review. *Journal of Environmental Management, 336*, Article 117662.

Dobrauz-Saldapenna, G., & Huras, N. (Eds.). (2023). *Circular economy: New solutions for a better tomorrow*. Stämpfli Verlag.

Doumkou, S. (2024). *Circular economy and sustainable agriculture: A comparative analysis of EU and MENA regions*.

Ekka, P., Patra, S., Upreti, M., Kumar, G., Kumar, A., & Saikia, P. (2023). Land degradation and its impacts on biodiversity and ecosystem services. *Land and Environmental Management through Forestry*, 77–101.

Elisha, O. D., & Felix, M. J. (2021). Destruction of coastal ecosystems and the vicious cycle of poverty in Niger Delta Region. *Journal of Global Agriculture and Ecology, 11*(2), 7–24.

Eze, C. G., Nwankwo, C. E., Dey, S., Sundaramurthy, S., & Okeke, E. S. (2024). Food chain microplastics contamination and impact on human health: A review. *Environmental Chemistry Letters, 22*(4), 1889–1927.

Ezeudu, O. B. (2024). Harnessing the drivers and barriers to implementation of extended producer responsibility for circular economy in Nigeria. *Circular Economy and Sustainability*, 1–26.

van Genuchten, E. (2024). How Amazon deforestation affects biodiversity. In *A Guide to a Healthier Planet, Volume 2: Scientific Insights and Actionable Steps to Help Resolve Climate, Pollution and Biodiversity Issues* (pp. 117–124). Springer Nature Switzerland.

Godfrey, L., Roman, H., Smout, S., Maserumule, R., Mpofu, A., Ryan, G., & Mokoena, K. (2021). Unlocking the opportunities of a circular economy in South Africa. In *Circular Economy: Recent Trends in Global Perspective* (pp. 145–180). Springer Nature Singapore.

Gorbatcheva, A., Gorbatchev, N., & Zenchanka, S. (2024). Belarus on the path to a green and circular economy: a comparison with the European Green Deal. *Academia Green Energy, 1*(2).

Hajam, Y. A., Kumar, R., & Neelam. (2024). Value addition to waste for circular economy and sustainable development. In *Zero Waste Management Technologies* (pp. 137–170). Springer Nature Switzerland.

Hartley, K., Schülzchen, S., Bakker, C. A., & Kirchherr, J. (2023). A policy framework for the circular economy: Lessons from the EU. *Journal of Cleaner Production, 412*, Article 137176.

Herrera Arango, J., Senent-De Frutos, J.A., & Molina, E.H. (2022). Murky waters: The impact of privatizing water use on environmental degradation and the exclusion of local communities in the Caribbean. *International Journal of Water Resources Development*, *38*(1), 152–172

Hossain, M. E., Ullah, E., Rehman, M. Z., Haseeb, M., & Esquivias, M. A. (2024). Going sustainable or going extinct: the consequences of clean technologies, green finance, and natural resources on the environment. *Sustainability, 16*(14), 5836.

Hussain, M., Butt, A. R., Uzma, F., Ahmed, R., Irshad, S., Rehman, A., & Yousaf, B. (2020). A comprehensive review of climate change impacts, adaptation, and mitigation on environmental and natural calamities in Pakistan. *Environmental Monitoring and Assessment, 192*, 1–20.

Hussain, M., Khan, M., & Saber, H. (2023). Thematic analysis of circular economy practices across closed-loop supply chains: An institutional theory perspective. *Sustainable Production and Consumption, 40*, 122–134.

Jackson, M., Lederwasch, A., & Giurco, D. (2014). Transitions in theory and practice: Managing metals in the circular economy. *Resources, 3*(3), 516–543.

Jahanger, A., Ozturk, I., Onwe, J. C., Joseph, T. E., & Hossain, M. R. (2023). Do technology and renewable energy contribute to energy efficiency and carbon neutrality? Evidence from top ten manufacturing countries. *Sustainable Energy Technologies and Assessments, 56*, Article 103084.

Jayathilake, H. M., Prescott, G. W., Carrasco, L. R., Rao, M., & Symes, W. S. (2021). Drivers of deforestation and degradation for 28 tropical conservation landscapes. *Ambio, 50*(1), 215–228.

Jimura, T. (2023). Introduction-sustainability and Japan's heritage and tourism. *Sustainability Management in Heritage and Tourism: The Concept and Practice of Mottainai in Japan* (pp. 3–32). Springer International Publishing.

John, C. K., Pu, J. H., Moruzzi, R., et al. (2023). Deposition and microbial analyses from roof-top sediments within different sanitation environments. *Topics in Catalysis*. https://doi.org/10.1007/s11244-023-01856-x

Johnson, J. E., Allain, V., Basel, B., Bell, J. D., Chin, A., Dutra, L. X., & Nicol, S. (2020). Impacts of climate change on marine resources in the Pacific Island region. *Climate change and impacts in the Pacific*, 359–402.

Jovine, R. (2021). *Light to life: The miracle of photosynthesis and how it can save the planet*. Hachette UK.

Kabir, M., Habiba, U. E., Khan, W., Shah, A., Rahim, S., Patricio, R., & Shafiq, M. (2023). Climate change due to increasing concentration of carbon dioxide and its impacts on environment in 21st century; a mini review. *Journal of King Saud University-Science, 35*(5), Article 102693.

Kanda, W., Klofsten, M., Bienkowska, D., Henry, M., & Hjelm, O. (2024). Challenges of circular new ventures: An empirical analysis of 70 cases. *Journal of Cleaner Production, 442*, Article 141103.

Kandasamy, J., Venkat, V., & Mani, R. S. (2023). Barriers to the adoption of digital technologies in a functional circular economy network. *Operations Management Research, 16*(3), 1541–1561.

Kandpal, V., Jaswal, A., Santibanez Gonzalez, E. D., & Agarwal, N. (2024). Circular economy principles: Shifting towards sustainable prosperity. In *Sustainable Energy Transition: Circular Economy and Sustainable Financing for Environmental, Social and Governance (ESG) Practices* (pp. 125–165). Springer Nature Switzerland.

Kattel, G. R. (2022). Climate warming in the Himalayas threatens biodiversity, ecosystem functioning and ecosystem services in the 21st century: Is there a better solution? *Biodiversity and Conservation, 31*(8), 2017–2044.

Kekana, H. N., Ruhiiga, T. M., Ndou, N. N., & Palamuleni, L. G. (2023). Environmental justice in South Africa: the dilemma of informal settlement residents. *GeoJournal, 88*(4), 3709–3725. Khoaele, K. K., Gbadeyan, O. J., Chunilall, V., & Sithole, B. (2023). The devastation of waste plastic on the environment and remediation processes: a critical review. *Sustainability, 15*(6), 5233.

Khan, Z. A., Chowdhury, S. R., Mitra, B., Mozumder, M. S., Elhaj, A. I., Salami, B. A., & Rahman, S. M. (2023). Analysis of industrial symbiosis case studies and its potential in Saudi Arabia. *Journal of Cleaner Production, 385*, Article 135536.

Kibria, M. G., Masuk, N. I., Safayet, R., Nguyen, H. Q., & Mourshed, M. (2023). Plastic waste: Challenges and opportunities to mitigate pollution and effective management. *International Journal of Environmental Research, 17*(1), 20.

Kirchherr, J., Yang, N. H. N., Schulze-Spüntrup, F., Heerink, M. J., & Hartley, K. (2023). Conceptualizing the circular economy (revisited): An analysis of 221 definitions. *Resources, Conservation and Recycling, 194*, Article 107001.

Kirikkaleli, D., & Ali, M. (2024). Resource efficiency, energy productivity, and environmental sustainability in Germany. *Environment, Development and Sustainability, 26*(5), 13139–13158.

Koide, R., Yamamoto, H., Kimita, K., Nishino, N., & Murakami, S. (2023). Circular business cannibalization: A hierarchical Bayes conjoint analysis on reuse, refurbishment, and subscription of home appliances. *Journal of Cleaner Production, 422*, Article 138580.

Kolawole, A. S., & Iyiola, A. O. (2023). Environmental pollution: Threats, impact on biodiversity, and protection strategies. In *Sustainable Utilization and Conservation of Africa's Biological Resources and eEnvironment* (pp. 377–409). Springer Nature Singapore.

Kumar, A., Singh, E., Mishra, R., Lo, S. L., & Kumar, S. (2023). Global trends in municipal solid waste treatment technologies through the lens of sustainable energy development opportunity. *Energy, 275*, Article 127471.

Kumar, B., Kumar, L., Kumar, A., Kumari, R., Tagar, U., & Sassanelli, C. (2024). Green finance in circular economy: A literature review. *Environment, Development and Sustainability, 26*(7), 16419–16459.

Kumar, M. S., & Donnelly, S. (2024). 'leave fossil fuels in the soil, halt deforestation': Stop threatening the planet. In *Eco-Politics and Global Climate Change* (pp. 239–255). Springer Nature Switzerland.

Kumar, A., Nagar, S., & Anand, S. (2021). Climate change and existential threats. In *Global Climate Change* (pp. 1–31). Elsevier.

Kumar, P., Kumar, A., Patil, M., Hussain, S., Sharma, S., Singh, S., & Singh, A. N. (2022). Biodiversity resources and ecosystem services in the context of the industrialized world. In *Biodiversity* (pp. 23–39). CRC Press.

Lamb, W. F., Grubb, M., Diluiso, F., & Minx, J. C. (2022). Countries with sustained greenhouse gas emissions reductions: An analysis of trends and progress by sector. *Climate Policy, 22*(1), 1–17.

Lu, C., & Wang, K. (2023). Natural resource conservation outpaces and climate change: Roles of reforestation, mineral extraction, and natural resources depletion. *Resources Policy, 86*, Article 104159.

Maina, J., Asamoah, E. F., D'Agata, S., Gamoyo, M., Rakoto, P., George, R., & Bosire, J. (2023). *Rising climate risk and loss and damage to coastal subsistence-oriented livelihoods*. Under Review at Nature Portfolio.

Majid, S., Zhang, X., Khaskheli, M. B., Hong, F., King, P. J. H., & Shamsi, I. H. (2023). Eco-efficiency, environmental and sustainable innovation in recycling energy and their effect on business performance: Evidence from European SMEs. *Sustainability, 15*(12), 9465.

Manić, E., & Lutovac, M. (2022). Natural resources and manufacturing sector. In *The Geography of Serbia: Nature, People, Economy* (pp. 207–220).

Marazziti, D., Cianconi, P., Mucci, F., Foresi, L., Chiarantini, I., & Della Vecchia, A. (2021). Climate change, environment pollution, COVID-19 pandemic and mental health. *Science of the Total Environment, 773*, Article 145182.

McKie, E. C., Saez de Tejada Cuenca, A., & Agrawal, V. (2023). Does less result in more? the role of information and rewards in take-back programs for clothing. In *The Role of Information and Rewards in Take-Back Programs for Clothing*.

Menton, M., Larrea, C., Latorre, S., Martinez-Alier, J., Peck, M., Temper, L., & Walter, M. (2020). Environmental justice and the SDGs: From synergies to gaps and contradictions. *Sustainability Science, 15*, 1621–1636.

Mondal, S., Singh, S., & Gupta, H. (2023). Assessing enablers of green entrepreneurship in circular economy: An integrated approach. *Journal of Cleaner Production, 388*, Article 135999.

Möslinger, M., Ulpiani, G., & Vetters, N. (2023). Circular economy and waste management to empower a climate-neutral urban future. *Journal of Cleaner Production, 421*, Article 138454.

Nair, K. P. (2023). Which are the threats to biodiversity? it's conservation and sustainability. In *Biodiversity in Agriculture: Sustainability of Soil, Soil Fauna and Soil Flora* (pp. 57–100). Springer Nature Switzerland.

Naqash, A. H., & Bhat, B. B. (2023). Environmental legislation and global initiatives. In *Geospatial Analytics for Environmental Pollution Modeling: Analysis, Control and Management* (pp. 415–433). Springer Nature Switzerland.

Nathaniel, S. P., Nwulu, N., & Bekun, F. (2021). Natural resource, globalization, urbanization, human capital, and environmental degradation in Latin American and Caribbean countries. *Environmental Science and Pollution Research, 28*, 6207–6221.

Nguyen, T. T., Grote, U., Neubacher, F., Do, M. H., & Paudel, G. P. (2023). Security risks from climate change and environmental degradation: Implications for sustainable land use transformation in the Global south. *Current Opinion in Environmental Sustainability, 63*, Article 101322.

Nguyen, P. M., Berrard, C., Daoud, N., Saillard, P., Peyroux, J., & Vitrac, O. (2024). Assessment of chemical risks and circular economy implications of recycled PET in food packaging with functional barriers. *Resources, Environment and Sustainability, 17*, Article 100163.

Nhemachena, C., Nhamo, L., Matchaya, G., Nhemachena, C. R., Muchara, B., Karuaihe, S. T., & Mpandeli, S. (2020). Climate change impacts on water and agriculture sectors in Southern Africa: Threats and opportunities for sustainable development. *Water, 12*(10), 2673.
Nighot, N. S., & Kumar, R. (2023). A comprehensive study on the synthesis and characterization of eco-cementitious binders using different kind of industrial wastes for sustainable development. *Developments in the Built Environment, 14*, Article 100135.
Njoku, A. U., & Sampson, N. R. (2023). Environment injustice and public health. *Handbook of Social Sciences and Global Public Health* (pp. 1987–2006). Springer International Publishing.
Nkonya, E., & Kato, E. (2020). Rethinking agro-food sector to combat land degradation and desertification. *Life on Land* (pp. 851–863). Springer International Publishing.
Novarlić, B., & Đurić, P. (2024). Enhancing comprehensive waste management in transition economies through green logistics: A case study of Bosnia and Herzegovina. *Journal of Intelligent Management Decision, 3*, 42–55.
Ogidi, O. I., & Akpan, U. M. (2022). Aquatic biodiversity loss: Impacts of pollution and anthropogenic activities and strategies for conservation. In *Biodiversity in Africa: Potentials, Threats and Conservation* (pp. 421–448). Springer Nature Singapore.
Okkerse, C., & Van Bekkum, H. (1999). From fossil to green. *Green Chemistry, 1*(2), 107–114.
Okoh, A. S., & Okoh, A. S. (2021). Political economy of fossil fuel exit. In *Oil Mortality in Post-Fossil Fuel Era Nigeria: Beyond the Oil Age* (pp. 47–88).
Padilla, L. A., & Padilla, L. A. (2021). The anthropocene: Are we in the midst of the sixth mass extinction? In *Sustainable Development in the Anthropocene: Towards a New Holistic and Cosmopolitan Paradigm* (pp. 93–167).
Palinkas, L. A., & Palinkas, L. A. (2020). Fleeing flooding: Asia and the Pacific. *Global Climate Change, Population Displacement, and Public Health: The Next Wave of Migration*, 99–125.
Paliwal, R. (2022). From waste to wealth: Stepping toward sustainability through circular economy. In *Handbook of Solid Waste Management: Sustainability through Circular Economy* (pp. 441–463). Springer Nature Singapore.
Parsons, E. S., Jowell, A., Veidis, E., Barry, M., & Israni, S. T. (2024). Climate change and inequality. *Pediatric Research*, 1–8.
Pastor, M., Cha, J. M., Méndez, M., & Morello-Frosch, R. (2024). California dreaming: Why environmental justice is integral to the success of climate change policy. *Proceedings of the National Academy of Sciences, 121*(32), Article e2310073121.
Pathan, M. S., Richardson, E., Galvan, E., & Mooney, P. (2023). The role of artificial intelligence within circular economy activities—a view from Ireland. *Sustainability, 15*(12), 9451.
Paudel, D., Neupane, R. C., Sigdel, S., Poudel, P., & Khanal, A. R. (2023). COVID-19 pandemic, climate change, and conflicts on agriculture: A trio of challenges to global food security. *Sustainability, 15*(10), 8280.
Prasad, R. D., Prasad, R. J., Shrivastav, R. K., Charmode, N., SP, R., Mamidpelliwar, P. M., & Sarvalkar, P. (2023). A review on concept of veterinary bio-technology and livestock products in medicine. *ES Food & Agroforestry, 14*, 1004.
Rani, G., Kaur, J., Kumar, A., & Yogalakshmi, K. N. (2020). Ecosystem health and dynamics: An indicator of global climate change. *Contemporary environmental issues and challenges in era of climate change*, 1–32.
Rankin, W. J. (2011). *Minerals, Metals and Sustainability: Meeting Future Material Needs*. CSIRO publishing.
Rashid, S., & Malik, S. H. (2023). Transition from a linear to a circular economy. *Renewable Energy in Circular Economy* (pp. 1–20). Springer International Publishing.
Reddy, V. V., Kumar, P., Rao, A. L. N., Kumar, R., Singh, S., Asha, V., & Kareem, S. H. (2023). Waste to wealth generation: Innovative methodologies in resource utilization and minimization in circular economy. In *E3S Web of Conferences* (vol. 453, p. 01035). EDP Sciences.
Reijnders, L. (2022). Life cycle assessment of greenhouse gas emissions. *Handbook of Climate Change Mitigation and Adaptation* (pp. 313–347). Springer International Publishing.

Rezania, S., Oryani, B., Nasrollahi, V. R., Darajeh, N., Lotfi Ghahroud, M., & Mehranzamir, K. (2023). Review on waste-to-energy approaches toward a circular economy in developed and developing countries. *Processes, 11*(9), 2566.
Roy, H., Islam, M. R., Tasnim, N., Roy, B. N., & Islam, M. S. (2024). Opportunities and challenges for establishing sustainable waste management. *Trash or Treasure: Entrepreneurial Opportunities in Waste Management*, 79–123.
Liow Li Sa, M., & Choon-Yin, S. (2023). Biophilic vision, regenerative sustainable urbanism and circular economy. In *Sustainable Urban Development in Singapore: Imagining Walkability in an Urban Concrete Jungle* (pp. 233–251). Springer Nature Singapore.
Salem, H. S., Pudza, M. Y., & Yihdego, Y. (2022). Water strategies and water–food Nexus: Challenges and opportunities towards sustainable development in various regions of the World. *Sustainable Water Resources Management, 8*(4), 114.
Santos, C., Coelho, A., & Marques, A. (2024). The greenwashing effects on corporate reputation and brand hate, through environmental performance and green perceived risk. *Asia-Pacific Journal of Business Administration, 16*(3), 655–676.
Sharma, K., Aswal, C., & Paul, J. (2023). Factors affecting green purchase behavior: A systematic literature review. *Business Strategy and the Environment, 32*(4), 2078–2092.
Sharma, D., Rao, K., & Ramanathan, A. L. (2021). A systematic review on the impact of urbanization and industrialization on Indian coastal mangrove ecosystem. In *Coastal Ecosystems: Environmental Importance, Current Challenges and Conservation Measures* (pp. 175–199).
Shemer, H., Wald, S., & Semiat, R. (2023). Challenges and solutions for global water scarcity. *Membranes, 13*(6), 612.
Shen, M., Huang, W., Chen, M., Song, B., Zeng, G., & Zhang, Y. (2020). (Micro) plastic crisis: Un-ignorable contribution to global greenhouse gas emissions and climate change. *Journal of Cleaner Production, 254*, Article 120138.
Short, J. R., & Short, J. R. (2021). *Hurricane Katrina, Infrastructure Deficit and the Costs of Climate Change* (pp. 37–86). Public Policy and Reaction to Disaster Events.
Siderius, T., & Zink, T. (2023). Markets and the future of the circular economy. *Circular Economy and Sustainability, 3*(3), 1569–1595.
Singh, M. (2024). Future perspectives on attaining zero waste: an empirical study of the contrasts in the municipal solid waste strategies of two municipalities in India. *Educational Administration: Theory and Practice, 30*(6), 4499–4506.
Singh, A., Mehta, S., Yadav, S., Nagar, G., Ghosh, R., Roy, A., & Singh, I. K. (2022). How to cope with the challenges of environmental stresses in the era of global climate change: An update on ROS stave off in plants. *International Journal of Molecular Sciences, 23*(4), 1995.
Smil, V. (2013). *Making the modern world: Materials and dematerialization*. John Wiley & Sons.
Srivastav, A. L., Dhyani, R., Ranjan, M., Madhav, S., & Sillanpää, M. (2021). Climate-resilient strategies for sustainable management of water resources and agriculture. *Environmental Science and Pollution Research, 28*(31), 41576–41595.
Srivastav, A. (2024). Assessing forest resilience amidst the challenges of climate change. In *Forest Policies, Laws, and Governance in India: Conservation Challenges in the Face of Climate Change* (pp. 233–280). Springer Nature Singapore.
Sun, L., Fang, S., Iqbal, S., & Bilal, A. R. (2022). Financial stability role on climate risks, and climate change mitigation: Implications for green economic recovery. *Environmental Science and Pollution Research, 29*(22), 33063–33074.
Tamizhdurai, P., Mangesh, V. L., Santhosh, S., Vedavalli, R., Kavitha, C., Bhutto, J. K., & Kumaran, R. (2024). A state-of-the-art review of multilayer packaging recycling: Challenges, alternatives, and outlook. *Journal of Cleaner Production*, 141403.
Tan, Y. L., Chen, J. E., Yiew, T. H., & Habibullah, M. S. (2022). Habitat change and biodiversity loss in South and Southeast Asian countries. *Environmental Science and Pollution Research, 29*(42), 63260–63276.

Tavares, T. M., Ganga, G. M. D., Godinho Filho, M., & Rodrigues, V. P. (2023). The benefits and barriers of additive manufacturing for circular economy: A framework proposal. *Sustainable Production and Consumption, 37*, 369−388.

Tiamgne, X. T., Kalaba, F. K., & Nyirenda, V. R. (2022). Mining and socio-ecological systems: A systematic review of Sub-Saharan Africa. *Resources Policy, 78*, Article 102947.

Turunen, T., Mateo, E. R., & Alaranta, J. (2023). Regulatory catalysts for the circular economy. In *The Routledge Handbook of Catalysts for a Sustainable Circular Economy* (pp. 169–186). Routledge.

Upreti, G. (2023). Climate change and its threat to humanity in the anthropocene. In *Ecosocio-centrism: The Earth First Paradigm for Sustainable Living* (pp. 137–162). Springer Nature Switzerland.

Ventura, V., La Monica, M., Bortolini, M., Cutaia, L., & Mora, C. (2024). Blockchain and industrial symbiosis: a preliminary two-step framework to green circular supply chains. *International Journal of Environmental Science and Technology*, 1–14.

Vidal-Ayuso, F., Akhmedova, A., & Jaca, C. (2023). The circular economy and consumer behaviour: Literature review and research directions. *Journal of Cleaner Production*, 137824.

Vitale, V., & Ziegler, M. (2024). *Transforming trash to treasure: Uncovering the social benefits of industrial symbiosis in Scandinavia.*

Wang, Y., Wu, R., Liu, L., Yuan, Y., Liu, C., Ho, S. S. H., & Cao, J. (2022). Differential health and economic impacts from the COVID-19 lockdown between the developed and developing countries: Perspective on air pollution. *Environmental Pollution, 293*, Article 118544.

Wassie, S. B. (2020). Natural resource degradation tendencies in Ethiopia: A review. *Environmental Systems Research, 9*(1), 1–29.

Weigend Rodríguez, R. (2023). *Circular futures: an interdisciplinary approach based on circular economy principles and futures studies methods to support small and medium-sized enterprises (SMEs) transitioning to circularity and a futures-oriented organisation (Doctoral dissertation).*

Wiebe, K. S., Norstebø, V. S., Aponte, F. R., Simas, M. S., Andersen, T., & Perez-Valdes, G. A. (2023). Circular economy and the triple bottom line in Norway. *Circular Economy and Sustainability, 3*(1), 1–33.

Wójcik-Karpacz, A., Karpacz, J., Brzeziński, P., Pietruszka-Ortyl, A., & Ziębicki, B. (2023). Barriers and drivers for changes in circular business models in a textile recycling sector: Results of qualitative empirical research. *Energies, 16*(1), 490.

Yadav, N., & Upadhyay, R. K. (2023). Global effect of climate change on seasonal cycles, vector population and rising challenges of communicable diseases: A review. *Journal of Atmospheric Science Research, 6*(1), 21–59.

Yang, M., Chen, L., Wang, J., Msigwa, G., Osman, A. I., Fawzy, S., & Yap, P. S. (2023). Circular economy strategies for combating climate change and other environmental issues. *Environmental Chemistry Letters, 21*(1), 55–80.

Zante, G., Elgar, C. E., Hartley, J., Mukherjee, R., Kettle, J. P., Horsfall, L., & Abbott, A. P. (2024). A toolbox for improved recycling of critical metals and materials in low-carbon technologies. *RSC Sustainability.*

Zhang, X., Lin, M., Wang, Z., & Jin, F. (2021). The impact of energy-intensive industries on air quality in China's industrial agglomerations. *Journal of Geographical Sciences, 31*, 584–602.

Zhang, X., Feng, X., Jiang, Z., Gong, Q., & Wang, Y. (2023). A blockchain-enabled framework for reverse supply chain management of power batteries. *Journal of Cleaner Production, 415*, Article 137823.

Zhang, Y., Liu, L., Zhang, L., Yu, C., Wang, X., Shi, Z., & Zhang, Y. (2022). Assessing short-term impacts of PM2. 5 constituents on cardiorespiratory hospitalizations: Multi-city evidence from China. *International Journal of Hygiene and Environmental Health, 240*, 113912.

Zorpas, A. A. (2024). The hidden concept and the beauty of multiple "R" in the framework of waste strategies development reflecting to circular economy principles. *Science of The Total Environment*, 175508.

# Unveiling Circularity: Challenges and Barriers in Transitioning to Circular Practices

**Seda Yıldırım**

**Abstract** Sustainable Development Goals guide both developed and developing countries to achieve a sustainable future. In particular, the 2030 Sustainable Development Goals clearly state what needs to be done or achieved by 2030 with 17 specific basic and sub-goals. In this respect, it can be considered a concrete guide for the 2030 SDGs. On the other hand, the 2030 SDGs or sustainable development can't progress together with the current economic system. At this point, the green transition has come into play. The green transition and circular economy system can make countries' sustainable transformations realistic. Adapting to green transition and circular economy practices is not equally easy for every country. The transition to and adaptation to circular economy practices may vary depending on the country's structure, population, economic power, size, and many other variables. This study explores the challenges that prevent adaptation to the green transition and circular economy in the long term. As it is purposed, this study will review current statistics, reports, and sources to conclude the main challenges and barriers in transitioning to circular practices.

**Keywords** Green transition · 2030 sustainable development goals · The COVID-19 · Russia-Ukraine war · Greenwashing · Cost of green transition

## 1 Introduction

The Earth is subject to some changes while continuing its existence. Climate change is also a part of this natural change. However, the effects of increasing industrial activity and city life of mankind with the industrial revolution have accelerated climate change (NASA, 2024; IPCC, 2022; Hulme et al., 1999; He & Silliman, 2019; Trenberth, 2018). The rising temperatures and changes in sea water temperature have created major changes in terms of ecosystems and biodiversity globally. For example, the effects of climate change experienced in the Mediterranean basin and the threat

S. Yıldırım (✉)
Tekirdag Namık Kemal University, Tekirdağ, Turkey
e-mail: sedayil1@gmail.com



P. Singh et al. (eds.), *Circular Economy and Environmental Resilience*,
https://doi.org/10.1007/978-3-031-93091-1_8

of local biodiversity in terrestrial and wetland areas of Mediterranean countries by invasive species are events that have left their mark on the twenty-first century. In particular, the habitat of invasive marine species such as puffer fish and lionfish in Mediterranean waters are among the biggest indicators of climate change (Kaplan & Yildirim, 2023; Kaplan et al., 2022; Yildirim & Kaplan, 2022; Yildirim et al., 2023).

In the global fight against climate change, serious measures have been taken to prevent global temperatures from increasing by another 1 degree. In 2015, with the United Nations' 2030 Sustainable Development Goals, developed and developing countries began working on 17 basic goals that must be achieved by 2030. The most important function of the 2030 Sustainable Development Goals is to give common goals to all UN member countries globally and to plan these goals according to concrete criteria (Yıldırım et al., 2024a, 2024b).

The 2030 Sustainable Development Goals include goals that are expected to be resolved globally. 17 themes have been determined for each goal. These goals and themes are summarized as in Table 1.

As seen in Table 1, 2030 SDGs have directly linked to three dimensions of sustainable development. Ensuring the balance between economic growth and social welfare is closely linked to Goal-1; Goal-2; Goal-3; Goal-4; Goal-5; Goal-8; Goal-9; Goal-10; Goal-11; Goal-12; Goal-16; Goal-17 and achieving Goal-6; Goal-7; Goal-13; Goal-14; Goal-15 is directly linked to the balance between environmental quality and economic development in the long term.

Adapting circular economy and green transition is not easy as task or plan comparing to current indicators based on real outcomes. Green transition is a big challenge but also the way of green transformation has many troubles and barriers too. For example, the COVID-19 pandemic shocked the world and stop each progress in 2030 SDGs (Farquharson, 2020; United Nations, Human Rights Office of The High Commissoner, n.d.; Yuan et al., 2023; Martín-Blanco et al., 2022). Then, the Russia-Ukraine conflict has created new global crises (Braimah, 2024; UNPD, 2023). This chapter aims to explore challenges and barriers to green transition in the long term. In this context, this chapter focuses on the below questions:

- How does the COVID-19 pandemic influence 2030 Sustainable Development Goals?
- What is the link between Ukraine-Russia Conflict and 2030 SDGs?
- What are the main challenges and barriers in the green transition?
- How does Greenwashing influence sustainable development?

## 2 The COVID-19 Pandemic and 2030 SDGs

Every pandemic affects the century it is in. Economic and social stagnation occurs globally during pandemic periods. Along with the deterioration of people's physical and mental health, many problems such as job loss, income loss, housing difficulties, and health accessibility problems occur during the pandemic. The COVID-19 pandemic in the twenty-first century has also shown that the same problems occur in global problems by equalizing countries in the world (Antipova, 2021).

It can be said that future generations will be more severely affected by the unemployment and employment problems caused by the COVID-19 crisis. Unfortunately, young people have been exposed to greater problems in terms of employment. According to ILO (2021) records, young people have experienced more employment problems than other age groups. According to global data, there was a 3.7% decrease in employment among adults in 2020. However, there was an 8.7% decrease in employment among young people in 2020.

Table 2 shows unemployment rates in OECD countries. During the COVID-19 period, according to the OECD, there was an 8% increase in the unemployment rate calculated for Colombia, Canada and the USA. However, in Europe (despite the greatest impact of COVID-19), unemployment increased by less than 1% (Buchholz, 2020).

Kretchmer (2020) has conducted some employment studies for the G7 during the COVID-19 period. Table 3 summarizes some data for the G7 for the COVID-19 pandemic and unemployment.

The most prominent problem areas because of the crisis caused by COVID-19 are given in Table 4. The rates of global focus issues are given in this table. COVID-19 is the biggest problem on a global scale. It is seen that unemployment is next. Due to COVID-19, health has entered the 3rd place among the important global issues (Chariton, 2020).

When combined with the COVID-19 pandemic and the effects of climate change, a sustainable future in the world has become a distant dream. The COVID-19 crisis has also affected some groups more. For example, unsustainable health problems have been triggered in the elderly and sick. Many people have lost their lives due to COVID-19. The COVID-19 crisis has created a huge obstacle in terms of sustainability and the problem of unsustainability in health has emerged. Both developed and developing countries were caught unprepared by the COVID-19 crisis (Moore, 2021).

The COVID-19 crisis has essentially revealed globality and the interdependence of countries. It has been seen that the world must act as a whole in response to global crises. As in the COVID-19 crisis, world leaders must make a joint plan to combat the climate crisis (Metzke, 2020).

The COVID-19 pandemic is seen as a major obstacle. As a global crisis, it has been a factor that prevents the establishment of a sustainable future. It has been seen that both developed and developing countries are not ready for a pandemic like COVID-19.

Table 1 2030 sustainable development goals and main issues

| 2030 SDGs and themes[a] | Issues[b] | Link to sustainable development[b] |
|---|---|---|
| Goal-1: poverty | Reducing poverty, ensuring that everyone has an income that meets the minimum living standards | Economic issues |
| Goal-2: hunger | Reducing the number of people exposed to hunger and ensuring food security and sustainability | Social welfare |
| Goal-3: health | Sustainability in the health and health conditions of individuals and societies | Social welfare |
| Goal-4: education | To ensure that girls and boys have access to equal and quality education opportunities | Social welfare |
| Goal-5: gender equity | Achieving gender equality in all areas of life | Social welfare |
| Goal-6: clean water | Ensuring people have access to clean drinking water and protecting freshwater resources | Environmental quality |
| Goal-7: clean energy | Providing people with access to clean/renewable energy at affordable prices | Environmental quality and economic issues |
| Goal-8: economic growth | Continuing economic growth and development by providing appropriate employment conditions | Economic issues |
| Goal-9: industry | Supporting growth and development in industries by adopting green and innovative technologies | Economic issues and environmental quality |
| Goal-10: inequalities | Reducing and ending inequalities globally | Social welfare |
| Goal-11: sustainable cities | Providing incentives for sustainable city planning and the formation of sustainable societies in city life | Environmental quality and social welfare |
| Goal-12: responsible consumption/production | Making the necessary arrangements for consumers to adopt responsible consumption behavior and for producers to switch to green production | Economic issues |
| Goal-13: climate change | Planning to combat climate change and developing strategies and policies | Environmental quality |

(continued)

**Table 1** (continued)

| 2030 SDGs and themes[a] | Issues[b] | Link to sustainable development[b] |
|---|---|---|
| Goal-14: biodiversity of sea/oceans | Taking necessary precautions for the sustainability of underwater life, ecosystem and biodiversity | Environmental quality |
| Goal-15: biodiversity of land | Taking necessary precautions for the sustainability of life on land, the ecosystem and biodiversity | Environmental quality |
| Goal-16: peace/justice/strong institutions | Establishing and protecting peace, justice and establishing strong institutional structures for a sustainable future | Social welfare |
| Goal-17: partnerships | Increasing cooperation between developed and developing countries and providing support to underdeveloped countries for sustainable policies | Social welfare |

*Source* Authors' Compilation
[a]United Nations, Department of Economic and Social Affairs Sustainable Development, (n.d.)
[b]Yıldırım et al. (2024)

**Table 2** OECD unemployment indicators (2020)

| Country[a] | Unemployment change rate (%)[a] |
|---|---|
| Colombia | + 10.3 |
| Spain | + 0.8 |
| Canada | + 8.1 |
| US | + 9.8 |
| OECD | + 3.2 |
| France | + 0.5 |
| South Korea | + 1.2 |
| Germany | + 0.4 |

[a] Based onBuchholz, (2020)
Change rate: February 2020–May 2020

## 3 Ukraine-Russia Conflict and 2030 SDGs

The Russo-Ukrainian War has brought energy and food security to the forefront after the COVID-19 period. In addition to health, food and energy security have become national issues (IPR, 2022).

According to the International Energy Agency, some opportunities may have emerged in energy. The Russia-Ukraine war has made it necessary for countries dependent on Russia for energy to invest in other renewable energy sources instead

**Table 3** Unemployment in G7 and COVID-19

| Country | Unemployment and COVID-19 pandemic term |
|---|---|
| USA | The US private sector saw more than 20 million job losses in April 2020 |
| Canada | In Canada, unemployment increased by 13% for the period April 2020 to March 2020. More than 7.2 million people applied for emergency unemployment benefits during the COVID-19 period (2020) |
| Japan | Japan's unemployment rate is rising at a slower pace than other G7 countries. On the other hand, there has been a greater increase in the number of people applying for emergency loans |
| UK | Official unemployment figures for the UK have been recorded later. However, unemployment has also been seen in the UK and has been significant in some sectors such as airlines |
| France | Job losses in France have been smaller than in other countries |
| Germany | The risk of unemployment due to COVID-19 has also been observed in Germany, but it is in a better position than other countries. Various support is being provided to people in Germany to manage the crisis |
| Italy | Although unemployment is low, the low number of people looking for work has also affected these figures |

**Table 4** COVID-19 and economic issues

| Global concern[a] | Importance of economic issues (%)[a] |
|---|---|
| COVID-19 pandemic | 55 |
| Unemployment | 42 |
| Poverty and social inequality | 31 |
| Healthcare | 25 |
| Financial/policial corruption | 24 |

[a] Based on Chariton, (2020)

of gas. However, according to the IEA, the Russia-Ukraine war primarily triggered the energy crisis (Buchholz, 2022).

Many researchers have stated that the European Union has returned to fossil fuel consumption in response to Russia's war against Ukraine. However, this is not true. According to gas and coal consumption figures, they have decreased in the second half of 2022 and are expected to decrease in the coming period. During the COVID-19 period, EU coal consumption has also decreased in 2020. On the other hand, investments in renewable energy in the EU have increased according to Russia's energy stance (Myllyvirta, 2023).

The energy impacts of the Russia-Ukraine war can be summarized as follows in the IEA's World Energy Outlook 2022 report (Thomson, 2022) (Table 5).

It is necessary to reduce greenhouse gas (GHG) emissions by 45% by 2030. At this point, green energy is a prominent sector. Emissions have decreased during the COVID-10 period (Bradshaw, 2022).

**Table 5** Six main effects of Russia-Ukraine war on energy

| Main effects | Issue |
|---|---|
| High progress in energy prices | One of the biggest effects of the Russia-Ukraine war in the field of energy is high prices. The effect of the Russia-Ukraine war has also been added to the increasing energy prices after the COVID-19 period. Millions of people cannot access energy |
| Changes in energy flow | Russia has reduced gas flow to the European Union by approximately 80%. On the other hand, Russia's oil production and exports in China, India and Turkey have increased |
| Energy and economy relationship | High energy prices have also been reflected in production costs and prices of goods and services have also increased. Therefore, inflation has been triggered |
| Emission status | It can be said that the energy crisis and emission rates are not clear in the future |
| Turning to renewable energy | Investment and turn to renewable energy sources instead of fossil-based energy sources may have increased |

* Thomson, (2022)

According to the World Bank, the economic crisis that began in the CIVD-10 period continued to grow with the Russia-Ukraine war. These countries and countries with close trade relations with these countries may be more affected by the economic crisis. In addition to the energy crisis, the food crisis is also one of the important effects of the Russia-Ukraine war. Problems with grain and seeds coming from Russia and Ukraine will also affect global food security (Masterson, 2022).

## 4 Greenwashing, Waste Problems, High Costs and Others

There can be many obstacles along the sustainability journey. When viewed at a micro level, many obstacles such as the lack of sustainable transition in consumer behavior, incorrect waste management, and inadequate government policies become barriers to the green transition. In developing countries with large populations such as India and China, people have high purchasing and spending rates. These people consume more, and the growth of the middle class means that consumption also grows. Especially in the fashion sector, as fast fashion continues, the consumption and waste problem will continue (Stanislaus, 2018).

Kirchherr et al. (2017) report also stated that there are 4 main obstacles to the transition to circular economy. These are: (Kirchherr et al., 2017):

- *Cultural obstacles*: Lack of climate change in the company culture, lack of environmental values, lack of environmental awareness among consumers, lack of business activities according to the value chain
- *Technological obstacles*: Lack of circular design, problems with the production of recycled products, lack of data, lack of sufficient application

- *Market-related obstacles*: High investment costs, insufficient funds, standardization obstacles
- *Legal regulatory obstacles*: Limited circular application and legal regulation, lack of a global legal infrastructure

Access to and investment in green technologies required for the green transition are high-cost areas (Söderholm, 2020). The costs of the green transition are high, but if we do not fight against climate change, there will be more difficult days in the future. Some costs in the transformation of the global economy to a green economy may be high and more complex. Because of the global crises, an increase in costs and prices has been observed in the last 5 years. However, there may be positive aspects such as the increase in the price of fossil fuels encouraging investment in green energy (Keller & O'Neal, 2023).

According to McKinsey, the green transition will be costly. They emphasized that countries would need to increase global spending by $3.5 trillion each year if net zero is to be achieved by 2050. Today, a 60 percent increase in investment rates would only be enough to pay for the green transition (Broom, 2022).

Currently, most countries do not have the green transition finance needed for climate change. Globally, the annual cost of $100 billion to be paid by 2020 has been achieved in four countries. For developing economies, the green transition is a slower process. Developed countries need to help other countries with the green transition (Zenghelis, 2021).

Huser (2024) determined main challenges in green transition as below:

- *Human-based constraints*: There may be some human-based obstacles within the organization in the transition to a circular economy. For example, what will be the reactions of employees towards the green transition?
- *Financial constraints*: New costs are a burden for businesses that operate and are established according to the traditional economy in the transition to a green economy. The return of these costs may be uncertain for businesses.
- *Technical constraints*: Inadequate technical and technological infrastructure is a significant obstacle. For example, how will materials be recycled? Can every material be recycled? How will waste management be?

Consumers' inadequate level of sustainable consumption behavior or a gap between consumers' cognitive attitudes and behavioral attitudes is a major obstacle to sustainable consumption and circular economy. Economic models shape consumer habits and consumers' behaviors do not change easily. It is expected that consumers will also change in the transition to circular economy. If this change is slow and unbalanced, the circular economy model will not yield successful results (Sustana, 2024).

Bain & Company research has revealed some findings regarding consumers' awareness of sustainable consumption as below: (Bain & Company, 2023):

- *Baby boomer versus Gen Z*: Generation Z is coded as more sensitive to sustainability than older people. However, it is seen that older people are as sensitive to the future and sustainability as Generation Z. It has been observed that baby boomer groups are extremely sensitive to sustainability in countries such as Japan, India and France.

- *Liberals versus conservatives*: Although it is thought that liberal people are more focused on environmental sensitivity than conservatives, a study conducted in the USA showed that everyone, regardless of political views, is concerned about sustainability.
- *High price and green products*: Environmentally friendly consumers are coded as willing to pay a high price for green products. However, this price has certain limits. It has been observed that they are generally willing to pay a high price of 11–12%. For example, in the USA, the highest 11% high price for consumers; In developing countries such as India, Indonesia, Brazil and China, consumers are willing to accept a price difference of 15% and 20% at most.
- *Consumer behavior and government policies*: Government regulations on sustainable policies and practices also affect consumer behavior. For example, government regulations play a leading role in practices such as plastic bag use and electric vehicle incentives.
- *Uncertainty about sustainability*: Consumers' doubts about green product information and uncertainties about sustainable lifestyles affect consumers' purchasing behavior.

Bain & Company research shows that government practices can be wrong due to incorrect target audience definition or incorrect selection of supposedly environmentalist groups. Both brand environmental campaigns and public service announcements that only target Generation Z can leave other generations alone in terms of the environment. For the entire society to successfully adapt to the green transition, all individuals and communities should be reached in the environmental economy.

According to Team Zorgle (2023), 2 main situations are stated as follows for the difficulties seen in the adaptation of sustainable consumption behavior:

- *Price and Accessibility problems:* The cost of green products can be considered as a primary obstacle in the adaptation of sustainable consumption behavior. Not everyone who wants to buy green or environmentally friendly products can afford the prices of these products. On the other hand, it may be possible to distribute green or environmentally friendly products at affordable prices and quantities all over the world. When there is an accessibility problem, a certain part of the consumers are deprived of green products.
- *Perceptual and Participation problems:* Perceptual constraints also create an obstacle for consumers to switch to green consumption and green purchasing behavior. Indifference to green products and not knowing which products are green are among the important problems.

The fact that some sectors continue to operate according to the traditional economy on a global scale can be considered as one of the important obstacles in the transition to circular economy. For example, the fast fashion industry is one of the big sectors that continues. However, fast fashion is a sector that has accelerating effects on climate change in terms of its own nature, waste production and pollution rates.

According to the United Nations, the fast fashion industry is responsible for at least 10% of carbon emissions. According to the World Bank, sales in the fashion sector will increase by 65% by 2030. In terms of environmental impact, the fashion sector is also inefficient in raw material use. On the other hand, there is a problem of excessive water use in both production and consumption (Stallard, 2022).

Fast fashion is among the sectors that consumers show great interest in in terms of consumption. It is a polluting sector with a high carbon footprint. Since the green transition is slow, the fast fashion sector continues to grow. Unfortunately, there are many problems that do not comply with sustainability. There are deficiencies in terms of wage payment and workers' rights. Excessive washing of clothes, damage to natural resources and water resources after production and consumption is also a significant loss (Cardona, 2024).

Greenwashing can also be seen as a new trap that prevents both consumers and producers from making a green transition. Greenwashing is a major obstacle to the green transition of businesses and brands. Brands or businesses respond and even deceive consumers about their green image and environmentalism with exaggerated advertising campaigns. It is seen that most big brands are turning to greenwashing practices under the name of sustainable practices and suspicions are increasing in this regard (Yildirim, 2023). It is also necessary to combat the negative effects of greenwashing. Greenwashing is considered a deception. It causes consumers to be misled (UN, n.d.; European CSFF, 2024; Parliament, 2024).

As greenwashing incidents are on the rise globally, the European Union and the UN have also started working on legal sanctions, as they will negatively affect sustainable future planning. Some values on social washing and greenwashing have been identified in RepRisk's study. Greenwashing goes one step ahead of consumer response and deepens the situation with certificates and some commitments. While the uncertainty and evidence of sustainability claims become more complex, consumers are also misled and skeptical about green products on the market (Fürer, 2023).

## 5 Conclusion

Global climate change and rising sea temperatures are changing ecosystems and biodiversity. Policy makers are trying to stop or slow down climate change by producing global sustainable policies. The green economy model, along with the 2030 Sustainable Development Goals, is included in the long-term planning of both developed and developing countries (Yıldırım and Yıldırım, 2020; Yıldırım & Vasilev, 2024; Yıldırım et al., 2024b).

The 2030 Sustainable Development Goals were presented to the world in 2015. Developed and developing countries have begun to develop new sustainable policies and strategies based on 17 basic goals expected to be implemented or achieved by 2030 (Yıldırım et al., 2023, 2024a). The sustainability journey is a long process. During this long green transformation journey, countries face various challenges.

Some of these challenges affect countries and institutions at micro levels, while others affect them at micro levels. It is expected that countries and institutions will achieve sustainability because of overcoming the challenges they face. The reason why each country and institution have different levels of sustainability or achieves different goals is that they continue their sustainability journey by giving different responses to the macro and micro problems they encounter.

Due to the literature, it can be said that the below challenges is seen as main barrier for adapting green economy in the long term:

- The challenges in the green transition of the producer.
- High costs in establishing green technological infrastructures.
- Uncertain time to pay back the installation costs based on green technology at the initial stage.
- Insufficient information on the purposes of using green technologies.
- Incorrect waste management.
- Increasing tendency towards greenwashing practices in brands and businesses and avoidance of sustainable transformation.
- Financial and patent problems in accessing green energy.
- Lack of sufficient incentive practices by the government.
- The Challenges in the green transition of the consumer.
- Lack of information of consumers about green products.
- Lack of environmental awareness among consumers.
- Exposure of consumers to greenwashing practices.
- Inability to access green and environmentally friendly products everywhere.
- High pricing of green and environmentally friendly products.
- Insufficient environmental awareness campaigns by governments.

As a result, although the 2030 SDGs are the first concrete guide for countries and institutions, they are expected to be revised and taken to a higher level within the next 5 years. Although there is very little time left until 2030, it is a fact that the various global crises felt since 2019 have prevented countries from achieving their sustainability goals with their subsequent micro effects. Therefore, when the 2030 SDGs are to be revised and updated, countries need to re-plan according to global crises and possible micro effects and determine new goals. Despite global campaigns and the 2030 SDGs, or the serious impacts of climate change, the green transition has not been as successful as desired. In this respect, examining the factors that hinder the green transition of countries or institutions can guide the search for more successful transition practices.

The 2030 SDGs are the first concrete target indicator given to UN member countries. It can be considered that the revised version of the 2030 SDGs will be updated with more concrete and realistic steps, customized for each country. Since each country has different shortcomings and advantages, it is expected that the shortcomings will be completed, and the advantages will be continued for a sustainable future.

## References


Antipova, A. (2021). Analysis of the COVID-19 impacts on employment and unemployment across the multi-dimensional social disadvantaged areas, *Social Sciences & Humanities Open*, *4*(1), 100224. https://doi.org/10.1016/j.ssaho.2021.100224

Bain & Company. (2023). *Consumers say their environmental concerns are increasing due to extreme weather; study shows they're willing to change behavior, pay 12% more for sustainable products*. https://www.wbcsd.org/news/consumers-say-their-environmental-concerns-are-increasing-due-to-extreme-weather-study-shows-theyre-willing-to-change-behavior-pay-12-more-for-sustainable-products/

Bradshaw, M. (2022). *How will Russia's war in Ukraine affect energy supplies and the planet?*. https://www.wbs.ac.uk/news/how-will-russias-war-in-ukraine-affect-energy-supplies-and-the-planet/

Braimah, A. I. (2024). *Russia-Ukraine war and sustainable development goals: Is arming Ukraine the answer to conflict resolution?*. In D. Mhlanga, E. Ndhlovu, (Eds.), *The Russia-Ukraine Conflict and Development in Africa. Contributions to Political Science*. Springer. https://doi.org/10.1007/978-3-031-63333-1_5

Broom, D (2022). *What's the price of a green economy? An extra $3.5 trillion a year*, https://www.weforum.org/stories/2022/01/net-zero-cost-3-5-trillion-a-year/

Buchholz, K. (2020). *Continental divide in COVID-19 unemployment Burden*. https://www.statista.com/chart/22223/oecd-labor-force-unemployment/

Buchholz, K. (2022). *Ukraine crisis expected to speed up shift to renewables*. https://www.statista.com/chart/28656/world-energy-supply-by-source-outlook/

Cardona, N. (2024). *Environmental impact of fast fashion statistics (2025)*. https://www.uniformmarket.com/statistics/fast-fashion-statistics

Chariton, E. (2020). *Unemployment worries spike around the world as coronavirus remains top global concern*. https://www.weforum.org/stories/2020/06/coronavirus-unemployment-top-global-worries-ipsos/

Cherevko, A. (2024). *Ukraine and the SDGs: How the war has influenced global development*. https://unu.edu/merit/news/ukraine-and-sdgs-how-war-has-influenced-global-development

CSFF. (2024). Greenwashing and green finance: The CSSF calls for vigilance and informs on www.letzfin.lu, https://www.cssf.lu/en/2024/11/greenwashing-and-green-finance-the-cssf-calls-for-vigilance-and-informs-on-www-letzfin-lu/

European Parliament. (2024). Stopping greenwashing: How the EU regulates green claims. https://www.europarl.europa.eu/topics/en/article/20240111STO16722/stopping-greenwashing-how-the-eu-regulates-green-claims

Farquharson, V. (2020). Sustainable development in the Year of COVID-19. https://www.iisd.org/articles/insight/sustainable-development-year-covid-19?gad_source=1&gclid=CjwKCAiAgoq7BhBxEiwAVcW0LPTNGzNK9105HbvHH-YlUydJ6QtU6ki-NJ2cKoRB78kUQAzh0lapBhoCmW8QAvD_BwE

Fürer, M. (2023). RepRisk data suggest increase in greenwashing and rise of social washing. https://www.reprisk.com/research-insights/news-and-media-coverage/reprisk-data-shows-increase-in-greenwashing-with-one-in-three-greenwashing-public-companies-also-linked-to-social-washing

He, Q. & Silliman, B. R. (2019). Climate change, human impacts, and coastal ecosystems in the anthropocene. *Current Biology*, *29*(19), R1021–R1035.

Hulme, M., Barrow, E., Arnell, N., Harrison, P. A., Johns, T. C. & Downing, T. E. (1999). Relative impacts of human-induced climate change and natural climate variability. *Nature*, *397*, 688–691 (1999). https://doi.org/10.1038/17789

Huser, P. (2024). *What are the circular economy limitations for companies and how to overcome them?*. https://www.manutan.com/blog/en/circular-economy/circular-economy-limitations-how-to-overcome-them#:~:text=A%20further%20key%20limitation%20of,still%20need%20to%20be%20perfected

ILO, (2021). *An update on the youth labour market impact of the COVID-19 crisis*. Statistical Brief. https://www.ilo.org/media/8151/download

Inevitable Policy Response (IPR). (2022). Ukraine war: The new geo-politics of energy and implications for climate policy. https://www.unpri.org/inevitable-policy-response/ukraine-war-the-new-geo-politics-of-energy-and-implications-for-climate-policy/9766.article

IPCC(The Intergovernmental Panel on Climate Change). (2022). *Climate change: A threat to human wellbeing and health of the planet*. Taking action now can secure our future. https://www.ipcc.ch/2022/02/28/pr-wgii-ar6/

Kaplan, M., Yildirim, S., & Yildirim, D. C. (2022). Pufferfish versus lionfish: Comparing risks for Turkish marine economics. *Marine Economics and Management, 5*(2), 173–187. https://doi.org/10.1108/MAEM-09-2022-0008

Kaplan, M. & Yildirim, S. (2023). The sustainability of seafood market under the attack of invasive alien species: A review of turkish marines. In A. Ben Salem, L. Rhazi, & A. Karmaoui (Eds.), *Climatic and Environmental Significance of Wetlands: Case Studies from Eurasia and North Africa* (pp. 74–95). IGI Global Scientific Publishing. https://doi.org/10.4018/978-1-7998-9289-2.ch005

Keller, C., O'Neal, M. (2023). *The green transition: Driving force behind a radical economic rethink*. https://www.ib.barclays/our-insights/green-transition-driving-force-behind-a-radical-economic-rethink.html#:~:text=The%20predicted%20costs%20of%20the,substantial%20but%20far%20from%20impossible

Kirchherr, J., Hekkert, M., Bour, R., Huijbrechtse-Truijens, A., Kostense-Smit, E., & Muller, J. (2017). *Breaking the barriers to the circular economy October 2017*, https://circulareconomy.europa.eu/platform/sites/default/files/171106_white_paper_breaking_the_barriers_to_the_circular_economy_white_paper_vweb-14021.pdf

Kretchmer, H. (2020). *How coronavirus has hit employment in G7 economies*. https://www.weforum.org/stories/2020/05/coronavirus-unemployment-jobs-work-impact-g7-pandemic/

Martín-Blanco, C., Zamorano, M., Lizárraga, C., & Molina-Moreno, V. (2022). The Impact of COVID-19 on the sustainable development goals: Achievements and expectations. *International Journal of Environmental Research and Public Health, 19*(23), 16266. https://doi.org/10.3390/ijerph192316266.PMID:36498340;PMCID:PMC9739062

Masterson, V. (2022). *This chart shows how much Ukraine and Russia export to the world*. https://www.weforum.org/stories/2022/04/world-bank-ukraine-food-energy-crisis/

Metzke, R., (2020). *Climate action must stay top of the global agenda as we emerge from COVID-19*. https://www.weforum.org/stories/2020/05/climate-action-top-global-agenda-covid-19/

Moore, S. (2021). Climate Change and COVID-19. News-Medical. Retrieved December 19, 2024, from https://www.news-medical.net/health/Climate-Change-and-COVID-19.aspx

Myllyvirta, L. (2023). *Russia–Ukraine war has not led to increased fossil fuel consumption in the EU*. https://energyandcleanair.org/russia-ukraine-war-has-not-led-to-increased-fossil-fuel-consumption-in-the-eu/

NASA (National Aeronautics and Space Administration). (2024). *Vidence*. https://science.nasa.gov/climate-change/evidence/

Söderholm, P. (2020). The green economy transition: The challenges of technological change for sustainability. *Sustain Earth, 3*, 6. https://doi.org/10.1186/s42055-020-00029-y

Stallard, E. (2022). *Fast fashion: How clothes are linked to climate change*. https://www.bbc.com/news/science-environment-60382624

Stanislaus, M. (2018). *Barriers to a circular economy: 5 reasons the world wastes so much stuff (and why it's not just the consumer's fault)*. https://www.wri.org/insights/barriers-circular-economy-5-reasons-world-wastes-so-much-stuff-and-why-its-not-just

Sustana. (2024). *How does consumer behavior challenge the circular economy?* https://sustainablebrands.com/read/consumer-behavior-challenge-circular-economy

Team Zorgle. (2023). Are consumers increasingly conscious of environmentally sound business?. https://zorgle.co.uk/are-consumers-increasingly-conscious-of-environmentally-sound-business/

Thomson, E. (2022). 6 ways Russia's invasion of Ukraine has reshaped the energy world. https://www.weforum.org/stories/2022/11/russia-ukraine-invasion-global-energy-crisis/

Trenberth, K. E. (2018). Climate change caused by human activities is happening and it already has major consequences. *Journal of Energy & Natural Resources Law, 36*(4), 463–481. https://doi.org/10.1080/02646811.2018.1450895

UN. (n.d.). Greenwashing–the deceptive tactics behind environmental claims. https://www.un.org/en/climatechange/science/climate-issues/greenwashing#:~:text=By%20misleading%20the%20public%20to,delay%20concrete%20and%20credible%20action

United Nations, Human Rights Office of The High Commissoner. (n.d.). *COVID-19 and the 2030 agenda for sustainable development.* https://www.ohchr.org/en/sdgs/covid-19-and-2030-agenda-sustainable-development

United Nations, Department of Economic and Social Affairs Sustainable Development. (n.d.). *THE 17 Goals.* https://sdgs.un.org/goals

UNPD. (2023). Sustainable development goals during the war: UNDP and the Government of Ukraine held a series of strategic sessions. https://www.undp.org/ukraine/press-releases/sustainable-development-goals-during-war-undp-and-government-ukraine-held-series-strategic-sessions

Yildirim, S. (2023). Greenwashing: A rapid escape from sustainability or a slow transition? *LBS Journal of Management & Research, 21*(1), 53–63. https://doi.org/10.1108/LBSJMR-11-2022-0077

Yildirim, S. & Kaplan, M. (2022). The threat of invasive alien marine species to the blue economy: The mediterranean case. In L. Raimi & J. Kah (Eds.), *Implications for Entrepreneurship and Enterprise Development in the Blue Economy* (pp. 50–80). IGI Global Scientific Publishing. https://doi.org/10.4018/978-1-6684-3393-5.ch004

Yıldırım, S., & Vasilev, V. (2024). The future of green collar workers: Green transition in employment. In S. Yıldırım, D. Yıldırım, I. Demirtaş & V. Kandpal (Eds.), *Green Transition Impacts on the Economy, Society, and Environment* (pp. 1–16). IGI Global Scientific Publishing. https://doi.org/10.4018/979-8-3693-3985-5.ch001

Yıldırım, S. & Yıldırım, D. Ç. (2020). Achieving sustainable development through a green economy approach. In S. Patti & G. Trizzino (Eds.), *Advanced Integrated Approaches to Environmental Economics and Policy: Emerging Research and Opportunities* (pp. 1–22). IGI Global Scientific Publishing. https://doi.org/10.4018/978-1-5225-9562-5.ch001

Yildirim, S., Kaplan, M., & Yildirim, D. C. (2023). climate change and the ocean economy: A review on OECD's ocean economy. In A. Pego (Ed.), *Handbook of Research on Bioeconomy and Economic Ecosystems* (pp. 212–232). IGI Global Scientific Publishing. https://doi.org/10.4018/978-1-6684-8879-9.ch012

Yıldırım, S., Bostancı, S. H., Yıldırım, D. Ç. (2023). Parameters for the study of climate refugees. In P. Singh, B. Ao, A. Yadav, (Eds.), *Global Climate Change and Environmental Refugees.* Springer. https://doi.org/10.1007/978-3-031-24833-7_11

Yıldırım, S., Sevik, N., Kandpal, V., & Yıldırım, D. C. (2024b). The role of green brands on achieving 2030 sustainable development goals (2030 SDGs). In K. Kankaew, S. Chaudhary & S. Widtayakornbundit (Eds.), *Contemporary Management and Global Leadership for Sustainability* (pp. 141–162). IGI Global Scientific Publishing. https://doi.org/10.4018/979-8-3693-1273-5.ch009

Yıldırım, D. Ç., Yıldırım, S., Bostancı, S. H. (2024). Anthropogenic impacts in the himalayas and the sustainable development goals (SDGs). In A. Borthakur & P. Singh (Eds.), *The Himalayas in the Anthropocene.* Springer. https://doi.org/10.1007/978-3-031-50101-2_12

Yuan, H., Wang, X., Gao, L., Wang, T., Liu, B., Fang, D., & Gao, Y. (2023). Progress towards the sustainable development goals has been slowed by indirect effects of the COVID-19 pandemic. *Communications Earth & Environment, 4,* 184. https://doi.org/10.1038/s43247-023-00846-x

Zenghelis, D. (2021). *What are the likely costs of the transition to a sustainable economy?.* https://www.economicsobservatory.com/what-are-the-likely-costs-of-the-transition-to-a-sustainable-economy

# Catalysts of Circular Change: Role Played by Institutions and Lobbyists in Promoting Sustainable Development

**Pasquale Marcello Falcone, Grazia Errichiello, and Patrizio Giganti**

**Abstract** This chapter explores the crucial role of institutions and leaders in promoting sustainable development through the principles of the circular economy (CE). We analyze how institutional quality can act as a catalyst for change, creating an enabling environment for the adoption of sustainable practices. Additionally, the chapter examines the influence of lobbying, often perceived negatively, highlighting how it can be strategically leveraged to support policies and regulations that favor CE initiatives. Through case studies and theoretical frameworks, we investigate the strategies employed by leaders and organizations to shape policymaking in favor of sustainability, shedding light on both successes and potential challenges. Our goal is to provide a comprehensive understanding of the mechanisms through which institutions and lobbying efforts can work together to accelerate the transition towards sustainable development. The chapter concludes with practical recommendations for policymakers and leaders on strengthening institutional frameworks and effectively utilizing lobbying to achieve long-term sustainability goals.

**Keywords** Circular economy · Institutional quality · Sustainable development · Lobbying · Sustainability policies

P. M. Falcone (✉) · G. Errichiello · P. Giganti
Department of Business and Economics, University of Naples Parthenope, Napoli, Italy
e-mail: pasquale.falcone@uniparthenope.it

G. Errichiello
e-mail: grazia.errichiello@collaboratore.uniparthenope.it

P. Giganti
e-mail: patrizio.giganti@collaboratore.uniparthenope.it

 
P. Singh et al. (eds.), *Circular Economy and Environmental Resilience*,
https://doi.org/10.1007/978-3-031-93091-1_9

# 1 Introduction

The transition to a circular economy (CE) is no longer an option but a necessity for achieving sustainable development. Moving beyond the traditional linear economic model, the CE focuses on the efficient use of resources, built upon three core principles: reducing waste, reusing materials, and recycling resources. This approach not only addresses pressing issues such as resource depletion and climate change but also generates economic and social benefits, fostering business resilience and greater equity.

As highlighted by (Geissdoerfer et al., 2017), the CE serves as a bridge between environmental sustainability and economic growth, unlocking new opportunities for innovation across various sectors. However, for this transition to succeed, institutions and policymakers must play a fundamental role in shaping the necessary regulatory and strategic framework.

Institutions and lobbyists are key drivers in advancing CE policies. Institutions—including governments, regulatory bodies, and international organizations—provide governance structures, policy incentives, and regulatory frameworks essential to facilitating this shift. A concrete example is the European Commission's Circular Economy Action Plan, which has demonstrated how institutional support can lead to legislative reforms and targeted investments (European Commission, 2020). At the same time, lobbying—often associated with private interests—can serve as a strategic tool to accelerate the adoption of sustainable models. Industry representatives, non-governmental organizations (NGOs), and other stakeholders act as intermediaries in policymaking, influencing decisions through data, expertise, and targeted proposals. The collaboration between institutions and lobbying efforts can foster the adoption of green technologies and the enhancement of extended producer responsibility, aligning industrial practices with CE objectives (Falcone & Tutore, 2025).

This chapter explores the dynamic interplay between institutions and lobbying in the context of the CE, analyzing how their collaboration influences policy innovation, stakeholder engagement, and the broader adoption of circular practices. Through case studies and empirical evidence, we highlight the potential of these synergies to accelerate the transition toward more sustainable economic models. However, we also address critical challenges such as power asymmetries, vested interests, and the risk of greenwashing, emphasizing the importance of greater transparency and inclusivity in public policies.

The contribution of this chapter lies in providing an integrated perspective on the institutional and political dynamics that support the CE. Beyond theoretical insights, we offer practical recommendations for policymakers, lobbyists, and other stakeholders on navigating the complexities of the transition, ensuring that regulatory frameworks are fair, transparent, and capable of fostering long-term economic and environmental resilience.

This chapter is organized as follows. Section 2 explores the role of institutions, focusing on institutional quality and governance frameworks that drive circular change, with particular attention to policy development and public–private partnerships. Section 3 examines the influence of lobbying on sustainable development, analyzing its role, strategies, challenges, and opportunities in advancing CE initiatives. Section 4 highlights the synergies between institutions and lobbying, emphasizing how collaboration can accelerate the adoption of CE principles. Finally, Sect. 5 concludes the chapter by discussing key policy implications and providing recommendations for strengthening institutional frameworks and leveraging lobbying to support long-term sustainability goals.

## 2 The Role of Institutions in Promoting Circular Economy

Institutions play a pivotal role in advancing the CE and sustainable development by establishing regulatory frameworks, fostering innovation, promoting collaboration, and implementing performance monitoring systems, as explored in the subsequent subsections.

### *2.1 Defining Institutional Quality*

Institutional quality is a complex, multidimensional concept that significantly influences economic, social, and environmental outcomes. It encompasses the effectiveness of both formal and informal institutions in creating, implementing, and sustaining policies that align with development goals, including the transition to a CE (Nguyen & Nguyen, 2023). Institutions, defined as the "rules of the game" (North, 1990), include the political, economic, and social systems that govern behavior and interactions. High-quality institutions are distinguished by their ability to promote stability, equity, efficiency, and accountability. A solid definition of institutional quality includes key dimensions such as governance, rule of law, transparency, and regulatory efficiency. Kaufmann et al. (1999), in their work on the Worldwide Governance Indicators (WGI), identify six essential governance elements that underlie institutional quality: voice and accountability, political stability, government effectiveness, regulatory quality, rule of law, and control of corruption. These indicators collectively assess the ability of institutions to serve the public good and foster an environment conducive to sustainable economic activities.

The role of institutions in economic development has been extensively discussed in the literature. Acemoglu and Robinson (2012), in their seminal book "Why Nations Fail", argue that inclusive institutions, which involve broad segments of society and provide opportunities for innovation and entrepreneurship, are key to long-term prosperity. In contrast, extractive institutions, which concentrate power and wealth in the hands of a few, hinder progress and exacerbate inequalities.

Transparency and accountability are critical components of institutional quality. According to Olson (1982), institutions that minimize corruption and ensure accountability build trust between citizens and business, thus fostering economic resilience. Studies such as those by Mauro (1995) and Rose-Ackerman (1999) highlight the negative effects of corruption on growth and institutional credibility, emphasizing the importance of strong governance mechanisms. The relationship between institutional quality and environmental governance is critical for promoting sustainable practices. Ostrom (1990) underscores the importance of well-structured institutions in managing common resources to avert the "tragedy of the commons". The study emphasizes that institutions must be tailored to local contexts, equipped with clear rules and enforcement mechanism, to effectively manage resources sustainability. The transition to a CE is significantly influenced by the quality of institutions. High-quality institutions are essential for creating and enforcing regulations that facilitate resource recycling, innovative product design, and sustainable business practices (Nguyen & Nguyen, 2023). Institutional quality plays a pivotal role in addressing systemic barriers to sustainability, particularly in the context of market failures and path dependencies. High-quality institutions with adaptive capacities are essential for navigating the complexity of sustainable transitions, as highlighted by Meadowcroft (2007). The quality and adaptability of institutions are pivotal in driving sustainability initiatives and achieving broader socio-economic and environmental goals (Barra & Falcone, 2024). The relationship between the CE and institutions is characterized by a continuous exchange that drives progress in both domains. Figure 1 describes how the CE and institutions interact with each other. The light green circles represent the components of the CE, and through its contact with institutions, it generates drives innovation, shapes new regulations, and promotes collaboration. The blue circles represent the components of institutions, and through their contact with the CE, they generate policy incentives, investment in technology, and education and awareness campaigns.

## 2.2 *Institutional Frameworks for Circular Change*

Institutional frameworks are essential for facilitating the transition to a CE, as they establish the necessary governance structures, regulatory mechanism, and incentives. These frameworks are characterized by several key features that support systemic change and align political objectives with practical strategies. An example of an effective institutional framework is the European Union's Circular Economy Action Plan (CEAP). The CEAP sets ambitious goals for recycling, waste reduction, and sustainable product design, supported by regulations like the EU Waste Framework Directive and the Single-Use Plastic Directive (European Commission, 2020). By fostering collaboration among EU members states and sectors, the CEAP has become a key driver for the EU's CE efforts, influencing both regional and global policies. Another example is the Global Environment Facility (GEF), managed by the UN

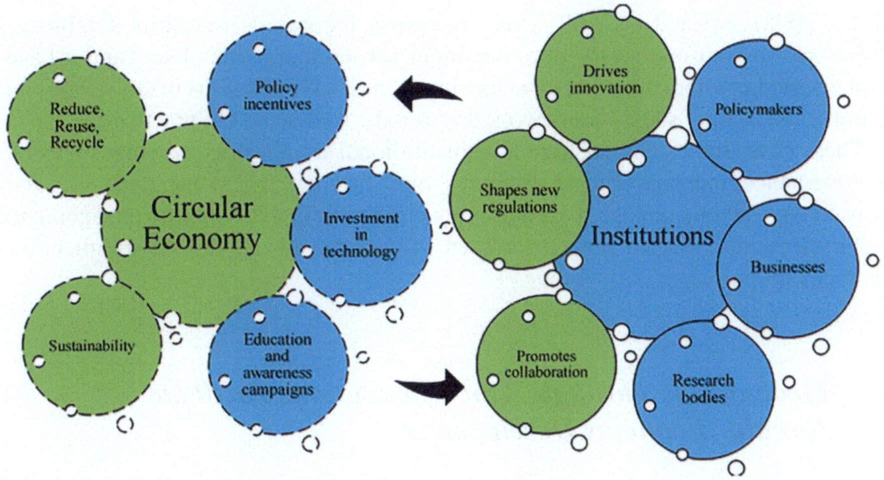


**Fig. 1** Interaction between CE and institutions. *Source* own elaboration

Environment Programme (UNEP). The GEF supports developing countries in implementing CE principles by funding projects that promote resource efficiency, waste reduction, and sustainable practices. With its global reach, the GEF helps align environmental goals with economic development, promoting a more sustainable global economy (UN Environment Programme, 1991). Furthermore, the Environmental Policy Committee (EPOC) of the OECD, provides strategic guidance on CE practices for member countries. Through its reports and recommendations, the OECD has shaped policies promoting waste management, resource recovery, and innovation in circular business models (OECD, 1971). These institutional frameworks are essential for fostering the systemic changes needed to promote CE practices at regional, national and global levels. Building on these examples, successful transitions to a CE rely heavily on institutional support tailored to local contexts. Several case studies from Europe and Asia highlight diverse approaches to fostering circularity. A prominent case is the Netherlands, often regarded as a leader in CE initiatives. The Dutch National Circular Economy Programme (NPCE) (The Ministry of Infrastructure and Water Management, 2023) aims to transition the Netherlands to a fully CE by 2050, with milestones set for 2030. The programme integrates national policies with regional and municipal efforts, fostering innovative practices such as urban mining and circular construction. This highlights the value of partnerships across governance levels to drive systemic change. In Sweden, the RE:Source programme (Energimyndigheten, 2016) exemplifies effective institutional support for industry. This initiative fosters collaboration among business, academia, and government to develop sustainable technologies. A key outcome has been progress in material recovery, reducing reliance on virgin resources. Sweden's approach shows how institutions can guide industrial practices towards sustainability. China offers another compelling example with its Circular Economy Promotion Law (President of the People's Republic of

China, 2008), enacted in 2009. This framework focuses on industrial symbiosis, where waste from one sector becomes input for another. Cities like Tianjin have implemented eco-industrial parks, achieving significant reductions in emissions and resource use. China's legal framework drives industrial-scale CE practices.

These case studies demonstrate how institutional frameworks, whether focusing on governance, industry or legal structures, must be adaptable, context-specific, and focused on systemic change. By combining clear policies, stakeholder engagement, and infrastructure investments, governments can create an enabling environment for CE innovations.

### *2.3 Mechanisms Through Which Institutions Facilitate Circular Economy Development*

Institutions are fundamental in formulating and enforcing policies that drive the transition to a CE. Effective policy frameworks create clear regulations, provide economic incentives, and set actionable goals that guide stakeholders towards sustainable practices (Kjaer & Vetterlein, 2018). Regulatory measures, such as waste management standards or resource efficiency targets, lay the normative foundation for industries and consumers, ensuring alignment with circularity principles while fostering accountability (Monciardini et al., 2023). At the same time, economic instruments including tax incentives, subsidies, and penalties, further motivate business to embrace practices like product redesign, material recovery, and recycling (Stabler et al., 2024). Beyond these regulatory and economic mechanisms, institutions also play a crucial role in driving innovation. By investing in research and development, they support the creation of advanced technologies for material recovery, sustainable product design, and industrial symbiosis (Abreu & Ceglia, 2018). Additionally, institutions contribute to the success of the transition by establishing robust performance indicators and reporting systems. These tools enable continuous monitoring of progress towards CE objectives and offer valuable insights into the effective of implemented policies (OECD, 2024). By tracking key metrics such as resource efficiency, waste reduction, and recycling rates, institutions can evaluate the impact of their initiatives, ensure alignment with broader sustainability goals, and enhance transparency. This enables stakeholders to make informed decisions and allows for adjustments to policies and strategies as needed (Nika et al., 2021). In addition to these policy and regulatory mechanisms, collaboration between public and private sector plays a vital in advancing the CE (World Economic Forum, 2024). Such partnerships help address systemic challenges like waste optimization, circular infrastructure development, and the adoption of sustainable business models. Eco-industrial parks, for example, unite companies and public entities to transform waste into shared resources, fostering sustainability and improving economic performance through innovative practices. In South Korea, the Eco-Industrial Park (EIP) program, initiated by the Korean National Cleaner Production Center (KNCPC) in 2003 with government support, demonstrates the effectiveness of industrial symbiosis for boosting resources efficiency and reducing waste (Park et al., 2016). At the urban level, partnerships are essential for fostering circular cities, with initiatives like the Circular Cities Declaration (ICLEI—Local Governments for Sustainability, 2024) encouraging sustainable urban practices such as shared mobility and renewable energy. Sweden's Hammarby Sjöstad, for instance, integrates renewable energy, advanced recycling, and water management through collaboration between municipalities and private companies (Solly, 2016). In the agri-food sector, public–private initiatives are transforming supply chains, as seen in the Netherlands's Green Deals program, which supports circular agriculture by prioritizing sustainability and resource efficiency (PBL Netherlands Environmental Assessment Agency, 2017). Similarly, Finland's Telaketju project unites governments, universities, and businesses to promote textile recycling and sustainable production (VTT Technical Research Centre of Finland Ltd, 2021).

These examples highlight how well-structured partnerships, backed by innovative policies, can effectively tackle environmental challenges and advance circular models across industries and scales.

## 3 Influence of Lobbying on Sustainable Development

### *3.1 Role of Lobbying in Circular Economy*

Lobbying is the process of interaction, communication, and influence exerted by individuals, companies, or organizations on policymakers with the goal of promoting specific interests by shaping regulations (Girard et al., 2023; Jiao & Lu, 2024; Zetter, 2014).

With the growing impact of climate change, the European Union places significant emphasis on climate-related policies[1] aimed at promoting sustainable development and the CE, reducing emissions, and protecting ecosystems.[2] These regulations often have profound implications for large multinational corporations, which must develop effective lobbying strategies to mitigate potential risks to their competitiveness (da Silva Hyldmo et al., 2024). Operating within a CE framework requires companies across the entire value chain, from sourcing raw materials to delivering final products, to align with its core principles. In this context, policymakers face the critical challenge of balancing diverse interests while maintaining transparency (Charłampowicz et al., 2025; Laboutková et al., 2020).

On one side, large multinational polluters must adapt their structures and production processes to comply with CE principles, often requiring substantial investments

---

[1] https://www.consilium.europa.eu/en/policies/paris-agreement-climate/.

[2] https://commission.europa.eu/strategy-and-policy/priorities-2019-2024/european-green-deal_en.

that could impact their competitiveness. On the other side, organizations and businesses, including start-ups, that already integrate CE practices gain a competitive edge from the implementation of new green regulations. This dynamic underscores the complexity of policymaking, where ensuring sustainability must be carefully weighed against economic and industrial realities.

Lobbyists employ various strategies to exert pressure, including direct meetings, report presentations, awareness campaigns, event organization, and media engagement. Lobbying efforts can take both positive and negative forms, either supporting or opposing regulations (Errichiello et al., 2025). In the sustainability sector, positive lobbying is primarily driven by citizen-led initiatives, organizations advocating for stronger environmental regulations, and innovative companies promoting policies that incentivize the CE. These efforts often focus on securing tax incentives for investments in sustainable technologies and fostering regulatory frameworks that support green innovation.

Conversely, negative lobbying emerges from resistance by traditional industries (such as petrochemicals, single-use plastics, and fast fashion) that seek to delay or obstruct the implementation of new sustainability regulations. In some cases, these industries engage in greenwashing campaigns to mislead the public about their actual commitment to environmental responsibility. Given these dynamics, ensuring transparency in lobbying activities is crucial. However, tracking the nature and extent of lobbying remains a challenge. One tool addressing this issue is LobbyFacts,[3] a platform that compiles data from the EU Transparency Register, which annually records the lobbying activities of various European and non-European organizations. These entities include corporations, professional consultants, law firms, in-house lobbyists, trade and business associations, NGOs, think tanks, research institutions, religious organizations, and public authorities. LobbyFacts not only tracks key meetings between lobbyists and the European Commission but also records the financial expenditures associated with lobbying efforts, offering valuable insights into the influence of different stakeholders on EU policymaking.

When examining the CE (particularly the 3Rs principle of recycling, reducing, and reusing) discussions on these issues began gaining traction in 2014. Between 2014 and 2023, a total of 883 meetings were held, emphasizing the concept of circularity.[4] These meetings saw the involvement of key stakeholders, including in-house lobbyists, business associations, NGOs, companies, and various industry groups (see Fig. 2).

[3] https://www.lobbyfacts.eu/.

[4] The meetings were selected by performing a textual analysis based on their titles, which provide information about the subject, specifically choosing those related to circularity and containing the keywords circular, reuse, reduce, and recycle.

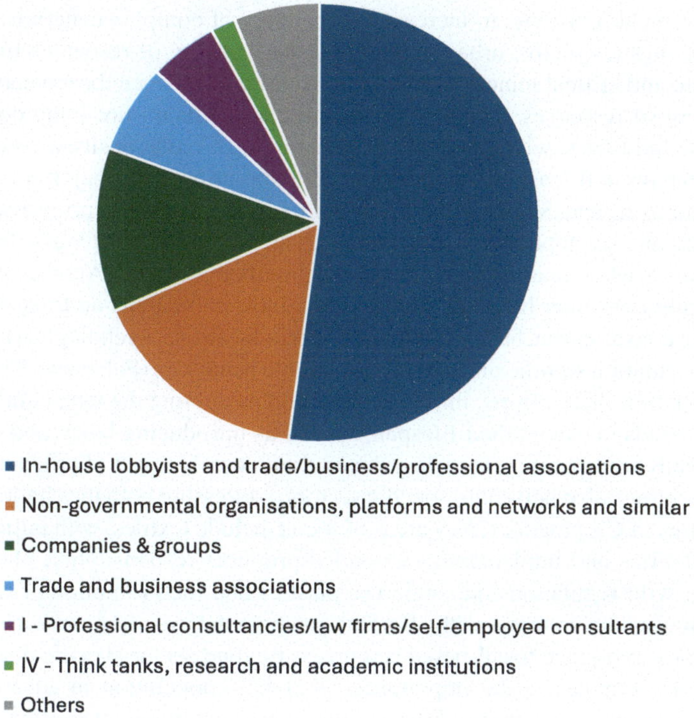


**Fig. 2** Lobbyist categories composition. *Source* own elaboration based on Lobby Facts

## 3.2 *Strategies for Effective Lobbying*

Tracking lobbyists' strategies is inherently challenging, as they apply pressure through both official and unofficial channels. However, participation in meetings with the European Commission offers valuable insight into the key issues organizations prioritize in their lobbying efforts. The number of meetings focused on the CE does not follow a clear trend. After an initial peak in 2015, surpassing 200 meetings, a fluctuation is recorded, with a notable increase in 2020 and 2022, each exceeding 100 meetings. These peaks coincide with periods of intensified legislative activity in the CE domain, particularly around the introduction and development of the Circular Economy Action Plan (CEAP).[5] Key topics discussed in these meetings include the implementation of the CEAP and the Green Deal, which aims to improve waste management by promoting recycling, establish rules for sustainable product design (eco-design), and introduce extended producer responsibility systems, particularly in sectors such as textiles and packaging. Innovation and sustainable technologies are central to lobbying discussions. Prominent areas in the field of circularity

[5] https://environment.ec.europa.eu/strategy/circular-economy-action-plan_en.

include chemical recycling, to increase the recovery of complex materials, particularly in the plastics sector; urban mining and the recovery of resources from electronic waste and critical minerals; and digitalization, to enhance the traceability and management of resources. There is also significant focus on involving companies and specific industries, with discussions on promoting circular business models such as reuse, repair, and leasing. This includes collaboration with major corporations such as Amazon, and Starbucks to align their policies with CE priorities. Efforts also include the development of protocols such as the "UK End of Waste Protocol," which clarifies when materials are no longer classified as waste. Another key focus is influencing consumer behavior through awareness campaigns on food waste and responsible consumption. Factors such as gender, education, psychological traits, and political orientation significantly shape sustainable behaviors (Falcone & Fiorentino, 2025). As part of these efforts, initiatives focus on promoting the repair and reuse of consumer goods to extend their lifespan, as well as introducing labels and certifications to ensure transparency regarding the environmental impact of products. In this context, meetings also delve into specific sectors, addressing both the challenges and opportunities of CE practices. Key areas of focus include textiles, with initiatives for recycling fabrics and implementing extended producer responsibility; plastics and packaging, with regulations on single-use plastics and the promotion of recycling; and the circular bioeconomy, with efforts to promote the use of biomass and biological materials to replace fossil-based resources. Beyond sector-specific discussions, meetings also emphasize the importance of global cooperation in advancing CE practices. A significant portion of these discussions centers on international collaboration, particularly with Latin America and Africa, to tackle e-waste recycling and management. Additionally, strategies to reduce greenhouse gas emissions and integrate CE principles into national climate contributions (NDCs)[6] are key topics of debate. The overall trend in total lobbying expenditure follows fluctuation similar to that observed in meeting activity. However, when considering average expenditures, a different pattern emerges. The entry of smaller lobbyists during periods of intense legislative activity leads to a decline in the mean expenditure, reflecting a shift in the composition of lobbying actors. During the period considered, organizations involved in CE lobbying spent an average of €735,423, with the maximum value for one organization reaching €12,300,000. Figure 3 represents the yearly lobbying expenditure in mean and in total.

### *3.3 Challenges and Opportunities*

Over the past decade, regulations on CE topics have increased, accompanied by a more diverse lobbying landscape. However, this rise does not necessarily indicate a growing consensus on the issue. Lobbying efforts can serve both to advance and

[6] https://www.consilium.europa.eu/en/press/press-releases/2023/10/16/paris-agreement-council-submits-updated-ndc-on-behalf-of-eu-and-member-states/.

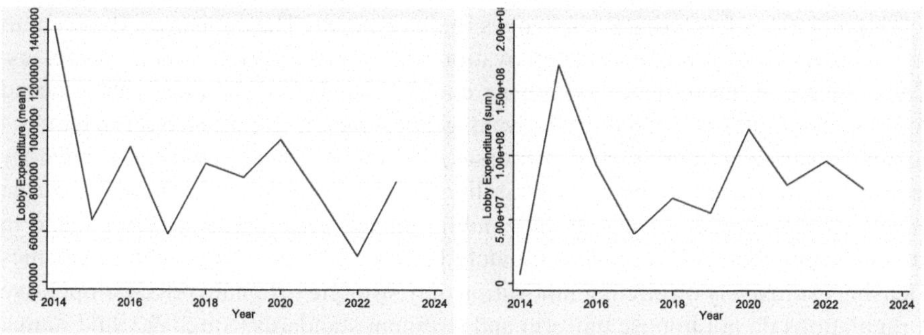


**Fig. 3** Trend of mean and sum lobbying of expenditure. *Source* own elaboration based on Lobby Facts

to hinder the effective implementation of CE regulations, as organizations employ strategies either to support or to delay policy measures.

Lobbying plays a crucial strategic role for organizations, given that these regulations are driving fundamental transformations in production processes and business models. Companies engaging in meetings on the CE and allocating greater resources to lobbying are primarily headquartered in Luxembourg, the United States, Belgium, China, Germany, and France (see Fig. 4). However, in terms of total lobbying expenditure, organizations based in Belgium account for the highest spending. This reflects not only Brussels' central role as a political hub but also the strong interest of industrialized economies in shaping EU regulatory decisions.

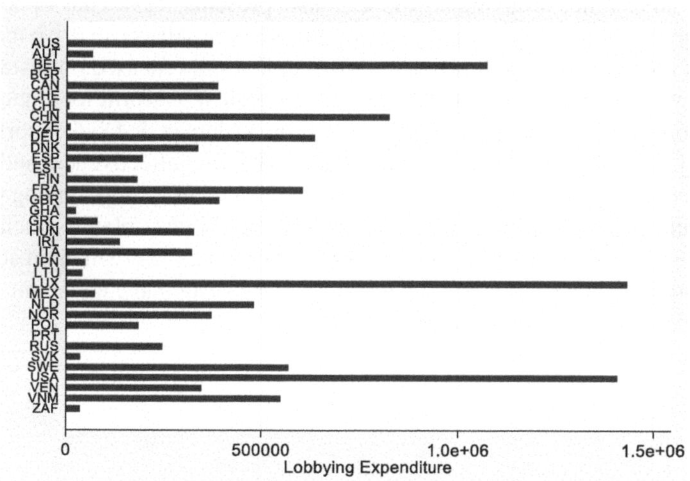


**Fig. 4** Mean of lobbying expenditure in different countries. *Source* own elaboration based on Lobby Facts

Organizations face critical challenges in this context. They must rapidly adapt to new regulations by implementing innovation and sustainability into their businesses. Companies that fail to do so risk losing competitiveness, while those that succeed could gain significant competitive advantages. However, compliance often requires costly adjustments to production processes, making it crucial to balance regulatory requirements with economic sustainability. Furthermore, organizations that resist these regulations risk damaging their public image, particularly if they engage in greenwashing tactics that could ultimately backfire. On the other hand, companies focusing exclusively on circular innovation may struggle to remain price-competitive if regulations do not impose uniform and stringent standards (Singh & Chudasama, 2021).

Measuring lobbying efforts in the CE presents additional complexities. There is no dedicated registry for lobbying expenditures specifically linked to CE policies. Limiting the analysis to recorded meetings, while valuable, could overlook related lobbying activities that indirectly support CE objectives. Additionally, considering only the organization's main headquarters to determine which countries invest more in lobbying could obscure some dynamics. For example, organizations and companies often operate across multiple locations and in transnational contexts. Moreover, legal headquarters might be based in countries with favorable tax regimes, altering perceptions of their actual influence. Moreover, the relationships between organizations and policymakers, as well as the networks among different lobbying entities, represent intangible yet critical elements of lobbying influence. These factors are difficult to quantify, making it challenging to capture the full scope of lobbying efforts. Nevertheless, lobbying expenditures and meeting records serve as useful proxies for estimating these activities.

At the same time, lobbying in the CE also presents significant opportunities. Organizations that actively engage in promoting CE initiatives can drive innovation, strengthen their competitive position, and attract investors focused on sustainability. Meetings with policymakers and stakeholders provide a platform for engaging with both supporters and opponents of CE regulations (Alemán & Dockendorff, 2024), facilitating strategic collaborations that enhance lobbying effectiveness and improve public perception. Additionally, the EU offers fiscal incentives and funding programs for companies investing in circular technologies. Lobbyists play a crucial role in maximizing access to these resources, not only advancing their own interests but also contributing to the broader transition toward a more sustainable economic model.

# 4 Synergies Between Institutions and Lobbying

## 4.1 How Institutions and Lobbyist Work Together

Collaboration between institutions and lobbyists is crucial for advancing the principles of a CE, as it fosters coordinated efforts across both public and private sectors (European Economic and Social Committee, 2019). By aligning institutional frameworks with industrial goals, this partnership facilitates the transition from a linear to a CE, creating a cohesive approach to sustainability. A widely recognized framework for such collaboration is the "Triple Helix" model, which emphasizes interactions between government, industry, and academia to enhance innovation and sustainability (Etzkowitz & Leydesdorff, 1995). This model not only supports the development of circular practices but also ensures that diverse stakeholders contribute to a shared vision of progress. CE initiatives are increasingly supported by lobbyists advocating for policies that promote green technologies, enhance waste management infrastructures, and provide regulatory incentives such as tax benefits or subsidies (Bocken et al., 2022). These efforts aim to foster a sustainable economic model that minimizes waste and maximizes resource efficiency, ensuring accountability and benefit-sharing among stakeholders.

Multi-stakeholder platforms, such as the Circular Economy 100 (CE100) program (Ellen MacArthur Foundation, 2019), exemplify how institutional and lobbying efforts can converge to promote CE initiatives. In these settings, lobbyists act as facilitators, translating industry needs into actionable policy recommendations and bridging gaps between legislative bodies and businesses. Moreover, as highlighted by Kirchherr et al. (2017), the success of CE initiatives relies heavily on trust-building and transparent communication channels. These elements are essential for fostering collaboration among various stakeholders, including institutions and lobbyists. The synergy created through these partnerships not only drives policy innovation but also enhances the social uptake of sustainable practices. Building on these collaborative frameworks, several successful case studies highlight the positive impact of collaboration between institutions and lobbyists in advancing the CE. A key example is the collaboration between the German government and industry associations in developing the Circular Economy Package (Federal Ministry for Environment, 2024). Industry groups, such as the German Association for Plastic Packaging and Films (IK), have advocated for policies promoting recycling and resource efficiency. IK, for example, has supported measures to improve plastic recyclability and extended producer responsibility frameworks (IK Industrievereinigung Kunststoffverpackungen, 2023). These efforts have led to stronger regulations and incentives aligned with CE principles, particularly in the plastic sector. Another case is the partnership between local governments and environmental NGOs in the UK, which significantly influenced the implementation of Extended Producer Responsibility (EPR) programs. Environmental lobbyists successfully shaped policy discussions, ensuring businesses bear greater responsibility for their products throughout their life cycle, thus promoting sustainability and waste reduction (Gupt & Sahay, 2015). This collaboration has enhanced recycling rates and reduced landfill dependency. Another noteworthy example is Kamikatsu, a small town in Japan, a model for the Zero Waste Japan initiative. This local effort showcases the potential of collaborative action between institutional actors and lobbyists in driving sustainable waste management practices (Shenyoputro & Jones, 2023). Local governments, supported by environmental lobbyists, pushed for policies emphasizing resource efficiency, product life extension, and waste reduction, contributing to Japan's leadership in CE practices. These case studies underscore the importance of multi-level cooperation and the circular role of lobbying in shaping effective CE policies. Policymakers have a crucial role: they must ensure transparency and balance the different interests involved to generate consensus around legislative designs and policy applications. This approach seeks to advance circularity objectives while minimizing negative impacts on national economies and mitigating the competitive disadvantages businesses may face due to stringent regulations. However, the application of CE regulations often requires monitoring, a significant challenge for authorities. Striking a balance among the various interests involved can facilitate the development of circular business models. In contrast, overly rigid regulations that push for a rapid transition to circularity may encounter resistance from key value chain actors. For example, value chain actors often adopt dynamic vigilance strategies (Day & Schoemaker, 2019), engaging with multiple similar business initiatives to identify and leverage competitive advantages. This approach keeps them agile and responsive to emerging business opportunities while minimizing disruptions to their existing operations. In the UK, research has shown that such strategies, combined with lobbying pressures, can delay the transition to a CE (Adelekan & Sharmina, 2024). This occurs because actors evaluate circular business models based on three key dimensions. First, Value Proposition, how the circular model offers value to customers and stakeholders. Second, Value Creation, how it generates value, including the processes, collaborations, and resources required for effective implementation. Third, Value Capture, how organizations retain value, such as through cost savings, increased revenue, or market differentiation. In this context, policymakers play a fundamental role in ensuring that circular business models are perceived as attractive by actors across the value chain. Non-governmental organizations (NGOs), startups, and innovative companies often have limited resources for lobbying compared to large multinational corporations. Policymakers must create mechanisms to ensure equitable involvement of lobbyists, which includes harmonizing regulations among countries with differing lobbying priorities and capacities. This is a complex challenge. For instance, the Single-Use Packaging and Disposable Industry has exerted pressure to influence the EU's decision-making process regarding the Packaging and Packaging Waste Regulation (PPWR). Their efforts aimed to reduce regulatory ambitions, achieve exemptions, and eliminate critical goals such as excluding paper and cardboard from restrictions and removing reuse targets for the takeaway sector. These actors leveraged scientific evidence, such as Life Cycle Assessment (LCA) studies, to support single-use packaging. They fostered debates by asserting the neutrality and objectivity of their evaluations while promoting technological innovations as solutions, often contrasting these with the EU's bureaucracy. This led to a reduction in regulatory ambitions, with the loss of key objectives such as promoting reuse and focusing on market-driven mechanisms. Additionally, alternative actors were excluded, reinforcing the power of incumbents, who shaped the regulatory content and institutional mechanisms to preserve the status quo (Yalçın et al., 2025). Lobbying strategies in the single-use packaging sector, based on instrumental use of science and institutional power, have hindered ambitious regulatory reforms. The outcome is an example how the lobbying can weaken circularity policies in favor of entrenched interests, with negative implications for the sustainability and inclusiveness of socio-ecological transitions. This evidence highlights the necessity of pursuing more democratic approaches to sustainability. Examples like Repair Cafés and Precious Plastic demonstrate how civil society initiatives can successfully implement CE practices. However, evidence of their political influence remains insufficient, primarily because these initiatives focus on applying existing circular practices rather than promoting new ones. As a result, they mitigate the impact of negative externalities without challenging legislative process (Spekkink et al., 2022). Well-balanced and transparent policies can position the EU as a global leader in sustainability, attracting investments and driving innovation. Incentivizing circular models such as reuse, repair, and leasing could create new markets and job opportunities. Moreover, implementing regulations that address citizens' needs, such as waste reduction and the promotion of responsible consumption, can increase public support for green policies.

In conclusion, the dialogue between lobbyists and policymakers is a complex yet crucial arena for shaping the future of sustainability and the CE. Transparency, cooperation, and innovation will be key elements to overcoming challenges and fully capitalizing on the opportunities in this sector, accelerating the transition to a green economy.

## 5 Conclusion and Implications

This chapter has examined the dynamic relationship between institutions and lobbyists, emphasizing their collaborative roles in advancing CE principles through policy innovation, stakeholder engagement, and the adoption of sustainable practices. By examining case studies and theoretical insights, we have identified the mechanisms and strategies that facilitate this transition while addressing the challenges posed by vested interests, power asymmetries, and risks of greenwashing.

Institutions, including governments, regulatory bodies, and international organizations, play a fundamental role in driving CE initiatives. Through policy incentives, regulatory frameworks, and investments in research and development, institutions create necessary condition for industries and consumers to adopt sustainable practices. Key mechanisms supporting this systemic transition include waste management standards, economic instruments such as incentives, and public–private collaborations.

Lobbyists, representing industry stakeholders, NGOs, and other interest groups, play a dual role in shaping CE policies. On the one hand, positive lobbying efforts have been instrumental in driving innovation, encouraging the adoption of green technologies, and aligning industrial practices with sustainability objectives. On the other hand, negative lobbying from traditional industries often seeks to delay regulatory advancements or promote greenwashing, thereby hindering meaningful progress. Despite these challenges, collaboration between institutions and lobbyists presents significant opportunities for developing comprehensive policies and fostering trust among stakeholders. Frameworks like the "Triple Helix" model and platforms such as CE100 demonstrate the potential of coordinated efforts in advancing CE principles. Moreover, successful case studies, such as Germany's Circular Economy Package and Japan's Zero Waste initiatives, highlight the tangible benefits of institutional-lobbyist partnerships in promoting circularity.

Policy implications for advancing the circular economy emphasize the need for transparent and inclusive policymaking that balances diverse stakeholder interests while ensuring regulations remain equitable, address power asymmetries, and mitigate greenwashing risks. Expanding economic instruments, such as tax incentives, subsidies, and penalties, can incentivize sustainable practices and technological innovation. To ensure accountability and track progress, these measures should be supported by standardized performance metrics and reporting systems. Additionally, promoting collaborative governance through public–private partnerships is crucial for tackling systemic challenges, like waste optimization and circular infrastructure development, with an emphasis on capacity building and knowledge-sharing.

Global cooperation is equally critical, particularly in tackling transboundary issues such as e-waste management and climate change. Leveraging international platforms to harmonize regulatory standards and share best practices, especially in developing regions, can enhance the effectiveness of circular economy initiatives on a broader scale.

Further research is necessary to address key areas fundamental to the circular economy's progress. One priority is the quantification of lobbying efforts and their impact on policymaking, which could help inform more balanced and equitable decisions. Exploring the dynamics of power asymmetries among stakeholders can contribute to more inclusive frameworks. Longitudinal studies on policy effectiveness across diverse sectors and regions will refine strategies and highlight best practices. Additionally, understanding consumer behavior is pivotal to designing targeted campaigns that encourage sustainable consumption. Leveraging digital technologies like blockchain for resource traceability and AI for waste optimization offers promising avenues to enhance CE implementation and efficiency.

The circular economy is not merely a strategy for sustainability but a pathway to resilience, innovation, and equity. When guided by transparency, inclusivity, and shared objectives, collaboration between institutions and lobbyists can serve as a powerful catalyst for systemic change. Policymakers, researchers, and stakeholders must work together to navigate the complexities of the CE transition, ensuring that regulatory frameworks are not only effective but also equitable and adaptable. By addressing existing challenges and embracing emerging opportunities, we can build a sustainable economic model that meets the needs of present and future generations.

## References

Acemoglu, D., & Robinson, J. A. (2012). *Why nations fail: The origins of power, prosperity and poverty*. Crown Publishers.

Adelekan, A., & Sharmina, M. (2024). Co-creating or confounding? Hybrid legitimacy evaluation of circular business models in the UK plastics sector. *Organization & Environment*.

Alemán, E., & Dockendorff, A. (2024). Seeking friends and influencers: Business and labor groups lobbying in the Chilean congress. *Business and Politics, 26*(4), 665–683.

Barra, C., & Falcone, P. M. (2024). Does institutional quality matter for bioeconomy performance? Insights from Italian regions. *Economic Change and Restructuring, 57*(6), 250. https://doi.org/10.1007/s10644-024-09834-3

Bocken, N. M. P., Niessen, L., & Short, S. W. (2022). The sufficiency-based circular economy—an analysis of 150 companies. *Frontiers in Sustainability*. https://doi.org/10.3389/frsus.2022.899289

Charłampowicz, J., Mathiyazhagan, K., Saikouk, T., & Gwalani, H. (2025). Strategic evaluation of European Inland transport companies for European green deal compliance: A multi-criteria decision making-analysis. *Journal of Environmental Management, 374*, 124105. https://doi.org/10.1016/j.jenvman.2025.124105

da Silva Hyldmo, H., Rye, S. A., & Vela-Almeida, D. (2024). A globally just and inclusive transition? Questioning policy representations of the European Green Deal. *Global Environmental Change, 89*, Article 102946.

Day, G. S., & Schoemaker, P. J. H. (2019). *See sooner, act faster: How vigilant leaders thrive in an era of digital turbulence*. Mit Press.

de Abreu, M. C. S., & Ceglia, D. (2018). On the implementation of a circular economy: The role of institutional capacity-building through industrial symbiosis. *Resources, Conservation and Recycling, 138*, 99–109. https://doi.org/10.1016/j.resconrec.2018.07.001

Ellen MacArthur Foundation. (2019). *The Circular Economy 100*.

Energimyndigheten. (2016). *RE:Source (Sweden)*.

Errichiello, G., Falcone, P. M., & Popoyan, L. (2025). Navigating climate policy: The influence of lobbying trends and narratives in Europe. *Environmental Science & Policy, 163*, Article 103974.

Etzkowitz, H., & Leydesdorff, L. (1995). The triple Helix—University-industry-government relations: A laboratory for knowledge based economic development. *EASST Review, 14*(1), 14–19.

European Commission. (2020). *A new circular economy action plan for a cleaner and more competitive Europe*.

European Economic and Social Committee. (2019). *Circular economy strategies and roadmaps in Europe Identifying synergies and the potential for cooperation and alliance building*.

Falcone, P. M., & Fiorentino, R. (2025). Nudging towards sustainability: Exploring the role of behavioral interventions in circular bio-economy development for the fashion industry. *Corporate Social Responsibility and Environmental Management, 32*(1), 661–678.

Falcone, P. M., & Tutore, I. (2025). Mapping the nexus: A bibliometric analysis and social network analysis of transformative innovation policies and sustainable development goals. *Business Strategy and the Environment, 34*(2), 2423–2435. https://doi.org/10.1002/bse.4104

Federal Ministry for Environment, N. C. N. S. and C. P. (2024). *The national circular economy strategy (Germany)*.

Geissdoerfer, M., Savaget, P., Bocken, N. M. P., & Hultink, E. J. (2017). The circular economy—a new sustainability paradigm? *Journal of Cleaner Production, 143*, 757–768. https://doi.org/10.1016/j.jclepro.2016.12.048

Girard, A., Gnabo, J.-Y., & Londoño van Rutten, R. (2023). Corporate lobbying and firm performance variability. *Finance Research Letters, 58*, 104524. https://doi.org/10.1016/j.frl.2023.104524

Gupt, Y., & Sahay, S. (2015). Review of extended producer responsibility: A case study approach. *Waste Management & Research: The Journal for a Sustainable Circular Economy, 33*(7), 595–611. https://doi.org/10.1177/0734242X15592275

ICLEI—Local Governments for Sustainability. (2024). *Circular cities declaration report 2024 insights on implementation, measurement, and nature.*

IK Industrievereinigung Kunststoffverpackungen. (2023). *Position of IK Industrievereinigung Kunststoffverpackungen on the proposal by the European Commission for a EU Packaging and Packaging Waste Regulation.*

Jiao, A., & Lu, J. (2024). Buffing firm innovation by lobbying. *Journal of Corporate Finance, 85*, 102557. https://doi.org/10.1016/j.jcorpfin.2024.102557

Kaufmann, D., Kraay, A., & Zoido-Lobatón, P. (1999). *Governance Matters* (2196).

Kirchherr, J., Reike, D., & Hekkert, M. (2017). Conceptualizing the circular economy: An analysis of 114 definitions. *Resources, Conservation and Recycling, 127*, 221–232. https://doi.org/10.1016/j.resconrec.2017.09.005

Kjaer, P. F., & Vetterlein, A. (2018). Regulatory governance: Rules, resistance and responsibility. *Contemporary Politics, 24*(5), 497–506. https://doi.org/10.1080/13569775.2018.1452527

Laboutková, Š, Šimral, V., & Vymětal, P. (2020). *Transparent lobbying and democracy*. Springer.

Mauro, P. (1995). Corruption and growth. *The Quarterly Journal of Economics, 110*(3), 681–712. https://doi.org/10.2307/2946696

Meadowcroft, J. (2007). Who is in charge here? Governance for sustainable development in a complex world*. *Journal of Environmental Policy & Planning, 9*(3–4), 299–314. https://doi.org/10.1080/15239080701631544

Monciardini, D., Maitre-Ekern, E., Dalhammar, C., & Malcolm, R. (2023). 13 circular economy regulation: An emerging research agenda. In *Handbook of the circular economy* (pp. 219–240). De Gruyter. https://doi.org/10.1515/9783110723373-016

Nguyen, T. C., & Nguyen, T. D. (2023). The role of institutional quality in the transition to a circular economy. *SSRN Electronic Journal*. https://doi.org/10.2139/ssrn.4643626

Nika, C.-E., Expósito, A., Kisser, J., Bertino, G., Oral, H. V., Dehghanian, K., Vasilaki, V., Iacovidou, E., Fatone, F., Atanasova, N., & Katsou, E. (2021). Validating circular performance indicators: the interface between circular economy and stakeholders. *Water, 13*(16), 2198. https://doi.org/10.3390/w13162198

North, D. C. (1990). *Institutions*. Cambridge University Press. https://doi.org/10.1017/CBO9780511808678

OECD. (1971). *Environmental policy committee (EPOC)* .

OECD. (2024). *Monitoring progress towards a resource-efficient and circular economy.*

Olson, M. (1982). *The rise and decline of nations: Economic growth.* Yale University Press.

Ostrom, E. (1990). *Governing the commons.* Cambridge University Press. https://doi.org/10.1017/CBO9780511807763

Park, J. M., Park, J. Y., & Park, H.-S. (2016). A review of the national eco-industrial park development program in Korea: Progress and achievements in the first phase, 2005–2010. *Journal of Cleaner Production, 114*, 33–44. https://doi.org/10.1016/j.jclepro.2015.08.115

PBL Netherlands Environmental Assessment Agency. (2017). *Evaluation of green deals for a circular economy.*

President of the People's Republic of China. (2008). *Circular economy promotion law (China).*

Rose-Ackerman, S. (1999). *Corruption and government.* Cambridge University Press. https://doi.org/10.1017/CBO9781139175098

Shenyoputro, K., & Jones, T. E. (2023). Reflections on a two-decade journey toward zero waste: A case study of Kamikatsu town, Japan. *Frontiers in Environmental Science.* https://doi.org/10.3389/fenvs.2023.1171379

Singh, P. K., & Chudasama, H. (2021). Conceptualizing and achieving industrial system transition for a dematerialized and decarbonized world. *Global Environmental Change, 70*, Article 102349.

Solly, A. (2016). From post-industrial wasteland to eco success: The innovative renewal of Hammarby Sjöstad. *NEWDIST*.

Spekkink, W., Rödl, M., & Charter, M. (2022). Repair Cafés and Precious Plastic as translocal networks for the circular economy. *Journal of Cleaner Production, 380*, Article 135125.

Stabler, D., Hakala, H., Huikkola, T., & Mention, A.-L. (2024). Aligning servitization and circularity: The role of institutional confluence in sustainable business models. *Journal of Cleaner Production, 462*, Article 142666. https://doi.org/10.1016/j.jclepro.2024.142666

The Ministry of Infrastructure and Water Management. (2023). *National circular economy programme 2023–2030.*

UN Environment Programme. (1991). *Global environment facility (GEF).*

VTT Technical Research Centre of Finland Ltd. (2021). *Telaketju—business from circularity of textiles.*

World Economic Forum. (2024). *Circular transformation of industries: The role of partnerships.*

Yalçın, N. G., Paredis, E., & Jaeger-Erben, M. (2025). Packaged science? Incumbent strategies of science capture from a power perspective. *Environmental Innovation and Societal Transitions, 54*, Article 100931.

Zetter, L. (2014). *Lobbying 3e: The art of political persuasion.* Harriman House Limited.

# Sustainable Synergy: Integrating Environmental Economics and Circular Economy

**Ameya Patil, Shantanu Saha, Rajeev Sengupta, Rakesh Yadav, and Kunal Gaurav**

**Abstract** The environment has been deteriorated at a rapid pace during the twentieth and twenty-first centuries. Circular economy and sustainability have become popular as possible answers to challenges of the environment and society. The article is based on the relationship between environmental economics and circular economy. Further, the fundamental principles that have to be understood for sustainable development have been explored. According to Environmental Economics, it is crucial to put prices for resources and, also, to include the costs that are given to the environment. Conversely, circular economy signifies a sustainable economic model. It seeks to minimize waste and maximize resource utilization. In this work, authors seek to examine whether it is possible to bridge these two fields and whether the principles of the two fields combined can be a basis for new approaches to tackle the global environmental problems, increase the economic stability and promote sustainable development. Case studies and review of literature have been used to demonstrate how this approach can be the first step to developing sustainable, circular and fair economies and hence assist in the creation of a sustainable and progressive thinking world economy. The findings of the current literature are utilized to develop a broad understanding of the association between environmental economics and circular economy. This has been demonstrated through studying the industrial symbiosis model in Denmark, the energy programme of Sweden, the circular economy model of Steelcase and electric mobility in China. Emphasis has been made on how these two concepts can be integrated so as to achieve sustainable development. Insights of

A. Patil (✉) · R. Sengupta · K. Gaurav
Faculty of School of Business, Dr. Vishwanath Karad MIT World Peace University, Pune, India
e-mail: ameyapatil0786@gmail.com

K. Gaurav
e-mail: kunal.gaurav@mitwpu.edu.in

S. Saha
Faculty of Ramcharan School of Leadership, Dr. Vishwanath Karad MIT World Peace University, Pune, India
e-mail: shantanu.saha@mitwpu.edu.in

R. Yadav
Faculty of S.P. Mandali, Prin. L.N. Welingkar Institute of Management Development and Research, Mumbai, India



P. Singh et al. (eds.), *Circular Economy and Environmental Resilience*,
https://doi.org/10.1007/978-3-031-93091-1_10

this work will be beneficial to policy makers, organizations and practitioners who are seeking for a sustainable integration of environmental economics and circular economy.

**Keywords** Circular economy · Environmental economics · Sustainability · Circular business models · Synergy · Amalgamation · Environmental benefits

## 1 Introduction

The present economy is centered on the linear "take-make-dispose" production and consumption model. Such a model has dire consequences on the environment. It is leading to resource depletion, pollution, and climate change (Vijai & Wisetsri, 2024). In response to these unyielding challenges, the circular economy model has been embraced as a sustainable approach. Such a style can help decouple growth from resource use. The core of circular economy lies in regenerative design, which focuses on designing products and services to enhance material use, decrease waste and keeping values of materials within the economic system. Thus, it leads to a more sustainable production and consumption pattern (Cheng & Chou, 2018). Circular economy models highlight the need to prolong the lifespan of products, components, and materials via strategies such as reuse, repair, remanufacturing, and recycling (Murray et al., 2017). Consequently, circular economy necessitates optimum utilization of resources.

Environmental economics offers a crucial perspective for evaluating the real costs associated with resource utilization. It also helps in understanding the impacts of waste generation on the environment and the economy. This branch of knowledge throws light on the impact of environmental degradation, thereby helping the populace in making wise decisions and encouraging sustainable management practices. The idea of circular economy has been rooted in environmental economics. By incorporating environmental economics and circular economy principles, businesses can gain a more comprehensive understanding of the economic and ecological implications of their activities (Ghisellini et al., 2016). Amalgamation of these two fields can lead to development of innovative business models, policies, and strategies that prioritize resource efficiency, waste reduction, and creation of closed-loop systems (Chen, 2020). A vital trait of amalgamation is the contemplation of consumer behavior and its part in circular economy. Consumer attitudes and consumption patterns play a critical part in driving the switch towards a more sustainable economic model (Cheng & Chou, 2018). Circular economy emphasizes closing resource loops through reuse, remanufacturing, and recycling. These practices align closely with the objectives of environmental economics. Thus, the combination of these two complementary fields can enable businesses and policymakers to unravel powerful synergies that drive more sustainable outcomes (Arthur et al., 2023). Environmental economics offers the enduring foundation for understanding and quantifying environmental influences of economic activities. Meanwhile, circular economy provides concrete strategies for achieving sustainable production and consumption paradigms.

Shift to a circular economy necessitates co-operation amongst all stakeholders, comprising industries, consumers and governments. Governments have a great opportunity to support the policies which encourage circularity and provide the necessary conditions for the transition. For instance, taxation of carbon emissions from industries makes the industries to adopt better production methods that are environmentally friendly. Companies, consecutively, must rethink their business models, embracing circular design principles and finding new ways to create value from waste and by-products (Islam & Iyer-Raniga, 2023). Eventually, the assimilation of environmental economics and circular economy upholds sustainable prospects. In such a world, economic prosperity ought to be accomplished in accord with ecological protection.

This article intends to offer a general understanding regarding ideas of circular economy and sustainability. The focus is predominantly on their relevance in arenas of economics and management. Furthermore, attribution, which focuses on factors that support the emergence of circular economy strategies is recognised as an important factor. Attribution helps in explaining the adoption of such strategies by businesses. Circular economy can be seen as a venue where usually competing priorities of three mainstays of sustainability, viz. economic, environmental, and social, can co-exist (Vayona & Demetriou, 2020). By synthesizing the insights from the prevailing research and case studies, this article presents a roadmap for the synergistic implementation of environmental economics and circular economy to achieve sustainable development (Nikolaou et al., 2021).

## 2 Different Angles of Circular Economy and Sustainability

The increasing importance of circular economy as a mode to attain sustainable development has invigorated scholars to propose different ways to comprehend the concept (Prieto-Sandoval et al., 2018). Circular economy has been designated as a "result-oriented approach". It encompasses amalgamation of several ideas from different fields. The major purpose is to develop a holistic approach to sustainability and management. Circular economy emphasizes the reduction of resource leakage and the maximization of resource utilization. Nevertheless, sustainable growth is the central objective. Circular economy is one of the most significant concepts that have been adopted in the pursuit of sustainability (Mondello et al., 2024), as it offers a different approach than the conventional linear economy model. The differentiating factor between the two is the pursuance of 3R strategy encompassing reduce, reuse, and recycle, in case of circular economy. The implications are that the resource to be used is reduced to a minimum, the reuse of products is expanded to a possible maximum, and the recycling of materials is maximized (Noman et al., 2022). Apart from 3 R's, circular economy also adopts other vital tactics. These include repurpose, wherein a discarded component can be used for another purpose and rethink, which implies enhancing product usage (Çimen, 2021). Design is the foremost element in circular economy. Accordingly, designers play a vital role in the switch towards a circular economy. These efforts involve redesigning products and optimizing manufacturing processes to encourage reuse. They also focus on implementing strategies that facilitate increased recycling. Further attention is on developing new business models to support sustainable economic development. The role of design in the circular economy goes further than product design, as it also implicates the development of new business models that embrace resource efficiency, longevity, and impression (Dam et al., 2020).

Sustainability is a larger notion that incorporates the economic, environmental, and social aspects of development (Barrier, 2017). There exists no doubt that circular economy can be perceived as a means to achieve sustainability (Mondello et al., 2024). However, these two concepts are not fully identical. The circular economy intends to transform the traditional "take-make-use-dispose" model into a further sustainable framework that underscores efficient resource utilization, waste reduction, and the protection of natural capital (Cheng & Chou, 2018). Notably, sustainable development is a multifaceted problem that entails large-scale societal changes (Wackernagel et al., 2017). Transitioning towards a circular economy is a multidimensional process that requires the active involvement and collaboration of various stakeholders like governments, businesses, and consumers, with each playing a critical role in driving this transformation (Ingemarsdotter et al., 2019; Jones & Comfort, 2021; Sahajwalla & Hossain, 2023).

Amalgamation of design thinking into circular economy stratagems help in navigating complications and obstacles related to sustainable development. It emphasizes understanding user needs and defining problems clearly. The focus is on developing innovative solutions tailored to address these problems effectively. Application of design thinking doctrines in the realm of circular economy provides opportunities for the practitioners to develop innovative and practical solutions for the move towards sustainable development. Key research studies in this arena call for fundamental systemic shifts to marshal evolution to circular economy (Corvellec et al., 2022). As per them, mere incremental changes are not enough to support this changeover. The execution of circular economy ideologies at the micro, meso, and macro levels can catalyse a systemic shift towards sustainability (Klein et al., 2020). At the micro level, businesses are adopting innovative business models that expedite the reuse, recycling, and remanufacturing of products and materials (Calvo-Porral & Lévy-Mangin, 2020). These circular business models are driving the development of circular supply chains that encompass several loops, such as closing, slowing, intensifying, narrowing, and dematerializing (Geissdoerfer et al., 2018). At the meso level, formation of industrial symbiosis networks and eco industrial parks helps to ensure effective exchange of materials, energy and waste between firms. Thus, it supports the more effective and cooperative management of resources. At the macro level, governments and policymakers are implementing regulatory frameworks and incentives to quicken the switch to a circular economy, creating enabling settings for sustainable development (Klein et al., 2020). Further, digital government initiatives can help a long way in this regard (Medaglia et al., 2024).

The literature in the arena of circular economy identifies its potential to act as a mitigation mechanism, providing a shared platform, wherein the conflicting priorities of the three pillars of sustainability can co-exist (Vayona & Demetriou, 2020). The integration of circular economy ideas such as waste reduction, resource recovery, and closed-loop systems, organizations can help combat the environmental and social challenges, while also ensuring economic viability (Prieto-Sandoval et al., 2018). Besides the social benefits of circular economy in terms of reduction in greenhouse gas emissions, economic benefits are estimated to be around $50 trillion by 2050 (Lacy & Rutqvist, 2015). This magnitude of economic benefits is even more than double the size of present global GDP.

## 2.1 *The Role of Attribution in Circular Economy Implementation*

The effective implementation of circular economy plans within organizations is profoundly influenced by the factors that drive and obstruct their adoption (Sarja et al., 2021). Attribution explores antecedents and elements influencing espousal of circular economy approaches. Thus, it is a crucial consideration in comprehending this process. According to attribution theory, individuals and establishments attribute the causes of their successes and failures to several internal and external factors (Vayona & Demetriou, 2020). These attribution factors are decisive in influencing perceptions, attitudes, and activities of organizational participants regarding espousal of circular economy strategies.

As per the prevailing literature, intra-organizational as well as inter-organizational factors can impact the attribution process, when seen from the perspective of circular economy implementation (Sarja et al., 2021). Intra-organizational factors such as organizational culture, leadership, and resource availability play a crucial role in shaping perceived feasibility of circular economy initiatives. These factors also impact the desirability of pursuing such endeavours within organizations. Likewise, inter-organizational factors such as industry dynamics, regulatory frameworks, and supply chain connections significantly influence the attribution of benefits and challenges associated with circular economy implementation. These factors shape how organizations perceive and approach circular economy strategies. The knowledge of attribution process and the influences that define it can help organizations to formulate the right strategies for the integration of circular economy concepts. Especially, it helps them to recognise the potential drivers and hurdles to circular economy adoption and modify their approaches accordingly (Ünal & Shao, 2019).

## 3 Observing Environmental Economics Through a Sustainability Lens

The field of environmental economics seeks to explore the complex association between economic activities and their effect on the natural environment. The way in which environmental-related things like resource depletion and climate change affect economic decision-making is the key focus area. The doctrines and perceptions that shape human understanding and thus, the approach to sustainability are the core of environmental economics (Pham & Ramiah, 2018). Environmental economics recognizes that economic activities are intensely connected with surrounding environmental systems. It emphasizes the intimate relationship between the economy and the environment. Environmental economists incorporate welfare economics as a branch of economics that analyses factors that affect the well-being of society to assess the impacts of environmental policies (Ando et al., 2024). Further, such policies ought to ascertain uncertain environmental costs (Pindyck, 2007). This entails adopting sustainable economic frameworks that prioritize a longstanding vision for nature conservation. They also emphasize maintaining ecosystem health for future generations. Such frameworks are vital for sustaining prevalent environmental conditions (Morseletto, 2022). Further, environmental economics seeks to take environmental externalities into account (Morrissey, 2020). Environmental externalities deal with inadvertent significances of economic activities that disturb the environment, but are not seen in market prices. Recognizing and accounting for environmental costs is important in view of their substantial bearing on the earning capacity and general economic performance of a business.

In the current world, environmental economics has evolved as a branch of sustainability science that emphasizes resources and resource efficiency in relation to the circular economy and its role (Halkos & Managi, 2023). Environmental economics has a notable contribution in promoting sustainability. Different stakeholders are now realizing an important fact that the purposes of environmental conservation and economic development are not contradictory. Rather, these are mutually reinforcing (Alfaro & Diaz, 2021). This understanding has led to demands for environmentally sustainable economic development, which entails balancing economic growth along with environmental protection. Indisputably, the task is arduous. Barbier (1987) acknowledges the impediments in describing sustainable development in a rigorous manner. Nevertheless, the idea of sustainability remains crucial owing to its ability to differentiate itself from other interpretations of development. Its emphasis on balancing economic, social, and environmental factors sets it apart as a comprehensive approach to long-term progress. Moreover, it offers a foundation for practical analysis and policymaking. The arena of environmental economics provides analytical tools and theoretical frameworks to address environmental challenges. It equips policymakers, organizations, and individuals with the information to adopt more sustainable practices. Morseletto (2022) emphasises 'continual environmental improvement', that entails unending effort to reduce the negative consequences of economic activity to improve environmental conditions. Environmental economics can thus be seen as a multidimensional arena which imparts treasured insights regarding intricate relationship between economic activities and environmental sustainability. Adoption of vital ideas of this field can steer towards a brighter future. In such a world, economic development and environmental stewardship would be viewed as complementary goals.

## 4 Review of Literature

The notion of circular economy has attained substantial interest in recent years as a promising approach to address the unrelenting environmental challenges posed by the traditional linear "take-make-dispose" economic model (Geissdoerfer et al., 2018). Practitioners and policymakers have recognized the urgency to move towards a more sustainable and regenerative system which decouples economic growth from the depletion of natural resources (Cheng & Chou, 2018). The review herein studies the connection of environmental economics and circular economy. An attempt is made to explore possible synergies. Thereby, the researchers seek to offer practical implications for organizations, governments, and the society.

The circular economy is engrained in the doctrine of maintaining the value of products, materials, and resources in the economy for as long as possible, minimizing waste and environmental impact (Klein et al., 2020). This model is strengthened by various practices, including oriented towards sustainability (Adam et al., 2017). The reduce, reuse, and recycle hierarchy, the idea of "one industry's waste is another's resource," and the regenerative principles of cradle-to-cradle design are the influential ones in this context. (Ranta & Saari, 2020) demonstrated the effectiveness of the rental model in the fashion industry to prolong product life through a number of uses. Environmental economics focuses on interrelationships between economic activities and the natural environment. It probes into the economic consequences regarding environmental issues as well as the part of policy interventions in that direction. Amalgamation of environmental economics and circular economy would unravel far-reaching welfare for the world.

The businesses can apply the principles of circular economy to decrease their dependence on virgin resources, minimize waste, and unearth new revenue streams (Cheng & Chou, 2018). Circular economy strategies are vital to environmental sustainability, as they help in reduction of carbon emissions (Hailemariam et al., 2023). This study used circular economy pointers such as recycling of municipal waste, biowaste, and packaging waste for 29 European nations. Circular practices can be applied for a sustainable and greener infrastructure development. Koohmishi et al. (2025) discussed the application of circularity for Ballasted tracks in railway infrastructure. Steel and fasteners can be recycled, whereas sleepers and ballast can be repurposed. Circularity, particularly product recovery and recycling, can be of vital use in the supply chain of the construction industry (David et al., 2025). This research further arouses interest of stakeholders in applying circularity principles in other sectors as well, including the service sector. Move to a circular economy can lead to local job creation and enhance energy independence. It also helps alleviate environmental degradation by promoting sustainable resource use. Governments play a critical part in facilitating the switch towards a circular economy through the implementation of supportive policies and incentives (Klein et al., 2020). Such policy interventions encompass producer responsibility programmes. They also involve creation of circular procurement guidelines to support sustainable practices. Additionally, investments in infrastructure and research and development are crucial. Further, digital initiatives and making conducive policies for green finance can help in advancement of circular economy (Kumar et al., 2024; Medaglia et al., 2024). The public sector enterprises can likewise lead the initiatives by embracing circular business models in their own operations, thereby setting an example for the private sector (Lewandowski, 2018). Circular economies will succeed when stakeholders, including industries, consumers, policymakers, and researchers, collaborate and commit to implementing them.

Environmental economics can assist policymakers in assessing the value of improvements in environmental quality and of programmes and schemes in this aspect (Cropper & Oates, 1992). Hence, they can utilise related research and tools to enhance environmental policies. Aligning economic incentives with environmental stewardship can encourage corporations and regimes to explore innovative solutions. This alignment promotes efficient resource use while enhancing societal well-being. On a macro level, switch to circular economy can create new economic avenues, foster domestic employment and improve the energy security of nations. Simultaneously, it can be useful in mitigating environmental degradation and resource depletion (Cheng & Chou, 2018). Governments and international organizations have announced policies and initiatives to fast-track the transition to a circular economy (Tátrai & Diófási-Kovács, 2021). One of the initiatives in this regard is European Green Deal, a broad directive embraced by European Union. It aims to drive sustainability across various sectors and promote a circular economy. A key element of this policy is circular procurement, which emphasizes role of public and private establishments in facilitating the shift to circular practices.

There are varied schools of thought in the arena of circular economy. Some of these are Regenerative Design, Performance Economy, Cradle to Cradle and Industrial Ecology (Smol, 2023). The perspectivies of these different schools converge on the fundamental norm of designing production and consumption processes to maximize ecosystem functioning and human well-being (Gibbons, 2020). Modern research has emphasized the need for a more integrated approach to environmental economics and the circular economy (OECD, 2020). Public sector, as one of the biggest purchaser and consumer of goods and services, has the capacity to drive the way forward and encourage the embrace of circular economy. Public sector organizations should formulate procurement policies and resource management strategies that align with circular economy. Additionally, they must adopt operational practices that support circular approaches. The existing literature shows that circular economy has capacity to resolve existing global environmental and social issues. However, its effective implementation necessitates a deep understanding of the factors that influence its adoption within societies (Centobelli et al., 2020).

## 5 Achieving Synergy: Bridging Environmental Economics and Circular Economy

Circular economy has the capability of linking environmental economics with sustainable development. The application of the circular economy concepts such as the principles of resource efficiency, waste management, and closed-loop systems can help businesses to address the environmental and social issues that are present in the current economy without compromising the financial returns (Patwa et al., 2021). Environmental taxes, especially taxation of raw materials, can encourage industries to move towards recycling (Hondroyiannis et al., 2024).

Such synergistic perspective to environmental economics and circular economy provide numerous advantages. First, it might assist organisations to reduce their negative effects on environment through the effective resource management. Additionally, it promotes waste minimization, contributing to sustainable practices. Second, it can open up new markets for products, services and business models. Last, it can result to social well-being of communities and enhanced quality of life through employment creation. To ensure that such a symbiotic implementation is realized, it is imperative that businesses contemplate on attribution process. Also, factors that affect adoption of circular economy concepts must be comprehended. From the understanding of the factors that promote or hinder the circular economy, the businesses will be able to come up with appropriate measures that can be taken in relation to the specific conditions of the given context (Sarja et al., 2021).

## 6 Real-World Success: Case Studies Integrating Circular Economy and Environmental Economics

The segment consists of some case studies that offer clarity to finest approaches towards amalgamation of circular economy and environmental economics. Such an unification is essential in order to realize sustainability objectives. From the combination of these two disciplines, the case studies demonstrate how effective approaches can be taken in order to support economic development while at the same time maintaining environmental sustainability.

### *6.1 Case Study 1: Kalundborg, Denmark – a Model of Industrial Symbiosis and Circular Economy*

Industrial symbiosis network in Kalundborg, Denmark, is testimony to effective amalgamation of environmental economics and circular economy philosophies. It can be seen as a truly innovative waste recovery model (Schmiedeknecht, 2024). Here, a set of businesses have formed brilliant system of resource exchange. Via this system, the waste products or by-products of one business concern are used as input for another, thereby creating a closed-loop system (Chen, 2020). For example, a power plant supplies steam to a refinery. Bio-sludge, a waste product from a pharma company is utilized in farms as a nutrient. Fly ash from a power plant are used in cement production. Such energy exchanges and material exchanges minimize waste and conserve resources (Paché, 2024).

The paybacks of this style are multidimensional. Viewing through the angle of environmental economics, this model reduces gross environmental impact of involved businesses. It achieves this by minimizing waste and limiting resource extraction. Thereby, cost savings are achieved for businesses in the network, satisfying their profitability agenda. The annual socio-economic cost savings have been assessed close to 15 million USD (Danielsson et al., 2018). Thus, a win–win sustainable business model is practised.

Furthermore, the Kalundborg model establishes the potential for collaboration and information sharing among diverse industries, which is a crucial feature of circular economy (Nassos & Avlonas, 2020). In this context, it can be seen that the businesses can cooperate in order to identify the areas of common interest and possible ways of resources utilization and new revenue generation (Branson, 2016). In the process, they can also support the overall goal of sustainable development.

## 6.2 *Case Study 2: Steelcase – Pioneering a Circular Economy in the Furniture Industry*

Businesses in furniture industry are looking at ways to extend product life through circular economy principles. Reuse and recycle ought to replace production and consumption. Steelcase, a furniture business has implemented a "product-service system" model. The company resorted to embedding design-for-disassembly in all its products (Zils et al., 2023). Rather than selling furniture, the company offers a leasing service, wherein customers pay a fee, for a particular period, to access the furniture (D'Itria et al., 2024). Such a model induces the company to design durable and maintenance-friendly products. These can be effortlessly disassembled and reused or recycled at the end of their useful life. By retaining ownership of these products, Steelcase ensures their proper maintenance as well as reparation. The furniture is eventually recycled or refurbished to extend its lifecycle. This process helps minimize waste while maximizing resource utilization.

It can be seen that Steelcase accepts the responsibility for the product's lifecycle. When viewed through the environmental economics lens, the Steelcase model helps to internalize the environmental costs related to production and disposal of furniture (Veleva, 2021). Moreover, costs over the long run as well as the environmental impact of furniture are closely associated to the business model. Hence, this approach bolsters the business concern to focus sustainability and resource efficiency in its design and manufacturing processes. At the same time, the circular business model creates new revenue streams and opportunities for the business (Monticelli & Costamagna, 2023). This case highlights possibility of win–win circumstances amid environmental and economic goals.

### *6.3 Case Study 3: Turning Trash into Treasure-Sweden's Waste-To-Energy Success*

Sweden is considered as a country with the best practices in waste management, which are the exemplar of sustainable and circular economy. The nation's waste-to-energy program has been a crucial component of this stratagem, effectively assimilating environmental economics and circular economy principles (Andersson & Stage, 2018). The basis of this waste management system exists in the nation's fidelity to circular economy. Constantly, Sweden has been highlighting the reuse, repair, and reintegration of materials into the production and consumption chain. Such initiatives have helped the country to minimize waste generation and maximize resource efficiency (Pluskal et al., 2021). This approach aligns with sustainable waste management systems that integrate feedback loops, emphasis on processes, and diversion of wastes from disposal (Seadon, 2010).

This program is a leading illustration of practical implementation of circular economy principles. The nation has effectively transformed its waste management sector into a resource recovery system, wherein the energy generated from the ignition of non-recyclable waste is utilized to power households and industries (Ranta & Saari, 2020). Biomass and waste is turned into syngas, which gets processed to biomethane, used for heating and as a fuel source (Al-Moftah et al., 2025). Integration of waste management and energy production minimize environmental impact of waste disposal. Moreover, it contributes to sustainable energy generation, creating a synergetic connection between the two (Ellacuriaga et al., 2021). Furthermore, this initiative has a substantial impact on the nation's overall sustainability. By diverting waste from landfills and converting it into a valuable energy resource, Sweden has reduced its greenhouse gas emissions, thereby playing a significant role in the fight against climate change (Shekdar, 2009). This initiative has equally created economic welfares. These include employment in waste management and energy sectors. Additionally, costs related to traditional waste disposal techniques have been minimized.

Nevertheless, the success of the program had its own set of challenges. The environmental impacts of waste ignition, including air pollution and the possible release of harmful substances, ought be prudently supervised and addressed. Furthermore, the program's dependence on the import of waste from other nations elevated apprehensions regarding the ethical implications of such practice (Miranda & Hale, 1997). In order to address these challenges, Sweden has further increased its investment on research and development. The resulting innovations help to improve performance of its waste-to-energy plants and to make them more effective. This way the nation has also supported the implementation of policies that are derived from circular economy strategy which aims to minimize waste at the source through recycling and reusing options.

In a nutshell, this initiative exemplifies a captivating case regarding amalgamation of environmental economics and circular economy. It prevails as a compelling case in arena of sustainable waste management. Successes and challenges associated with this initiative offer valuable insights for other nations looking to formulate effective and sustainable waste management systems (Hoang & Fogarassy, 2020; Rajaeifar et al., 2017).

## 6.4 Case Study 4: Driving the Future- Shenzhen's Transformation Through Electric Mobility

Shenzhen, a city in China, has made a momentous shift towards electric mobility. The switch is a key component of its circular economy strategy aimed at reducing noise pollution. It also focuses on improving air quality for a healthier environment. Shenzhen replaced its entire fleet of internal combustion engine buses by electric buses (Hou et al., 2023). To accommodate this change, the city made huge investments in the necessary charging infrastructure and supporting systems (Tsoi et al., 2022). For business concerns, an innovative model of renting e-buses from manufacturers over a long-term helped city to avoid huge upfront capital expenditures related with outright purchase of vehicles (Li et al., 2020). Such an approach ensured the longer usage of buses, prompting regular upgrades to batteries and drivetrains. This went a long way to enhancing sustainability and efficiency of whole system. Due to this large scale electrification project Shenzhen has been able to make annual reductions of more than 4 million tons of particulate matter emitted into the air. Also, it has been able to decrease greenhouse gas emissions by 40% for every km travelled by the electric buses. Nevertheless, Shenzhen acknowledges the necessity for a shift in energy mix used in the charging stations towards greater utilization of renewable sources, to achieve a real circular operation. At present, merely 1 percent of energy utilised comes from renewable sources. However, Shenzhen believes that creating large-scale demand for electric mobility will drive a crucial shift in the energy mix. This shift aligns with the sustainability and environmental goals of circular economy system. The electric bus endeavour of city is a great example for other cities which are seeking to make changes towards using eco friendly transport systems. This shows how large scale electrification can lead to significant reduction in emissions and enhancement of air quality in cities. The electric bus initiative has been well received and implemented in the city thus reducing the cities carbon footprint. Furthermore, it has paved way for other cities across the globe to adopt the same vision for attaining sustainable and environmental friendly city development.

## 7 Challenges and Barriers

In order to successfully implement circular economy practices, a number of crucial issues needs to be addressed (Corvellec et al., 2022). The transition to circular economy faces various challenges and barriers, at the micro-level of individual businesses, at the meso-level of business networks and the macro-level of policy and regulations (Klein et al., 2020). It is true that the circular economy holds numerous benefits for society. However, its evasion of scrutiny has prevented it from exhibiting as an inclusive solution to environmental problems (Buchmann-Duck & Beazley, 2020). Additional obstacle to implementation of circular economy is the heterogeneous effect of government policies (Arthur et al., 2023; Ferriz-Papi et al., 2024). Complexity of integrating circular economy into existing business structures is yet another issue (Tuboalabo et al., 2024). Tecchnical knowledge is other impediment towards effective initiatives in this direction (Agyekum et al., 2024), especially in the case of small and medium enterprises (Kondala et al., 2024). Governments around the world have formulated numerous policies in order to mitigate environmental impact of human activities. However, the efficacy of these policies in promoting circular economy objectives has been mixed. Moreover, the economic feasibility of implementing circular economy practices can be hindered due to the high costs of collection, sorting, and processing of waste streams (Van Beukering et al., 2014). Policymakers must carefully consider the cost implications of new circular economy initiatives. They should also ensure these initiatives are financially viable for both businesses and consumers. This is because financial constraints might emerge as an impediment in such initiatives (Pacheco et al., 2024).

Economic barriers to circular economy can be tackled through financial incentives and support mechanisms (Tura et al., 2019). Governments across the globe are implementing policies and tactics to endorse resource efficiency and further move to a circular economy. Some of these policies include tax incentives, subsidies, and extended producer responsibility schemes (OECD, 2020). Enterprises and consumers can leverage such financial incentives and support mechanisms, which make circular economy practices economically attractive as well. Another strategic challenge is the requirement to resolve the misalignment between industrial and environmental policies. Further, poorly aligned waste characterization and biomass competition can lead to conflicts between the purposes of bio-economy and circular economy (Stegmann et al., 2020). Policymakers ought to make an effort to harmonize these policies in order to ensure a smooth and economically viable transition towards circular economy. Scholars and practitioners have identified some policy elements that can fast-track the transition to circular economy. These include further robust standards and norms in production, expansion of circular procurement, tax relief for circular products, liberalization of waste trading, and awareness campaigns (Hartley et al., 2020).

## 8 Conclusion

Increasing resource scarcity and mounting environmental concerns raise doubts on sustainability of traditional linear "take-make-dispose" model of production and consumption (Korhonen et al, 2018). To resolve these challenges pertaining to environment owing to industrial production, notion of circular economy has gained traction. It is perceived as a practical approach to endorse resource efficiency, waste reduction, as well as the formation of closed-loop systems. This article presents the practical implications and benefits of the circular economy through the exploration of a few case studies. These cases highlight the potential for circular economy strategies to generate novel business prospects, in addition to generating environmental benefits.

The amalgamation of environmental economics and circular economy doctrines provides a synergistic vision for a sustainable future. Environmental economics offer treasured insights into the valuation of natural capital, the pricing of externalities, and the design of market-based mechanisms to acclerate circular practices (Calvo-Porral & Lévy-Mangin, 2020; Geissdoerfer et al., 2018). Equally, the circular economy agenda can contribute to the operationalization of environmental economics through practical strategies and business models for decoupling economic growth from resource extraction and environmental degradation (Cheng & Chou, 2018; Klein et al., 2020). Ultimately, the synergy between environmental economics and circular economy can facilitate the development of a more resilient and sustainable economic system that respects the limits of natural resources, promotes the efficient use of materials, and fosters a collaborative and regenerative approach to economic growth (Chen, 2020; Klein et al., 2020).

## 9 Future Directions

The concept of the circular economy has drawn increased attention as a solution to resolve the environmental consequences of linear production and consumption processes (Aguilar-Hernandez et al., 2021). Even the Governments have put in place several policy measures to support circular economy. This emphasize the importance of taking into account future strategies for enhancing integration of environmental economics with circular economy principles.

Next area for further research is development of metrics to assess circularity at the product and company levels. Additionally, advancing frameworks for measuring circular economy practices stand paramount. The current ones quantify circularity taking economic value into account. These frameworks should incorporate additional dimensions such as environmental impact and social considerations. (García-Barragán et al., 2019; Klein et al., 2020). Another possible research can be conducted with regards to analysis of consumer behavior within circular economy framework. Comprehending the influence of cultural differences and consumer preferences on the adoption of circular economy practices can enlighten towards the development of effective policies and business models that encourage a shift away from the "take-make-dispose" mentality (Cheng & Chou, 2018). Another key area should be regarding leveraging technology and innovation to drive greater integration between environmental economics and circular economy principles (Maier et al., 2020; Parida & Wincent, 2019). As businesses grapple with the challenges of sustainability, they must look to novel solutions and business models that can unlock new value streams while mitigating environmental impact (Oe et al., 2022). This might involve investments in green technologies, the development of closed-loop supply chains, or the adoption of advanced data analytics to optimize resource usage and waste reduction (Bocken et al., 2014).

## References

Agyekum, K., & Amudjie, J. (2024). Challenges to implementing circular economy: Empirical evidence among built environment firms in Ghana. *International Journal of Construction Management, 24*(3), 281–297. https://doi.org/10.1080/15623599.2023.2222993

Aguilar-Hernandez, G. A., Rodrigues, J. F. D., & Tukker, A. (2021). Macroeconomic, social and environmental impacts of a circular economy up to 2050: A meta-analysis of prospective studies. *Journal of Cleaner Production, 278*, Article 123421. https://doi.org/10.1016/j.jclepro.2020.123421

Ando, A. W., Awokuse, T. O., Chan, N. W., González-Ramírez, J., Gulati, S., Interis, M. G., ... & Stolper, S. (2024). Environmental and natural resource economics and systemic racism. *Review of Environmental Economics and Policy, 18*(1), 143–164.

Al-Moftah, A. M. S., Alnajideen, M., Alafifi, F., Czyzewski, P., Shi, H., Alherbawi, M., ... & Valera-Medina, A. (2025). Economic feasibility of using municipal solid waste and date palm waste for clean energy production in Qatar. *Energies, 18*(4), 988.https://doi.org/10.3390/en18040988.

Andersson, C., & Stage, J. (2018). Direct and indirect effects of waste management policies on household waste behaviour: The case of Sweden. *Waste Management, 76*, 19–27. https://doi.org/10.1016/j.wasman.2018.03.038

Alfaro, M. T. B., & Diaz, R. A. (2021). Impact of green initiatives on the financial performance of small and medium enterprises: The case of manufacturing firm in Central Luzon. *WPOM-Working Papers on Operations Management, 12*(1), 28–41.

Arthur, E. E., Gyamfi, S., Gerstlberger, W., Stejskal, J., & Prokop, V. (2023). Towards circular economy: unveiling heterogeneous effects of government policy stringency, environmentally related innovation, and human capital within OECD countries. *Sustainability, 15*(6), 4959. https://doi.org/10.3390/su15064959

Adam, S., Bucker, C., Desguin, S., Vaage, N., & Saebi, T. (2017). Taking part in the circular economy: four ways to designing circular business models. Available at SSRN: 2908107. https://doi.org/10.2139/ssrn.2908107

Barrier, E. B. (2017). The concept of sustainable economic development. In *The Economics of Sustainability*, pp. 87–96. Routledge.

Bocken, N. M., Short, S. W., Rana, P., & Evans, S. (2014). A literature and practice review to develop sustainable business model archetypes. *Journal of Cleaner Production, 65*, 42–56. https://doi.org/10.1016/j.jclepro.2013.11.039

Branson, R. (2016). Re-constructing Kalundborg: The reality of bilateral symbiosis and other insights. *Journal of Cleaner Production, 112*, 4344–4352. https://doi.org/10.1016/j.jclepro.2015.07.069
Buchmann-Duck, J., & Beazley, K. F. (2020). An urgent call for circular economy advocates to acknowledge its limitations in conserving biodiversity. *Science of the Total Environment, 727*, Article 138602. https://doi.org/10.1016/j.scitotenv.2020.138602
Corvellec, H., Stowell, A. F., & Johansson, N. (2022). Critiques of the circular economy. *Journal of Industrial Ecology, 26*(2), 421–432. https://doi.org/10.1111/jiec.13187
Calvo-Porral, C., & Lévy-Mangin, J. P. (2020). The circular economy business model: Examining consumers' acceptance of recycled goods. *Administrative Sciences, 10*(2), 28. https://doi.org/10.3390/admsci10020028
Centobelli, P., Cerchione, R., Chiaroni, D., Del Vecchio, P., & Urbinati, A. (2020). Designing business models in circular economy: A systematic literature review and research agenda. *Business Strategy and the Environment, 29*(4), 1734–1749. https://doi.org/10.1002/bse.2466
Chen, C. W. (2020). Improving circular economy business models: opportunities for business and innovation: A new framework for businesses to create a truly circular economy. *Johnson Matthey Technology Review, 64*(1), 48–58. https://doi.org/10.1595/205651320X15710564137538
Cheng, C. C., & Chou, H. M. (2018, April). Applying the concept of circular economy—Using the cultural difference of European consumers as an example. In *2018 IEEE International Conference on Applied System Invention (ICASI)*, pp. 449–452. IEEE. https://doi.org/10.1109/icasi.2018.8394281.
Cropper, M. L., & Oates, W. E. (1992). Environmental economics: A survey. *Journal of Economic Literature, 30*(2), 675–740.
Çimen, Ö. (2021). Construction and built environment in circular economy: A comprehensive literature review. *Journal of Cleaner Production, 305*, Article 127180. https://doi.org/10.1016/j.jclepro.2021.127180
David, L. O., Aigbavboa, C., Adepoju, O., & Nnamdi, N. (2025). Applying circular economy strategies in mitigating the perfect storm: The built environment context. *Sustainable Futures*, 100444.https://doi.org/10.1016/j.sftr.2025.100444
D'Itria, E., Pei, X., & Bertola, P. (2024). Designing sustainability today: An analytical framework for a design for sustainability model in European fashion and furniture industries. *Sustainability, 16*(8), 3240. https://doi.org/10.3390/su16083240
Dam, K. V., Simeone, L., Keskin, D., Baldassarre, B., Niero, M., & Morelli, N. (2020). Circular economy in industrial design research: A review. *Sustainability, 12*(24), 10279. https://doi.org/10.3390/su122410279
Danielsson, S., Møller, P., & Randers, L. (2018). Modelling $CO_2$ savings and economic benefits for the Kalundborg symbiosis. Retrieved from http://www.symbiosis.dk/wp-content/uploads/2018/06/KalundborgSymbiosisPaperLCA2018-06.pdf
Ellacuriaga, M., García-Cascallana, J., & Gómez, X. (2021). Biogas production from organic wastes: Integrating concepts of circular economy. *Fuels, 2*(2), 144–167. https://doi.org/10.3390/fuels2020009
Ferriz-Papi, J. A., Lee, A., & Alhawamdeh, M. (2024). Examining the challenges for circular economy implementation in construction and demolition waste management: A comprehensive review using systematic methods. *Buildings, 14*(5), 1237. https://doi.org/10.3390/buildings14051237
García-Barragán, J. F., Eyckmans, J., & Rousseau, S. (2019). Defining and measuring the circular economy: A mathematical approach. *Ecological Economics, 157*, 369–372. https://doi.org/10.1016/j.ecolecon.2018.12.003
Geissdoerfer, M., Morioka, S. N., de Carvalho, M. M., & Evans, S. (2018). Business models and supply chains for the circular economy. *Journal of Cleaner Production, 190*, 712–721. https://doi.org/10.1016/j.jclepro.2018.04.159

Ghisellini, P., Cialani, C., & Ulgiati, S. (2016). A review on circular economy: The expected transition to a balanced interplay of environmental and economic systems. *Journal of Cleaner Production, 114*, 11–32. https://doi.org/10.1016/j.jclepro.2015.09.007

Gibbons, L. V. (2020). Regenerative—The new sustainable? *Sustainability, 12*(13), 5483. https://doi.org/10.3390/su12135483

Halkos, G., & Managi, S. (2023). New developments in the disciplines of environmental and resource economics. *Economic Analysis and Policy, 77*, 513–522. https://doi.org/10.1016/j.eap.2022.12.008

Hailemariam, A., & Erdiaw-Kwasie, M. O. (2023). Towards a circular economy: Implications for emission reduction and environmental sustainability. *Business Strategy and the Environment, 32*(4), 1951–1965. https://doi.org/10.1002/bse.3229

Hoang, N. H., & Fogarassy, C. (2020). Sustainability evaluation of municipal solid waste management system for Hanoi (Vietnam)—Why to choose the 'Waste-to-Energy' concept. *Sustainability, 12*(3), 1085. https://doi.org/10.3390/su12031085

Hartley, K., van Santen, R., & Kirchherr, J. (2020). Policies for transitioning towards a circular economy: Expectations from the European Union (EU). *Resources, Conservation and Recycling, 155*, Article 104634. https://doi.org/10.1016/j.resconrec.2019.104634

Hondroyiannis, G., Sardianou, E., Nikou, V., Evangelinos, K., & Nikolaou, I. (2024). Circular economy and macroeconomic performance: Evidence across 28 European countries. *Ecological Economics, 215*, Article 108002. https://doi.org/10.1016/j.ecolecon.2023.108002

Hou, X., Lv, T., Xu, J., Deng, X., Liu, F., & Lam, J. S. L. (2023). Electrification transition and carbon emission reduction of urban passenger transportation systems—A case study of Shenzhen. *China. Sustainable Cities and Society, 93*, Article 104511. https://doi.org/10.1016/j.scs.2023.104511

Islam, M. T., & Iyer-Raniga, U. (2023). Circular business model value dimension canvas: Tool redesign for innovation and validation through an Australian case study. *Sustainability, 15*(15), 11553. https://doi.org/10.3390/su151511553

Ingemarsdotter, E., Jamsin, E., Kortuem, G., & Balkenende, R. (2019). Circular strategies enabled by the internet of things—A framework and analysis of current practice. *Sustainability, 11*(20), 5689. https://doi.org/10.3390/su11205689

Jones, P., & Comfort, D. (2021). The circular economy: an exploratory case study from the paper and retail industries. *Athens Journal of Business and Economics, 7*(4), 379–394. https://doi.org/10.30958/ajbe.7-4-5.

Klein, N., Ramos, T. B., & Deutz, P. (2020). Circular economy practices and strategies in public sector organizations: An integrative review. *Sustainability, 12*(10), 4181. https://doi.org/10.3390/su12104181

Korhonen, J., Honkasalo, A., & Seppälä, J. (2018). Circular economy: The concept and its limitations. *Ecological Economics, 143*, 37–46. https://doi.org/10.1016/j.ecolecon.2017.06.041

Koohmishi, M., Kaewunruen, S., He, X., & Guo, Y. (2025). Advancing railway sustainability: Strategic integration of circular economy principles in ballasted track systems. *Journal of Cleaner Production, 144713*,. https://doi.org/10.1016/j.jclepro.2025.144713

Kondala, M., Nudurupati, S. S., & Pappu, R. P. (2024). The challenges in adoption of circular economy in SMEs–a research agenda and way forward. *Benchmarking: An International Journal, 31*(5), 1667–1699. https://doi.org/10.1108/BIJ-04-2023-0272.

Kumar, B., Kumar, L., & Kumar, A. (2024). Green finance in circular economy: A literature review. *Environment, Development and Sustainability, 26*, 16419–16459. https://doi.org/10.1007/s10668-023-03361-3

Lewandowski, M. (2018). Public sector and circular business models: From public support towards implementation through design. *Sustainable Business Models: Principles, Promise, and Practice, 85–101*,. https://doi.org/10.1007/978-3-319-73503-0_5

Lacy, P., & Rutqvist, J. (2015). *Waste to wealth: The circular economy advantage* (Vol. 91). Palgrave Macmillan. https://doi.org/10.1057/9781137530707

Li, M., Ye, H., Liao, X., Ji, J., & Ma, X. (2020). How Shenzhen, China pioneered the widespread adoption of electric vehicles in a major city: Implications for global implementation. *Wiley Interdisciplinary Reviews: Energy and Environment, 9*(4), Article e373. https://doi.org/10.1002/wene.373

Medaglia, R., Rukanova, B., & Zhang, Z. (2024). Digital government and the circular economy transition: An analytical framework and a research agenda. *Government Information Quarterly, 41*(1), 101904. https://doi.org/10.1016/j.giq.2023.101904Monticelli; A., & Costamagna, M. (2023). Environmental assessment of the rental business model: A case study for formal wear. *Environment, Development and Sustainability, 25*(8), 7625–7643. https://doi.org/10.1007/s10668-022-02363-x.

Miranda, M. L., & Hale, B. (1997). Waste not, want not: The private and social costs of waste-to-energy production. *Energy Policy, 25*(6), 587–600. https://doi.org/10.1016/s0301-4215(97)00050-5

Maier, D., Maier, A., Aşchilean, I., Anastasiu, L., & Gavriş, O. (2020). The relationship between innovation and sustainability: A bibliometric review of the literature. *Sustainability, 12*(10), 4083. https://doi.org/10.3390/su12104083

Mondello, A., Salomone, R., & Mondello, G. (2024). Exploring circular economy in the cosmetic industry: Insights from a literature review. *Environmental Impact Assessment Review, 105*, Article 107443. https://doi.org/10.1016/j.eiar.2024.107443

Murray, A., Skene, K., & Haynes, K. (2017). The circular economy: An interdisciplinary exploration of the concept and application in a global context. *Journal of Business Ethics, 140*, 369–380. https://doi.org/10.1007/s10551-015-2693-2

Morseletto, P. (2022). Environmental principles for modern sustainable economic frameworks including the circular economy. *Sustainability Science, 17*(5), 2165–2171. https://doi.org/10.1007/s11625-022-01208-w

Morrissey, K. (2020). *Resource and environmental economics*.https://doi.org/10.1016/B978-0-08-102295-5.10755-3

Nikolaou, I. E., Jones, N., & Stefanakis, A. (2021). Circular economy and sustainability: The past, the present and the future directions. *Circular Economy and Sustainability, 1*, 1–20. https://doi.org/10.1007/s43615-021-00030-3

Nassos, G., & Avlonas, N. (2020) The need and growth of a circular economy. In N. Avlonas, & G. Nassos, (Eds.), *Practical Sustainability Strategies: How to Gain a Competitive Advantage.* John Wiley & Sons: New York, NY, USA, 2020; pp. 127–136. https://doi.org/10.1002/9781119561132.ch10

Noman, A. A., Akter, U. H., Pranto, T. H., & Haque, A. K. M. (2022). Machine learning and artificial intelligence in circular economy: A bibliometric analysis and systematic literature review. *Annals of Emerging Technologies in Computing (AETiC), 6*(2), 13–40. https://doi.org/10.33166/aetic.2022.02.002

OECD. (2020). Improving resource efficiency and the circularity of economies for a greener world. OECD Environment Policy Papers, No. 20, OECD Publishing, Paris, https://doi.org/10.1787/1b38a38f-en.

Oe, H., Yamaoka, Y., & Duda, K. (2022). How to sustain businesses in the post-COVID-19 era: A focus on innovation, sustainability and leadership. *Business Ethics and Leadership, 6*(4), 1–9. https://doi.org/10.21272/bel.6(4).1-9.2022

Pacheco, D. A. D. J., Rampasso, I. S., Michels, G. S., Ali, S. M., & Hunt, J. D. (2024). From linear to circular economy: The role of BS 8001:2017 for green transition in small businesses in developing economies. *Journal of Cleaner Production, 439*, 140787. https://doi.org/10.1016/j.jclepro.2024.140787

Parida, V., & Wincent, J. (2019). Why and how to compete through sustainability: A review and outline of trends influencing firm and network-level transformation. *International Entrepreneurship and Management Journal, 15*, 1–19. https://doi.org/10.1007/s11365-019-00558-9

Patwa, N., Seetharaman, A., Arora, A., Agrawal, R., & Mandalia, H. (2021). Circular economy: Bridging the gap in sustainable manufacturing. *The Journal of Developing Areas, 55*(1). https://doi.org/10.1353/jda.2021.0012

Paché, G. (2024). Kalundborg industrial symbiosis: Circular strategy in the light of mutualism. *Journal of Strategic Innovation and Sustainability, 19*(3), 89–97.

Pindyck, R. S. (2007). Uncertainty in environmental economics. *Review of Environmental Economics and Policy, 1*(1), 45–65.

Pluskal, J., Šomplák, R., Nevrlý, V., Smejkalová, V., & Pavlas, M. (2021). Strategic decisions leading to sustainable waste management: Separation, sorting and recycling possibilities. *Journal of Cleaner Production, 278*, Article 123359. https://doi.org/10.1016/j.jclepro.2020.123359

Prieto-Sandoval, V., Jaca, C., & Ormazabal, M. (2018). Towards a consensus on the circular economy. *Journal of Cleaner Production, 179*, 605–615. https://doi.org/10.1016/j.jclepro.2017.12.224

Ranta, V., & Saari, U. A. (2020). Circular economy: Enabling the transition towards sustainable consumption and production. *Responsible Consumption and Production, 78–89*,. https://doi.org/10.1007/978-3-319-95726-5_3

Rajaeifar, M. A., Ghanavati, H., Dashti, B. B., Heijungs, R., Aghbashlo, M., & Tabatabaei, M. (2017). Electricity generation and GHG emission reduction potentials through different municipal solid waste management technologies: A comparative review. *Renewable and Sustainable Energy Reviews, 79*, 414–439. https://doi.org/10.1016/j.rser.2017.04.109

Sahajwalla, V., & Hossain, R. (2023). Rethinking circular economy for electronics, energy storage, and solar photovoltaics with long product life cycles. *MRS Bulletin, 48*(4), 375–385. https://doi.org/10.1557/s43577-023-00519-2

Sarja, M., Onkila, T., & Mäkelä, M. (2021). A systematic literature review of the transition to the circular economy in business organizations: Obstacles, catalysts and ambivalences. *Journal of Cleaner Production, 286*, Article 125492. https://doi.org/10.1016/j.jclepro.2020.125492

Schmiedeknecht, M. H. (2024). The principle of recovery in circular economy: Generating energy. In E. Fröhlich, K. Steinbiß, M. H. Schmiedeknecht, & C. R. Loza Adaui (Eds.), *Circular economy in sustainable supply chains*. CSR, sustainability, ethics & governance. Springer, Cham. https://doi.org/10.1007/978-3-031-70749-0_13

Seadon, J. K. (2010). Sustainable waste management systems. *Journal of Cleaner Production, 18*(16–17), 1639–1651. https://doi.org/10.1016/j.jclepro.2010.07.009

Shekdar, A. V. (2009). Sustainable solid waste management: An integrated approach for Asian countries. *Waste Management, 29*(4), 1438–1448. https://doi.org/10.1016/j.wasman.2008.08.025

Smol, M. (2023). Inventory and comparison of performance indicators in circular economy roadmaps of the European countries. *Circular Economy and Sustainability, 3*(1), 557–584. https://doi.org/10.1007/s43615-021-00127-9

Stegmann, P., Londo, M., & Junginger, M. (2020). The circular bioeconomy: Its elements and role in European bioeconomy clusters. *Resources, Conservation & Recycling: X, 6*, Article 100029. https://doi.org/10.1016/j.rcrx.2019.100029

Tuboalabo, A., Buinwi, U., Okatta, C. G., Johnson, E., & Buinwi, J. A. (2024). Circular economy integration in traditional business models: Strategies and outcomes. *Finance & Accounting Research Journal, 6*(6), 1105–1123.

Tátrai, T., & Diófási-Kovács, O. (2021). European green deal–the way to circular public procurement. In ERA Forum (Vol. 22, No. 3, pp. 523–539). Berlin/Heidelberg: Springer. https://doi.org/10.1007/s12027-021-00678-2

Tura, N., Hanski, J., Ahola, T., Ståhle, M., Piiparinen, S., & Valkokari, P. (2019). Unlocking circular business: A framework of barriers and drivers. *Journal of Cleaner Production, 212*, 90–98.https://doi.org/10.1016/j.jclepro.2018.11.202

Tsoi, K. H., Loo, B. P., Tal, G., & Sperling, D. (2022). Pioneers of electric mobility: Lessons about transport decarbonisation from two bay areas. *Journal of Cleaner Production, 330*, Article 129866. https://doi.org/10.1016/j.jclepro.2021.129866

Ünal, E., & Shao, J. (2019). A taxonomy of circular economy implementation strategies for manufacturing firms: Analysis of 391 cradle-to-cradle products. *Journal of Cleaner Production, 212*, 754–765. https://doi.org/10.1016/j.jclepro.2018.11.291

Veleva, V. (2021). The role of entrepreneurs in advancing sustainable lifestyles: Challenges, impacts, and future opportunities. *Journal of Cleaner Production, 283*, Article 124658. https://doi.org/10.1016/j.jclepro.2020.124658

Vijai, & Wisetsri, W. (2024). Circular economy for sustainable development in India. In S. Kulkarni & A. K. Haghi (Eds.), *Global sustainability*. World Sustainability Series. Springer. https://doi.org/10.1007/978-3-031-57456-6_4

Vayona, A., & Demetriou, G. (2020). Towards an operating model for attribution in circular economy. In *2020 16th International Conference on Distributed Computing in Sensor Systems* (DCOSS), pp. 490–495. IEEE. https://doi.org/10.1109/dcoss49796.2020.00082

Van Beukering, P., Kuik, O., & Oosterhuis, F. (2014). The economics of recycling. In M. L. Letcher & D. A. Vallero (Eds.), *Handbook of recycling*, pp. 479–489. Elsevier. https://doi.org/10.1016/B978-0-12-396459-5.00031-3

Wackernagel, M., Hanscom, L., & Lin, D. (2017). Making the sustainable development goals consistent with sustainability. *Frontiers in Energy Research, 5*, 18. https://doi.org/10.3389/fenrg.2017.00018

Zils, M., Howard, M., & Hopkinson, P. (2023). Circular economy implementation in operations & supply chain management: Building a pathway to business transformation. *Production Planning & Control, 36*(4), 501–520. https://doi.org/10.1080/09537287.2023.2280907

# Adopting Digital Technologies in the Circular Economy: Strategies to Improve Environmental Resilience

**Ernesto D. R. Santibanez Gonzalez, Luciano Ferreira da Silva, Diego de Melo Conti, and Marcos Ferasso**

**Abstract** This chapter explores how digital technologies can drive the transition from a linear to a circular economy, addressing the gap in understanding their role in enhancing environmental resilience. The study examines the adoption of technologies such as AI (Artificial Intelligence), Cloud Computing, Big Data, IoT (Internet of Things), and Blockchain within circular economy strategies. The arguments presented in this chapter highlight how these technologies optimize resource use, minimize waste, and foster sustainable innovation. Key findings reveal that blockchain improves traceability, AI enhances resource management, and IoT supports innovative business models, contributing to the synergy between biological and technical cycles. By integrating production chains and keeping materials in circulation longer, these technologies enhance the resilience of economic ecosystems. This chapter also demonstrates how digital transformation could overcome environmental challenges and creates competitive advantages and new business opportunities, contributing to a more sustainable global society.

**Keywords** Sustainability · Circular economy · Supply chains · Digital technologies · Environmental resilience

E. D. R. Santibanez Gonzalez (✉)
Industrial Engineering Department, CES4.0, Faculty of Engineering, University of Talca, Curico, Chile
e-mail: santibanez.ernesto@gmail.com

L. F. da Silva
UNINOVE, and American Institute of Applied Sciences (AUS), São Paulo, Switzerland
e-mail: lf.silvabr@uni9.pro.br

D. de Melo Conti
PUC-Campinas, São Paulo, Brazil
e-mail: diego.conti@puc-campinas.edu.br

M. Ferasso
Grupo de Investigación de Estudios Organizacionales Sostenibles, Universidad Autónoma de Chile, Santiago, Chile
e-mail: marcos.ferasso@uautonoma.cl

 
P. Singh et al. (eds.), *Circular Economy and Environmental Resilience*,
https://doi.org/10.1007/978-3-031-93091-1_11

# 1 Introduction

To meet the demand of approximately 9 billion people by 2050, production systems will require around 140 billion tons of minerals, fossil fuels, and biomass annually, while global food production must increase by at least 50% (Grafton et al., 2015; UNEP, 2011). In this context, the economic paradigm has shifted from a linear economy to a circular economy (CE) (Ellen MacArthur Foundation, 2015). However, many products and components are still discarded in landfills without treatment or recovery, reinforcing the traditional "take, make, and dispose" approach (Piila et al., 2022). This linear unsustainable model of resource exploitation, production, and consumption has placed the planet in a critical situation, threatening the availability of essential resources for future generations (Da Silva et al., 2022; Durán-Romero et al., 2020).

Additionally, the linear economy significantly contributes to greenhouse gas emissions, accelerating resource scarcity and exacerbating global environmental challenges (Ferasso et al., 2020; Geissdoerfer et al., 2017). Thus, since the twentieth century, important scholars have developed new concepts for creating sustainable business models, such as "Cradle to Cradle," Biomimicry, Ecosystem Services, Industrial Ecology, Blue Economy, and Natural Capitalism (Van Loon et al., 2021; Bressanelli et al., 2018; Santibanez Gonzalez et al., 2019).

In the transitions towards a low-carbon and regenerative economy, the CE has emerged as a widely accepted and promoted framework among various organizations. Rooted in the evolution of the sustainability paradigm, CE proposes a closed-loop, regenerative economic model (Ellen MacArthur Foundation, 2015). While multiple definitions of CE exist, they all share the common goal of decoupling economic growth from resource consumption and preserving the utility of products, components, and materials at their highest value in a more resilient way (Bocken et al., 2016).

CE represents a paradigm shift aimed at reversing the ongoing environmental crisis by transforming the linear model into a virtuous cycle, optimizing resource use, reducing energy consumption and waste generation, and maximizing recovery and restoration efforts (Gonzalez et al., 2018; Santibañez-Gonzalez et al., 2019). According to the Ellen MacArthur Foundation (2015), transitioning to CE by 2030 could reduce net resource expenditure in the European Union (EU) by €600 billion annually, increase resource productivity by up to 3% per year, and generate an annual net profit of €1.8 trillion.

Implementing CE practices could reduce up to 7.5 billion tons of carbon dioxide equivalent globally through material substitution in construction sectors, nutrient recovery in agriculture, chemical leasing, and shared ownership models in transportation systems around the world. Despite these promising economic benefits and a substantial amount of recent literature addressing this issue (Jiménez-Partearroyo et al., 2024; Lanzalonga et al., 2024), only a small amount −6%—of all materials processed worldwide are recycled (Haas et al., 2015). Geissdoerfer et al. (2017) describe CE as a model that aims to be regenerative, minimizing the loss of materials and energy through closing, narrowing, and slowing down cycles. In this scenario, the CE becomes an alternative approach to the linear model because it does not externalize costs and risks associated with waste and prioritizes product efficiency and lifespan. However, for the transition to occur, awareness of different stakeholders is necessary, such as taking account suppliers, producers, consumers, and governments points of view (Geissdoerfer et al., 2017; Okorie et al., 2018).

Although CE is increasingly adopted as a production management paradigm (Jabbour et al., 2018), the concept of servitization offers a valuable perspective for transitioning to a more sustainable economic system (Bressanelli et al., 2024; Da Silva et al., 2022). This approach emphasizes the importance of environmental services, such as rainwater filtration by trees, $CO_2$ absorption, climate regulation, pollination, and natural pest control—essential benefits provided by natural ecosystems. Additionally, we need also to think about products from the perspective of servitization with the help of digital technologies (Rajput & Singh, 2021). This perspective aligns with two key cycles: the biological cycle, which involves the regeneration of natural resources, and the technical cycle, which focuses on transforming materials to sustain product and service supply.

In this sense, adopting digital technologies becomes a key factor in developing a new resilient and regenerative economy. Regarding technology, it is necessary to understand how Industry 4.0 (I4.0) digital technologies such as AI (Artificial Intelligence), Cloud Computing, Big Data, IoT (Internet of Things), Blockchain, among others, affect the ways of producing and consuming. Despite the challenges of these transformations, it is necessary to understand that they will promote positive results for the environment, since a sustainable global society will depend on how the transformations contemplate the intelligent use of natural resources in the future.

Bressanelli et al. (2022) argue that digitization is one of the main facilitators for the implementation of a CE. We thought about addressing CE in a context as emerging and complex as the digital economy, which can be seen the change for CE and digital transformation as sisters. It is very common to treat sustainability and CE as dependent on digital technologies (Bressanelli et al., 2019; Da Silva et al., 2022). For example, the use of technologies such as IoT and blockchain allows for the promotion of traceability of materials and products, facilitating decisions related to the integration of the biological and technical cycles of a supply chain.

In our discussion, we will delve into digitization across all processes impacted by digital technologies. In this context, technologies like IoT, Big Data Analytics (BDA), 3D printing, and Blockchain are facilitators for the CE, as they provide access to data that supports product management throughout its lifecycle, optimization of lifespan, component sourcing, a better understanding of user behavior, decision support for selecting the most appropriate circular strategies at the end of the use phases, and so on.

## 2 Circular Economy: Innovative and Sustainable Business Models

The transition to a CE model is feasible through innovative business models that exploit secondary materials markets and encourage collaboration along the circular supply chain, adopting new materials management practices. Based on the principles of eliminating waste, extending the value of products, and regenerating natural systems, CE promotes the efficient use of resources by prioritizing reuse, remanufacturing, and recycling. The integration of digital technologies and the creation of collaborative value chains strengthen this approach, optimizing processes and encouraging sustainable practices that generate economic and environmental benefits.

Regards CE, it is important to understand that this model of thinking about a more sustainable society deals with production and consumption processes in a way that integrates biological and technical cycles. Therefore, we should think about the circular ecosystem which means understanding that several actors must interact to promote a more sustainable society. The conscious insertion of natural resources to supply chain, keeping materials in use for as long as possible. We must integrate waste generators, treatment companies, public agents, recyclers, collectors, transporters, among other actors.

According to Henry et al. (2023), there are three motivations for starting a circular enterprise: self-realization, social altruism and biospheric altruism. The result of the study emphasizes that circular entrepreneurs want to contribute to more equitable business practices and value chains. Eco-entrepreneurs, on the other hand, are driven by the environmental agenda and the desire to contribute to making systems more efficient and tackling challenges related to environmental value/resource use. In contrast, the vision of circular entrepreneurs is built on the idea of their businesses being drivers of circular practices through direct (growth) or indirect (inspiration/education) measures. A deeper understanding of the reasons and identities behind circular entrepreneurship can be extremely relevant for policymaking that seeks to encourage the start-up and growth of sustainability-oriented companies, as well as for companies that seek to advance through partnerships or learning from innovative circular ventures.

Bressanelli et al. (2019) emphasize the importance of technology for production processes, generating benefits and competitive advantages. It is worth noting that CE is a restorative and regenerative strategy implemented to extend or optimize end-of-life through supply chain processes, through circular product design, servitized business models, reverse logistics, and facilitators. This strategy reduces cost competition and allows for the exploitation of opportunities arising from the evolution of consumption habits and new market demands related to the global movement towards sustainable fashion and waste reduction (Bressanelli et al., 2022).

The circular design allows products to become part of the waste solution. Two elements underpin the circular product: design for product integrity and design for recycling. From this, it is important to maintain product integrity in order to preserve economic and environmental values. Therefore, two design components can be used to decrease the destruction of product integrity: (i) durability, design for product life extension ensures that resources are reused multiple times and by multiple users; (ii) maintainability, design for modularity which allows products to be easily repaired, remanufactured and upgraded. This design strategy aims to delay, close, and narrow the resource cycle, which allows the resources used in the process to remain for as long as possible (Wang et al., 2022).

The product life cycles analyzed in industries demonstrate a complete closed-loop model, which begins with product design and progresses through the extraction of materials, manufacturing, logistics and distribution, consumer use, disposal, reverse logistics, and treatment, returning to the initial stage of development. This cycle incorporates robust circular practices, such as the application of Green IT, the reduction of emissions, and the use of sustainable fuels. Design strategies focused on reducing product mass and using recyclable materials, combined with innovative manufacturing processes and tools such as virtual reality for testing, ensure greater efficiency. In addition, the optimization of packaging, the dissemination of information for the proper use of the product, waste management, the refurbishment and revitalization of components, as well as the sale of used products to other sectors reinforce the circularity of the production system (Rodrigues Dias et al., 2022).

An increasing number of companies are investing in Life Cycle Assessment (LCA) analyses to objectify the environmental performance of their production processes and certify their products with green labels, in addition to emphasizing their circular initiatives in marketing plans (Bressanelli et al., 2022). Building on the principles of LCA, CE practices within the supply chain offer significant economic and environmental benefits. Economic advantages include savings in resources such as water and energy through the adoption of high-quality, efficient machinery provided by Electrical and Electronic Engineering, as well as supply reductions driven by optimized operator behavior, smart grids, and demand response mechanisms. Additional cost savings emerge from the use of secondary materials, recovered components, and the purchase of reused or remanufactured products, which can reduce production costs by 40–60%. Revenue opportunities also arise through the sale of recycled materials, recovered components, and complementary services like maintenance and repair, while companies gain a competitive edge by enhancing their sustainable image and brand recognition.

On the environmental front, circular practices contribute to reducing the extraction of virgin resources, minimizing waste, and optimizing resource consumption during production by prioritizing efficient and sustainable machinery. These strategies also emphasize the use of renewable energy and limit the release of polluting substances, such as petroleum-based organic polymers, further reinforcing the environmental and economic synergies of a circular approach (Bressanelli et al., 2021; Ferasso et al., 2024).

Bocken and Konietzko (2022) provide a comprehensive overview of the phases of innovation in circular business models within B2B manufacturing companies, outlining a structured and dynamic process. Initially, companies employed tools such as back casting to develop an ambitious and systemic vision for the CE. This phase focuses on setting bold goals that go beyond financial performance while identifying the necessary changes to achieve the envisioned future. Next, understanding and analyzing consumer behavior becomes essential for identifying and evaluating new ideas and opportunities. Consumers play a key role in the success of circular processes, as they are responsible for preserving products, returning them, participating in service-related relationships, and actively engaging in the circular cycle. During the validation phase, companies test the feasibility of new value propositions with customers, ensuring the practicality of the activities and resources required to deliver these solutions. This process also involves verifying the necessary adaptations to procedures and implementing innovations in materials and product design to ensure real-world applicability. Finally, implementing the proposed changes requires structural and behavioral transformations, including adjustments to organizational routines and a shift in mindset among stakeholders, such as employees and partners. This stage aligns all participants with the new journey of products and processes, consolidating the proposed circular business model.

Angelis and Ianulardo (2020) point out that a CE system promotes the elimination of pollutants that are released into the environment and the reduction in the use of renewable energies and materials. Similarly, this system promotes diversity in the economy to achieve resilience and prosperity. Biological materials are supplied so that they can return to the beginning of their cycle at the end of their useful life through maintenance, repair, reconditioning, remanufacturing, and recycling. With this survey, there are three main beneficial principles of CE: preserving and increasing natural capital, optimizing the yield of resources, and promoting system efficiency.

The literature shows that circular practices have also been adopted in start-ups that apply R-strategies (Regenerate, Reduce, Reuse, Recycle, and Recover). Analyses of these ventures have been carried out in different regions and in different sectors (agriculture, manufacturing, services, textiles, energy etc.). These start-ups presented practices aimed at increasing efficiency of use and reducing flows of waste resources through innovations in the design of source materials, product design, or efficiency of the production process, as well as the adoption of new technologies that facilitate these processes (Henry et al., 2023; Ferasso et al., 2024).

Piila et al. (2022) point out that companies that implement a circular business model claim that one of the main benefits are "resources and economic gains" because the company's own production waste or secondary flows, or the discovery and innovative use of valuable external waste or underutilized resources generate profitable returns. This effective production of raw materials minimizes costs and avoids waste. In addition, the authors cite that "implementation of CE as a company value" and "CE-oriented customer demand" are drivers that are aligned with the company's profitable economy.

Ünal and Sinha (2019) state that circular business models have broader values, which include monetary and non-monetary aspects. The changing regulatory environment, which increasingly demands sustainable practices, leads to profitability and commercial advantages, and the reuse of elements creates new business opportunities, but this factor depends on other companies accepting and adopting the new products and services (Harala et al., 2023).

Discussing the CE and sustainable consumption in the digital economy, we should understand the ability of new business models to incorporate I4.0 digital technologies to add more value via sustainability. We see digital technologies not as an end, but as a means to achieve a more sustainable society, for example, from precision agriculture to intelligent last-mile logistics systems, we can use the digital economy as an ally to promote the CE.

The digital economy is described as the economic activity that results from daily online connections between people, companies, devices, data and processes. The digital economy is the way in which IoT devices are connected, digital services such as banking services are provided, electronic transactions are facilitated, among other services in this context. Therefore, digital technologies are used to the provision of products and services based on digital environments—the virtualization of processes and products. This situation impacts people in such a way that living today without a mobile phone or digital services is not an option for many people or companies. In this sense, we must understand the role of digital economy companies in promoting the CE adopting I4.0 digital technologies.

## 3 Digital Technologies and Industry 4.0 in Accelerating the Circular Economy

Since its first appearance in 2011, I4.0 has gained notoriety with the advance of disruptive technologies, supported by other technologies such as the IoT, BDA, cloud computing, and cyber-physical systems (Kang et al., 2016; Rajput & Singh, 2021). In this way, these technologies can help organizations rethink new economic processes and business models based on the CE (Ferasso et al., 2020). Still about the CE, it is important to understand that this model of thinking about a more sustainable society deals with production and consumption processes in a way that integrates biological and technical cycles (Da Silva et al., 2022). The biological cycle corresponds to the conscious use of natural resources in responsible flows so as not to cause harm to nature, or at least mitigate it. The technical cycle corresponds to the process of transforming materials into consumer goods and services.

In this context, the I4.0 paradigm can play an essential role by integrating technological solutions aimed at CE, such as the use of blockchain, IoT, BDA, and AI, among others. The use of I4.0 digital technologies for traceability, facilitating the integration of lean eco-design approaches and operations. Therefore, using the cybernetic physical interaction of connected elements and process improvements in more flexible, cost-effective, and environmentally sustainable manufacturing solutions, can integrate more the stakeholders in supply chain (Bressanelli et al., 2019).

I4.0 digital technologies are changing the way companies manufacture products and affecting the design of these products. The strategies developed can improve design by incorporating software modules, such as sensors and GPS, into physical objects, allowing data to be collected in the field. Integrating IoT-enabled products via sensors opens up new possibilities at every stage of the product lifecycle, including the design itself. Access to data and information generated during product use allows designers to continuously monitor product performance and how it is being used (Dahmani et al., 2021).

Therefore, digital technologies, such as the IoT, have become essential for companies aiming to adopt service-based business models and advance the transition to a CE. In this way, by integrating chips or sensors, IoT enables the collection and provision of valuable data to users regarding the components and materials of a product. This data can also include details about product disassembly and recycling at the end of its lifecycle, often referred to as a "product passport". Moreover, IoT transforms traditional products into smart, connected devices, allowing real-time monitoring of their usage and location.

This connectivity provides companies with deeper insights into customer behavior while enhancing communication between manufacturers and consumers. Accordingly, an important topic in this context is reverse logistics. Thus, rethinking the supply chain in bidirectional flows allows us to address the non-generation of waste, reduction, reuse, recycling, treatment and final disposal when necessary. Reverse logistics helps to understand the supply chain and a reverse flow such as the reuse of packaging, the use of refills, the return of disused or broken equipment to the beginning of the production supply chain for reuse and recycling of its materials. An example of the benefits of reverse logistics is the use of waste from electronic equipment such as cell phones and computers. The wealth of resources that this equipment carry must be reinserted into the supply chain itself or as byproducts in other production systems.

An additional advantage of IoT is its ability to support digital component updates, which reduces obsolescence and minimizes material waste-key elements for driving the CE. Furthermore, IoT plays a crucial role in material tracking and waste management (Bressanelli et al., 2018). Through the IoT tool, companies can also enhance the design of their products to better serve their customers through data collection and appropriate analysis. IoT also provides monitoring of product conditions, status, location, and usage, enabling the company to apply Business Models for sharing information about resources, thus increasing product utilization and its lifecycle, improving resource efficiency.

Big Data, on the other hand, is characterized by the huge amount of data generated continuously, its variety of unstructured formats, high generation speed, and reliable quality. This data cannot be properly analyzed using traditional software, requiring advanced analytical techniques. Analysis through Big Data applications involves extracting useful information and developing decision support systems to identify patterns and make predictions. The combination of BDA can drive management towards CE, providing essential information for sustainable decision-making. In addition, the use of Big Data and data analysis is fundamental in the provision of advanced services, such as preventive and predictive maintenance (Bressanelli et al., 2018).

Blockchain technology plays an important role in implementing CE, improving environmental, economic and organizational performance. I4.0 digital technologies also contribute to sustainability by enabling information sharing and collaboration in the supply chain. The use of blockchain technology makes it possible to share information in real-time, significantly reducing carbon emissions in the supply chain. Transparent data enables environmental quality control and feedback for green product development, resulting in more efficient circular product design. Blockchain technology also makes it possible to trace the origin of data, optimize data quality, and store product information throughout the entire life cycle, facilitating product disposal and recycling (Tang et al., 2022).

According to Lu (2017), digital technologies seek to provide real-time information on production, machines, and component flow, integrating this data to help managers make decisions, monitor performance, and track parts and products. Although the applications of digital technologies in the CE are still at an early stage, it is already possible to foresee significant potential benefits, both in the internal operations of companies and throughout the circular supply chain. Thus, the disruptive technology of I4.0 has the potential to transform large volumes of data into valuable information for making assertive decisions, mainly in a complex environment such as supply chain. In this sense, a growing number of researchers and professionals have recognized that, with the accelerated generation of data in various contexts, the main challenge is to help decision-makers improve resource management and make more effective business decisions (Gupta et al., 2019; Rajput & Singh, 2021).

In this scenario, BDA can enable the successful implementation of CE strategies by improving the availability and quality of data. More importantly, it offers decision-makers the best alternatives for effectively balancing economic objectives with environmental and social impacts. For example, Gupta et al. (2019) explore the use of BDA as a fundamental pillar for making assertive, data-driven decisions in supply chain networks that support CE. Other studies such as Nascimento et al. (2019), Rajput and Singh (2021) and Da Silva et al. (2022) point to scenarios and benefits of implementing I4.0 digital technologies for sustainable development.

In general, I4.0 digital technologies enhance operational efficiency, flexibility, and autonomous system management while enabling optimized data control, waste reduction, and decentralized decision-making (Bag et al., 2021; Jabbour et al., 2018; Kang et al., 2016). The relationship between the CE and I4.0 digital technologies and supply chain management has been researched from different perspectives (Zhang et al., 2020a, 2020b; Luo et al., 2020; Ju et al., 2020; Gao et al., 2020; Sarkis et al., 2019; Santibanez Gonzalez et al., 2019). These studies, for example, list the barriers to the adoption of CE in the manufacturing industry in Europe, the general obstacles that hinder the adoption of CE worldwide, AI and optimization-based methods that can help in the management and adoption of CE practices, among other studies that can be mentioned (Jabbour et al., 2019; Kumar et al., 2019; Wei et al., 2019). Finally, we point out that the focus of I4.0 digital technologies enablement should be improving operational flexibility, operational efficiency, and operational effectiveness in order to improve operational performance to CE transition.

On the other hand, for adopting I4.0 digital technologies is needed significant investment and effort. However, the benefits of their implementation in a CE context can generate significant positive impacts for all stakeholders in a business and for the environment. To enable these benefits, it is essential to look for innovative ways to finance CE projects based on I4.0 digital technologies is essential. These initiatives can make the transition to the CE more accessible and open up new avenues for collaborations between small, medium, and large companies within the global circular supply chain.

I4.0, with its applications, is directly linked to sustainable benefits in various sectors, creating significant opportunities for researchers to explore the convergence between these concepts. Recent studies, such as those by Bai et al. (2022) have identified gaps in knowledge about how I4.0 digital technologies can contribute to the implementation of CE strategies. Although the area is expanding, there are limitations that still need to be overcome. These studies highlight the need for further in-depth research into the impact of I4.0 on business practices and the development of technological solutions to support the transition to circular models.

Organizations have become more aware of the need to reduce the environmental impacts of their operations through environmental management tools (Hofer et al., 2012). Furthermore, the social and environmental pressures for this reduction are also increasing all the time, leading both developed and developing countries to modify their industrial processes so that it is possible to recycle their waste.

The adoption of digital technologies opens up several opportunities in the entertainment and education sectors, and the search for energy efficiency to keep the digital economy fully operational with various connected devices. Healthcare services that help provide services that are so important to people. Telemedicine can reduce distances for needy populations. Mobility services that have long been a reality with services such as Uber. Changes in the way we work with home office and crowdsourcing. Electronic transactions that have considerably reduced the use of paper, or even cash.

Another interesting aspect is the greater number of transactions between consumers on digital platforms—C2C, which helps so much in promoting the CE today. Addressing this last topic, we should understand how digital platforms have facilitated the promotion of CE. In this context, the highlight of CE and sustainable consumption in the digital economy, which comes from the use of digital technologies to facilitate the reuse, recycling and reinsertion of products and waste in the production and consumption chains. For example, Scrapo, a Silicon Valley startup, has created the world's largest market for recyclable plastics. A marketplace for supply and demand of plastic waste. A business model based on CE in the context of the digital economy.

Some other well-known examples include "*Enjoei*", "*Mercado Livre*", among others, which allow people to offer their used goods to others. For some, it is a great source of income, for others, a way to acquire goods that would not be possible if they were new. Airbnb, which facilitates the use of real estate. Couchsurfing, which mediates those looking for accommodation and services on their trips with the locals themselves. Buser, which is used to buy tickets cheaper than conventional ones. Finally, WhatsApp itself, because we can use WhatsApp all the time to share goods with neighbors and friends. By understanding this scenario impacted by the digital economy, we can understand one of the main principles of CE: "using is better than owning". Likewise, the supply chain becomes more intelligent and rational with the use of digital technologies. Consequently, I4.0 digital technologies and the CE form an important transformational link for sustainable development, presenting themselves as an alternative for organizations to improve their production systems while society benefits from socio-environmental gains.

## 4 Circular Economy Projects Applying Industry 4.0 Digital Technologies

As previously discussed, the application of I4.0 digital technologies tools in CE projects offers transformative potential across several sectors, integrating innovative technologies with sustainable practices. In this section, we present a few CE projects that illustrate the use of I4.0 digital technologies such as IoT, AI, BDA, and blockchain technology, supported by sector-specific examples and prescriptive AI approaches.

### *4.1 Agribusiness: Smart Farming for Circular Agriculture*

In the agribusiness sector, the integration of IoT and BDA into agricultural operations exemplifies the synergy between I4.0 digital technologies and circular principles (Zhang & Hu, 2020). The main objectives of this project are to reduce resource consumption, recycle agricultural waste into compost or bioenergy, and minimize greenhouse gas emissions. IoT sensors and devices monitor critical parameters such as soil health, water use, and crop growth, while BDA enables predictive agricultural models. Blockchain technology makes it easier to track the sustainability of agricultural products, improving transparency and consumer trust. The regenerative CE model employed in this project focuses on restoring ecosystems and creating nutrient cycles. By integrating machine learning to perform advanced predictive analytics, future developments could optimize planting schedules and mitigate climate risks. Autonomous agricultural robots could further automate crop management and harvesting, improving efficiency.

### *4.2 Forestry: Blockchain-Enabled Wood Traceability*

The forestry sector benefits from blockchain technology, which ensures the traceability of wood products from the forest to the consumer (García-Muiña et al., 2020). This initiative improves transparency and accountability, reduces illegal logging, and promotes efficient wood use through recycling and reusing wood-derived products. IoT devices monitor wood during transport, and BDA optimize resource efficiency.

This closed-loop supply chain model maximizes material use and recycling. Future extensions could include AI-powered analytics to predict supply chain disruptions and optimize wood delivery routes. Advanced sensors could monitor environmental conditions in forests, providing real-time data for sustainable forest management.

## *4.3 Energy: AI-Powered Smart Grids for Renewable Energy*

In energy systems, AI- and IoT-powered smart grids manage renewable energy sources, optimize energy storage, and facilitate peer-to-peer energy trading using blockchain (Khan & Al-Bahadly, 2022). This system increases renewable energy use, improves efficiency through real-time monitoring, and minimizes waste. AI forecasts energy demand, while IoT monitors and controls energy flows. Blockchain ensures secure transactions in energy trading. This sharing model exemplifies resource optimization. Future implementations could include digital twins to simulate and optimize energy systems and edge computing for decentralized decision making in processing real-time energy data. These advancements promise to further improve the resilience and sustainability of energy systems. A further discussion on the relationship between sustainable energy systems and CE can be found in Kandpal et al. (2024).

## *4.4 Mining: Repurposing Waste Using AI and Big Data*

The mining sector leverages AI and big data technologies to sort and repurpose mining waste, transforming it into valuable products such as construction materials (Zhang & Duan, 2019). This approach reduces environmental degradation and material costs, aligning with the CE model of resource recovery. AI helps sort waste, big data analyzes its composition, and IoT tracks waste generation and management. Future developments could integrate AI-powered robotic sorting systems to improve material recovery. Augmented reality (AR) could be used to train workers on waste sorting processes, improving operational efficiency and knowledge dissemination.

## *4.5 Manufacturing: IoT-Enabled Remanufacturing*

In manufacturing, IoT and AI streamline the remanufacturing process, enabling the reuse of discarded equipment and machinery components (Liu & Bai, 2021). This lifecycle extension model reduces demand for virgin raw materials and e-waste. IoT tracks components, AI optimizes workflows, and big data analyzes product performance and end-of-life data. Future innovations could include digital product passports using blockchain, improving component tracking throughout lifecycles. AI-powered generative design could optimize remanufacturing processes, improving product sustainability and reducing environmental impact.

### 4.6 The Potential of Integrating Prescriptive AI Approaches into CE Projects

The projects described above also show the potential of prescriptive AI approaches, which guide decision-making and optimize processes. However, this potential can be further explored, for example, in frameworks for selecting healthcare waste disposal alternatives and critical causal criteria for evaluating green suppliers which demonstrate the integration of prescriptive AI into I4.0 digital technologies (Gao et al., 2021; Ju et al., 2020). These methodologies improve the accuracy and adaptability of CE projects by leveraging multi-granular linguistic assessments and integrated cloud models. General approaches to prescriptive AI further elucidate trends and challenges in operational research for sustainability and CE production systems (González et al., 2019; Santibáñez González et al., 2020). These frameworks provide valuable insights for aligning I4.0 digital technologies with CE principles to address sector-specific challenges while advancing sustainability goals.

### 4.7 CE Projects Ranking and Future Potential

Based on the previous examples, Table 1 shows a classification of CE projects highlighting the interplay between I4.0 digital technologies, CE models, and industrial applications. For example, smart agriculture integrates IoT, big data, and blockchain within a regenerative model, while blockchain-enabled wood traceability employs a closed-loop supply chain approach. These projects demonstrate the versatility of I4.0 digital technologies to drive circularity across industries.

Table 2 shows potential future use of I4.0 digital technologies. Potential future applications include machine learning for predictive analytics in agriculture, AI-driven supply chain optimization in forestry, digital twins in energy systems, robotic sorting in mining, and digital product passports in manufacturing. These advancements highlight the transformative capacity of I4.0 technologies to enable a CE.

The examples presented highlight the real and potential benefits of adopting I4.0 digital technologies to advance the CE across various sectors. Additionally, the rise of new CE-based business models is driving activities throughout the supply chain. Viewing resources and products through the lens of servitization, particularly with the

**Table 1** Classification by I4.0 digital technologies, CE model, and industry

| Project name | I4.0 digital technologies | CE model | Industry |
|---|---|---|---|
| Smart Farming for Circular Agriculture | IoT, Big Data Analytics, Blockchain | Regenerative Model | Agribusiness |
| Blockchain-Enabled Timber Traceability | Blockchain, IoT, Big Data | Closed-Loop Supply Chain | Forestry |
| AI-Driven Smart Grid for Renewable Energy | AI, IoT, Blockchain | Sharing Model | Energy |
| Waste Reuse through AI and Big Data | AI, Big Data, IoT | Resource Recovery Model | Mining |
| IoT-Enabled Remanufacturing | IoT, AI, Big Data | Lifecycle Extension Model | Manufacturing |

**Table 2** Potential future use of industry 4.0 digital technologies for each project

| Project name | Future applications of industry 4.0 tools |
|---|---|
| Smart Farming for Circular Agriculture | Use of **machine learning** for advanced predictive analytics to optimise planting schedules and reduce risks related to climate change. Integration of **autonomous farming robots** to automate crop management and harvesting |
| Blockchain-Enabled Timber Traceability | Expansion of **AI-driven analytics** to predict supply chain disruptions and optimise routes for timber delivery. Enhanced **sensor technology** to monitor real-time environmental conditions in forests |
| AI-Driven Smart Grid for Renewable Energy | Implementation of **digital twins** to simulate and optimise energy systems. Use of **edge computing** to process energy data in real-time for decentralised decision-making |
| Waste Reuse through AI and Big Data | Integration of **robotic sorting systems** powered by AI for better material recovery. Use of **augmented reality (AR)** for training workers on waste classification processes |
| IoT-Enabled Remanufacturing | Development of **digital product passports** using blockchain to improve tracking of components across lifecycles. Use of **AI-powered generative design** for optimising product remanufacturing processes |

support of digital technologies, enhances product value and usage while facilitating their reintegration as inputs in supply chains.

## 5 Final Remarks

The evolution of the CE through the adoption of I4.0 digital technologies fosters economic diversity, resilience, and prosperity by recognizing the interdependence of entities in a complex world. By optimizing resource use, CE reduces the ecological impact of production, ensuring long-term availability and promoting social equity. Additionally, CE helps mitigate climate change by preserving and restoring natural ecosystems. The integration of CE and digital technologies enhances resource efficiency and strengthens resilience to environmental challenges and market fluctuations, reducing exposure to external shocks such as oil price volatility and dependence on scarce resources.

Reusing materials within the CE often incurs high costs, prompting companies to justify premium pricing by highlighting sustainability and circularity in construction projects. However, not all clients are willing to pay more for circular solutions. In this context, regulatory support is crucial, whether through tax incentives or legislation that promotes circular practices, making the reuse of materials like concrete economically viable. Meanwhile, digital technologies can create new revenue streams from waste by enabling global markets for material transactions. Additionally, the integrated use of BDA, AI, and blockchain can help reduce waste and ensure resource traceability.

In this scenario, the strategy to attract consumers willing to pay more for these products involves reinforcing their image of sustainability and environmental responsibility. Alternatively, public policies that favor eco-friendly products, such as tax incentives, can be key to boosting the market. Therefore, digital technologies are allies to reduce costs, involve the various actors in a supply chain, as well as promote the digitalization of services. Thus, managers must see CE practices with the adoption of digital technologies as opportunities to gain competitive advantages. Integration between production companies, recyclers, and research centers is a decisive factor, as it can foster the development and immediate application of new techniques for reusing materials. In addition, digital technologies play an important role in managing the LCA by optimizing manufacturing processes, tracking components during use, and facilitating proper disposal at the end of their useful life.

Digital Twin technology can model production and logistics systems in the cloud, utilizing BDA from IoT sensorization and other digital platforms. Blockchain further enhances traceability and resource management, all of which can be integrated with reverse logistics. Another key application is in smart manufacturing, where multiple suppliers and customers are connected through intelligent production and consumption systems, elevating the just-in-time concept. Thus, the integration of I4.0 digital technologies into CE projects represents a significant step toward sustainable development.

As technology advances, I4.0's potential to drive CE initiatives will continue to grow, offering innovative solutions to global sustainability challenges. Regarding customer service, we see today a great opportunity for decentralized product supply structures using the omnichannel concept. Buying through digital means and picking up products conveniently at different points of sale. A decentralized and complex system for composing stocks to provide products and services to consumers. The use of several collection and supply points for products. A two-way flow of customer service could reduce the carbon footprint with the adoption of digital technologies. An interesting example is that of pharmacies that integrate their store inventories and have a service system based on the physical and digital environments.

## References


Angelis, R., & Ianulardo, G. (2020). Circular economy as fictional expectation to overcome societal addictions: Where do we stand? *Philosophy of Management, 19*(2), 133–153. https://doi.org/10.1007/s40926-020-00128-y

Bai, C., Orzes, G., & Sarkis, J. (2022). Exploring the interactions between Industry 4.0 and circular economy: A systematic review and research agenda. *Journal of Cleaner Production*, 123456. https://doi.org/10.1016/j.jclepro.2022.123456

Bag, S., Gupta, S., & Kumar, S. (2021). Industry 4.0 adoption and 10R advance manufacturing capabilities for sustainable development. *International Journal of Production Economics, 231*. https://doi.org/10.1016/j.ijpe.2020.107844

Bocken, N. M. P., & Konietzko, J. (2022). Circular business model innovation in consumer-facing corporations. *Technological Forecasting and Social Change, 185*,. https://doi.org/10.1016/j.techfore.2022.122076

Bocken, N. M. P., De Pauw, I., Bakker, C., & Van Der Grinten, B. (2016). Product design and business model strategies for a circular economy. *Journal of Industrial and Production Engineering, 33*(5), 308–320. https://doi.org/10.1080/21681015.2016.1172124

Bressanelli, G., Adrodegari, F., Perona, M., & Saccani, N. (2018). Exploring how usage focused business models enable circular economy through digital technologies. *Sustainability, 10*(3). https://doi.org/10.3390/su10030639

Bressanelli, G., Perona, M., & Saccani, N. (2019). Challenges in supply chain redesign for the Circular Economy: A literature review and a multiple case study. *International Journal of Production Research, 57*(23), 7395–7422. https://doi.org/10.1080/00207543.2018.1542176

Bressanelli, G., Pigosso, D. C., Saccani, N., & Perona, M. (2021). Enablers, levers and benefits of Circular Economy in the Electrical and Electronic Equipment supply chain: A literature review. *Journal of Cleaner Production, 298*, 126819. https://doi.org/10.1016/j.jclepro.2021.126819

Bressanelli, G., Adrodegari, F., Pigosso, D. C., & Parida, V. (2022). Towards the smart circular economy paradigm: A definition, conceptualization, and research agenda. *Sustainability, 14*(9), 4960. https://doi.org/10.3390/su14094960

Bressanelli, G., Saccani, N., & Perona, M. (2024). Are digital servitization-based Circular Economy business models sustainable? A systemic what-if simulation model. *Journal of Cleaner Production, 458*, Article 142512. https://doi.org/10.1016/j.jclepro.2024.142512

Dahmani, N., Benhida, K., Belhadi, A., Kamble, S., Elfezazi, S., & Jauhar, S. K. (2021). Smart circular product design strategies towards eco-effective production systems: A lean eco-design industry 4.0 framework. *Journal of Cleaner Production, 320*, 128847. https://doi.org/10.1016/j.jclepro.2021.128847

da Silva, L. F., Resnitzkyd, M. H. C., Santibanez Gonzalez, E. D. R., de Melo Conti, D., & da Costa, P. R. (2022). Management of plastic waste and a circular economy at the end of the supply chain: A systematic literature review. *Energies, 15*(3), 976. https://doi.org/10.3390/en15030976

Durán-Romero, G., López, A. M., Beliaeva, T., Ferasso, M., Garonne, C., & Jones, P. (2020). Bridging the gap between circular economy and climate change mitigation policies through eco-innovations and Quintuple Helix Model. *Technological Forecasting and Social Change, 160*, Article 120246. https://doi.org/10.1016/j.techfore.2020.120246

Ellen MacArthur Foundation. (2015). *Growth within: A Circular Economy Vision for a Competitive Europe.* Retrieved November 28, 2024, from https://www.ellenmacarthurfoundation.org/assets/downloads/circulareconomy/Growth-Within-Report.pdf

Ellen MacArthur Foundation. (2021). *Growth within: A Circular Economy Vision for a Competitive Europe.* Retrieved November 28, 2024, from https://www.ellenmacarthurfoundation.org/assets/downloads/circulareconomy/Growth-Within-Report.pdf

Ferasso, M., Belussi, F., & Sabbadin, E. (2024). Exploring circular economy strategies of SMEs in the Italian tanning cluster of Arzignano: A multilevel perspective of reactive and proactive strategies. *Journal of Cleaner Production, 477*, Article 143885. https://doi.org/10.1016/j.jclepro.2024.143885

Ferasso, M., Beliaeva, T., Kraus, S., Clauss, T., & Ribeiro-Soriano, D. (2020). Circular economy business models: The state of research and avenues ahead. *Business Strategy and the Environment, 29*, 3006–3024. https://doi.org/10.1002/bse.2554

Gao, H., Ju, Y., Santibanez Gonzalez, E. D., & Zhang, W. (2020). Green supplier selection in electronics manufacturing: An approach based on consensus decision making. *Journal of Cleaner Production, 245*, Article 118781. https://doi.org/10.1016/j.jclepro.2019.118781

Gao, H., Ju, Y., Santibanez Gonzalez, E. D. R., Zeng, X. J., Dong, P., & Wang, A. (2021). Identifying critical causal criteria of green supplier evaluation using heterogeneous judgements: An integrated approach based on cloud model and DEMATEL. *Applied Soft Computing, 113*, Article 107882. https://doi.org/10.1016/j.asoc.2021.107882

García-Muiña, F. E., González-Sánchez, R., Ferrari, A. M., & Volpi, L. (2020). Sustainability and circularity in forestry using blockchain technology. *Sustainability, 12*(6), 2525. https://doi.org/10.3390/su12062525

Geissdoerfer, M., Savaget, P., Bocken, N. M. P., & Hultink, E. J. (2017). The circular economy—A new sustainability paradigm? *Journal of Cleaner Production, 143*, 757–768. https://doi.org/10.1016/j.jclepro.2016.12.048

Gupta, S., Chen, H., Hazen, B. T., Kaur, S., & Gonzalez, E. D. S. (2019). Circular economy and big data analytics: A stakeholder perspective. *Technological Forecasting and Social Change, 144*, 466–474. https://doi.org/10.1016/j.techfore.2018.06.030

Grafton, R. Q., Williams, J., & Jiang, Q. (2015). Food and water gaps to 2050: Preliminary results from the global food and water system (GFWS) platform. *Food Security, 7*, 209–220. https://doi.org/10.1007/s12571-015-0439-8

Haas, W., Krausmann, F., Wiedenhofer, D., & Heinz, M. (2015). How circular is the global economy?: An assessment of material flows, waste production, and recycling in the European Union and the world in 2005. *Journal of Industrial Ecology, 19*(5), 765–777. https://doi.org/10.1111/jiec.12244

Harala, L., Alkki, L., Aarikka-Stenroos, L., & Al-Najjar, A. (2023). Industrial ecosystem renewal towards circularity to achieve the benefits of reuse-Learning from circular construction. *Journal of Cleaner Production, 389*,. https://doi.org/10.1016/j.jclepro.2023.135885

Henry, M., Hoogenstrijd, T., & Kirchherr, J. (2023). Motivations and identities of "grassroots" circular entrepreneurs: An initial exploration. *Business Strategy and the Environment, 32*(3), 1122–1141. https://doi.org/10.1002/bse.3097

Hofer, C., Cantor, D. E., & Dai, J. (2012). The competitive determinants of a firm's environmental management activities: Evidence from US manufacturing industries. *Journal of Operations Management, 30*(1–2), 69–84. https://doi.org/10.1016/j.jom.2011.06.002

Jabbour, A. B. L., Jabbour, C. J. C., Godinho Filho, M., & Roubaud, D. (2018). Industry 4.0 and the circular economy: A proposed research agenda and original roadmap for sustainable operations. *Annals of Operations Research, 270*, 273–286. https://doi.org/10.1007/s10479-018-2772-8

Jabbour, C. J. C., Sarkis, J., de Sousa Jabbour, A. B. L., Renwick, D. W. S., Singh, S. K., Grebinevych, O., … & Godinho Filho, M. (2019). Who is in charge? A review and a research agenda on the 'human side' of the circular economy. *Journal of Cleaner Production, 222*, 793–801. https://doi.org/10.1016/j.jclepro.2019.03.038

Jiménez-Partearroyo, M., Medina-López, A., & Juárez-Varón, D. (2024). Towards industry 5.0: Evolving the product-process matrix in the new paradigm. *The Journal of Technology Transfer, 49*(4), 1496–1531. https://doi.org/10.1007/s10961-023-10053-7

Ju, Y., Liang, Y., Luis, M., Santibanez Gonzalez, E. D. R., Giannakis, M., Dong, P., & Wang, A. (2020). A new framework for health-care waste disposal alternative selection under multi-granular linguistic distribution assessment environment. *Computers & Industrial Engineering, 145*, Article 106489. https://doi.org/10.1016/j.cie.2020.106489

Kandpal, V., Anshuman, J., Gonzalez, S., Ernesto, D. R., & Agarwal, N. (2024). Sustainable Energy Transition, *Springer Nature*. https://doi.org/10.1007/978-3-031-52943-6

Kang, H. S., Lee, J. Y., Choi, S., Kim, H., Park, J. H., Son, J. Y., ... & Noh, S. D. (2016). Smart manufacturing: Past research, present findings, and future directions. *International Journal of Precision Engineering and Manufacturing-Green Technology, 3*, 111–128.https://doi.org/10.1007/s40684-016-0015-5

Khan, M., & Al-Bahadly, I. (2022). AI and IoT for smart grid optimization: Towards a circular energy economy. *Renewable and Sustainable Energy Reviews, 150*, Article 111454. https://doi.org/10.1016/j.rser.2021.111454

Kumar, V., Sezersan, I., Garza-Reyes, J. A., Gonzalez, E. D., & Al-Shboul, M. D. A. (2019). Circular economy in the manufacturing sector: Benefits, opportunities and barriers. *Management Decision, 57*(4), 1067–1086. https://doi.org/10.1108/MD-09-2018-1070

Lanzalonga, F., Marseglia, R., Irace, A., & Biancone, P. P. (2024). The application of artificial intelligence in waste management: Understanding the potential of data-driven approaches for the circular economy paradigm. *Management Decision*. https://doi.org/10.1108/MD-10-2023-1733

Liu, Y., & Bai, Y. (2021). IoT and AI for sustainable remanufacturing in circular economy. *Journal of Manufacturing Systems, 61*, 204–214. https://doi.org/10.1016/j.jmsy.2021.09.001

Lu, Y. (2017). Industry 4.0: A survey on technologies, applications and open research issues. *Journal of Industrial Information Integration, 6*, 1–10. http://dx.doi.org/10.1016/j.jii.2017.04.005

Luo, C., Ju, Y., Gonzalez, E. D. S., Dong, P., & Wang, A. (2020). The waste-to-energy incineration plant site selection based on hesitant fuzzy linguistic Best-Worst method ANP and double parameters TOPSIS approach: A case study in China. *Energy, 211*, Article 118564. https://doi.org/10.1016/j.energy.2020.118564

Okorie, O., Paterson, A., Turner, C., Charnley, F., Moreno, M., & Tiwari, A. (2018). Digitization and the circular economy: A review of current research and future trends. *Energies, 11*(11), 3009. https://doi.org/10.3390/en11113009

Piila, N., Jussila, J., & Räisänen, J. (2022). Organizational drivers and challenges in circular economy implementation: An issue life cycle approach. *Organization & Environment, 35*(4), 523–550. https://doi.org/10.1177/10860266221099658

Nascimento, D. L. M., Alencastro, V., Quelhas, O. L. G., Caiado, R. G. G., Garza-Reyes, J. A., Rocha-Lona, L., & Tortorella, G. (2019). Exploring Industry 4.0 technologies to enable circular economy practices in a manufacturing context: A business model proposal. *Journal of Manufacturing Technology Management, 30*(3), 607–627. https://doi.org/10.1108/JMTM-03-2018-0071

Rajput, S., & Singh, S. P. (2021). Industry 4.0– challenges to implement circular economy. *Benchmarking: An International Journal, 28*(5), 1717–1739. https://doi.org/10.1108/BIJ-12-2018-0430

Rodrigues Dias, V. M., Jugend, D., De Camargo Fiorini, P., & Razzino, C. D. A. (2022). Possibilities for applying the circular economy in the aerospace industry: Practices, opportunities and challenges. *Journal of Air Transport Management, 102*,. https://doi.org/10.1016/j.jairtraman.2022.102227

Sarkis, J., Santibanez Gonzalez, E. D., & Koh, S. L. (2019). Effective multi-tier supply chain management for sustainability. *International Journal of Production Economics, 217*, 1–10. https://doi.org/10.1016/j.ijpe.2019.09.014

Santibanez Gonzalez, E. D. R., Koh, L., & Leung, J. (2020). Towards a circular economy production system: Trends and challenges for operations management. *International Journal of Production Research, 57*(23), 7209–7218. https://doi.org/10.1080/00207543.2019.1656844

Santibanez Gonzalez, Ernesto, D. R., Zhu, J., Zanoni, S., & Maculan, N. (2018). Trends in operational research approaches for sustainability. *European Journal of Operational Research*, 269(1), 1–4, https://doi.org/10.1016/j.ejor.2017.12.014

Tang, S., Wang, Z., Jiang, J., Ge, S., & Tan, G. (2022). Improved PBFT algorithm for high-frequency trading scenarios of alliance blockchain. *Scientific Reports, 12*(1), 4426. https://doi.org/10.1038/s41598-022-08587-1

Ünal, E., & Sinha, V. K. (2023). Sustainability trade-offs in the circular economy: A maturity-based framework. *Business Strategy and the Environment.*

UNEP. (2011). *Decoupling natural resource use and environmental impacts from economic growth: A Report of the Working Group on Decoupling to the International Resource Panel.* United Nations Environment Programme. Retrieved from: https://www.resourcepanel.org/reports/decoupling-natural-resource-use-and-environmental-impacts-economic-growth

Van Loon, P., Diener, D., & Harris, S. (2021). Circular products and business models and environmental impact reductions: Current knowledge and knowledge gaps. *Journal of Cleaner Production, 288*,. https://doi.org/10.1016/j.jclepro.2020.125627

Wang, Q., Feng, X., Liu, Y., Cui, W., Sun, Y., Zhang, S., & Wang, F. (2022). Effects of microplastics and carbon nanotubes on soil geochemical properties and bacterial communities. *Journal of Hazardous Materials, 433*, 128826. https://doi.org/10.1016/j.jhazmat.2022.128826

Wei, J., Wang, Y., Zhao, J., & Santibanez Gonzalez, E. D. (2019). Analyzing the performance of a two-period remanufacturing supply chain with dual collecting channels. *Computers & Industrial Engineering, 135*, 1188–1202. https://doi.org/10.1016/j.cie.2018.12.063

Zhang, F., Ju, Y., Dong, P., Wang, A., & Santibanez Gonzalez, E. D. (2020). Multi-period evaluation and selection of rural wastewater treatment technologies: a case study. *Environmental Science and Pollution Research, 27*, 45897–45910. https://doi.org/10.1007/s11356-020-10307-z

Zhang, F., Ju, Y., Dong, P., & Santibanez Gonzalez, E. D. (2020b). A fuzzy evaluation and selection of construction and demolition waste utilization modes in Xi'an. *China. Waste Management & Research, 38*(7), 792–801. https://doi.org/10.1177/0734242X20908925

Zhang, X., & Duan, Z. (2019). Circular economy strategies in mining: A big data approach. *Resources Policy, 63*, Article 101426. https://doi.org/10.1016/j.resourpol.2019.101426

Zhang, Y., & Hu, L. (2020). Application of IoT and big data in sustainable agriculture. *Journal of Cleaner Production, 259*, Article 120901. https://doi.org/10.1016/j.jclepro.2020.120901

# Circular Economy: Transition to Sustainable Development or a New Model of Economic Transformation

**Venelin Terziev, Marusya Lyubcheva, and Marin Georgiev**

**Abstract** The present study and research aim to briefly analyse the development of the idea of realizing a circular economy in Bulgaria. A comprehensive review of the basic strategic documents developed in Bulgaria, such as a national strategy and an action plan in this direction, is made, by defining and analyzing the goals and objectives in detail. Emphasis is placed on the problems that have arisen during the implementation and the author's analysis of the current state is made, which is considered as part of the general state of the country and the European Union. Examples are given to support the author's theses, as well as those that serve to support the conclusions and conclusions. The circular economy is examined in both planes as a revolutionary model or a model that exists and is being upgraded. The main emphasis is placed on the approaches and relevant sectors in which the principles of the circular economy can and are successfully applied. The development is supported by a project survey and statistical data from Eurostat. Bulgaria is significantly lagging behind in implementing the strategy and planned activities, which is a result of the political crisis of the last few years and the inability to elect a regular government. All this and a complex of economic and other social factors further complicate the situation and create obstacles to the implementation of the national strategy.

**Keywords** Circular economy · Priorities · Goals · Business models · Sustainable development

V. Terziev (✉) · M. Lyubcheva · M. Georgiev
Black Sea Institute, Bourgas, Bulgaria
e-mail: vkterziev@gmail.com

M. Georgiev
e-mail: clementon@abv.bg

 
P. Singh et al. (eds.), *Circular Economy and Environmental Resilience*,
https://doi.org/10.1007/978-3-031-93091-1_12

# 1 Introduction

The circular economy is a model aimed at extending the life cycle of products. In practice, this means sharing, borrowing, reusing, repairing, and recycling existing materials and products for as long as possible (Fig. 1). When a product reaches the end of its life, the materials it is made of continue to be used in other ways, thus minimizing waste generation. The demand for raw materials and energy resources is growing, but many of them are limited in quantity. Resources are often not available within the European Union and European countries are becoming dependent on imports from other countries. The effect on the environment should not be underestimated—the extraction and consumption of raw materials increases energy consumption and greenhouse gas emissions. More reasonable use of raw materials is also a measure against climate change (Terziev et al., 2025a).

The creation and destruction of value inherent in the linear economy leads to an increasing amount of waste, which means resource and financial losses, as well as

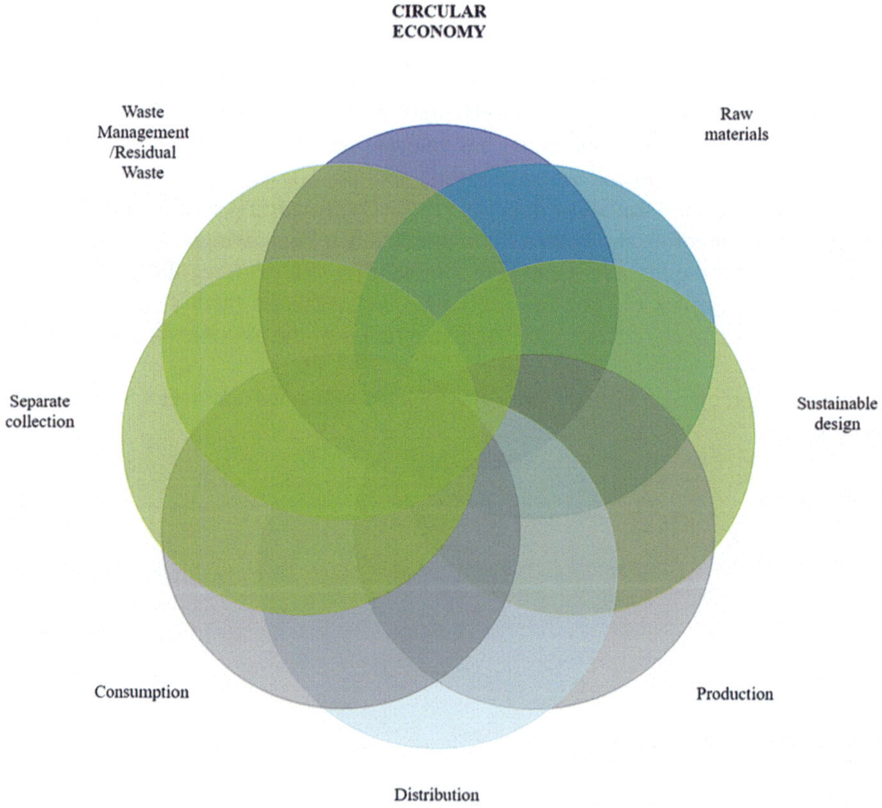


**Fig. 1** The circular economy model: fewer raw materials, less waste, fewer emissions. *Source* https://www.europarl.europa.eu/portal/bg

the depletion of non-renewable and scarce resources and the deterioration of the state of the environment. The transition from a linear economy to a circular economy is a significant change in the economic order. This change must encompass the entire production cycle—product design, the movement of material flows, and the creation of the right incentives and legal structures to stimulate business ideas. A change in the way consumers think is essential for these business ideas to be realized. This change requires the joining of efforts of the executive branch, business, financial, and educational institutions. In essence, the circular economy is a systemic change that builds sustainability in the long term, stimulates business, generates economic opportunities, and creates social benefits with a neutral footprint on nature. Breaking the link between economic growth and waste generation is the goal of Bulgaria's policy and the adopted National Strategy (Terziev, 2022a, 2022b).

The circular economy is based on the following guiding principles: Designing and manufacturing products in a way that does not lead to waste and pollution; Most products are designed to become obsolete too quickly. The efforts of the European Union countries are aimed at changing this model by regulating the design and manufacturing of products in a way that increases their durability and makes them reusable, repairable, and recyclable. Products in the circular economy contain a significant proportion of recycled materials, which in turn creates conditions for secondary production and high-quality recycling; Extending the life cycle of products and materials; The way products are designed and manufactured can prevent up to 80% of their environmental impact, but some products can remain in the economy instead of being landfilled; Restoring natural systems; Applying a circular approach will allow the restoration of natural systems and the achievement of a harmonious relationship between humans and nature.

The new business models that are implementing the circular economy are deployed through: design—products should be designed in such a way that they allow for longer use, and be easy to maintain, repair, overhaul, remanufacture, or recycle. The new design model should include the production of renewable materials, less resource-intensive or fully recyclable; optimal use—increase the value and use of the product during its life cycle. Such models represent the provision of a service instead of selling a product, as well as taking responsibility for the product throughout its useful life; value recovery—maximum recovery and recycling of products and materials after their use in new products or useful resources. The aim is to reduce material losses and preserve resources for new production; Circular support—the coordination of networks with circular value and resource flows. These include the development or implementation of enabling technologies that facilitate other business models. (Terziev & Georgiev, 2024a, 2024b, 2024c, 2024d).

## 2 Literature Review

The study of social transformations, the development of the idea of realizing a circular economy in Bulgaria, social processes and social changes is particularly important at the current stage of social development. Particularly significant in this regard are the works of Prof. Venelin Terziev, published in various scientific and other publications in different languages. Their knowledge and study is important both for the scientific community and from the point of view of the management of social systems (Terziev, 2023a; Terziev & Georgiev, 2023b; Terziev et al., 2025a; Georgiev & Terziev, 2023c; Terziev et al., 2025b; 2025c).

## 3 European Union Policies for the Circular Economy

The European Green Deal (2019) focuses on taking action to ensure that environmental and climate-related issues are addressed, thus ensuring a sustainable future for future generations. Achieving climate neutrality by 2050 is the most important priority.

By moving towards a circular economy, which is one of the focuses of the Pact, the aim is to strengthen the competitiveness of the economy and increase the gross domestic product of the community by 0.5% by 2030. In March 2020, the European Commission published a new Circular Economy Action Plan with measures to accelerate the transition to a sustainable model and the recovery of natural resources, thus reducing the impact on the environment and society. The new action plan aims to: establish a framework for sustainable product policy; key product value chains; a more effective waste policy aimed at preventing, reusing, and recovering waste; creating circular models for the benefit of people, regions, and cities.

Key strategic documents at the European level supporting the circular economy include: The Plastics Strategy in a Circular Economy (2018) guides the use of plastics, highlighting the need to increase the recycling and recovery of plastic materials. By 2030, all plastic packaging must be recyclable or reusable. Member States also have targets to achieve a 25% recycled content of plastic in PET bottles by 2025, and all bottles must be made of at least 30% recycled materials by 2030. In addition, plastic products such as forks, knives, spoons, plates, and straws are banned; The European Industrial Strategy (2020) envisages reorganising industry, supporting small and medium-sized enterprises in the transition, preserving Europe's resilience and competitiveness, based on the green transition, digitalisation, and setting global standards; From Farm to Fork (2020) is a strategy to build sustainable food chains to protect nature, ensure healthy food, support farmers, and reduce the use of chemicals, pesticides, and fertilisers in growing crops. Among the main objectives of the Strategy is the reduction of food loss and waste. The Commission is committed to halving food waste per capita at the retail and consumer level by 2030.

## 4 National Context for Implementing the Circular Economy Strategy

Bulgaria has set the circular economy as a long-term priority of its development policy. The National Development Program: Bulgaria 2030 prioritizes the "circular and low-carbon economy". The Strategy for the transition to a circular economy is the first and most important step in this direction. The transition to a circular economy will provide the country with economic growth and a better environment. The necessary institutional, financial, and human resources have been mobilized for this purpose.

In April 2020, an Advisory Council was established in Bulgaria in connection with the European Green Deal. Among the main roles of the Council is to advise and assist the Council of Ministers in achieving a balance and reconciling national priorities in the field of energy security, accelerated economic development, social justice, and environmental protection.

With the Strategy, the Government of Bulgaria takes decisive action in this direction, committing to providing resources and funds for their implementation. The measures set out in this document are aimed at increasing resource productivity and leading to connectivity in the economy by introducing new business models. Waste is beginning to be considered as a resource for the economy. The aim is to keep materials in production for as long as possible and recycle them with high quality, to produce better quality raw materials, and to reduce the country's economy's dependence on raw material imports.

The strategy is also in line with the National Strategy for Small and Medium-sized Enterprises 2021–2027 (Environment priority), the Innovation and Competitiveness Program 2021–2027 (Circular Economy priority), the National Waste Management Plan 2021–2028, and the Environment Program 2021–2027.

The circular economy needs a new and more active role for consumers. Consumer awareness and interest in the circular economy are crucial for the success of the transition to a circular economy. Consumers need to be provided with the most objective information possible for decision-making, including the social and environmental footprint of products. Consumers also play an important role in waste prevention and recycling. The measures in the Strategy are aimed at an informed, responsible, and active society. The high public attitudes towards the need to protect the environment can be used to increase engagement in such activities leading to a circular economy, such as buying second-hand products, repairing, sharing, or renting products.

The transition of production to a circular economy is assessed by a set of indicators such as resource productivity, the contribution of recycled materials to the demand for raw materials, the amount of investment and added value in sectors related to the circular economy, innovation, eco-innovation, and their results, etc., and small and medium-sized enterprises are the basis of the circular economy. In Bulgaria, there is significant potential for extraction and recycling of many of the raw materials of critical importance for the European Union, in addition to the development of innovation and particular eco-innovation, recycling and repair and reuse, environmentally friendly products or services, higher quality of eco-labeled products and industrial symbiosis are among the key actions that will transform entire sectors of the economy (Terziev et al., 2025a; 2025e).

The public sector, through the application of "green" criteria in public procurement procedures, could stimulate the market for ecological products in the country. To increase knowledge in this area, training can be organized for central and local government employees regarding the inclusion of "green" criteria by product groups in public procurement, as well as the creation of a virtual space for the exchange of good practices.

Separate waste collection is also included in the Strategy, as studies show that citizens have positive attitudes towards separate waste collection, but their actual actions do not confirm them. The weak motivation of citizens can be corrected through financial instruments if the household waste fee is tied to the actual quantity and volume of the waste disposed of. Effective separate collection can also be achieved by improving the infrastructure, providing an opportunity for the separation of biological waste, and promoting the activities of separate collection sites (Table 1) (Terziev et al., 2025e).

## 5 Priority Areas

### *5.1 Polymers*

The production of plastics in primary form in Bulgaria in 2019 amounted to 175 thousand tons, while the production of plastic products is significantly larger and exceeds 500 thousand tons. The significant production of plastic products and articles in the country provides an opportunity for the use of recycled plastics. The available recycling capacity is estimated at over 150,000 tons, which allows for the recycling of the entire amount of plastic waste collected in the country and at the same time provides opportunities for the processing of the main categories of materials—LDPE, HDPE, PET, PE. Additional efforts should be made to reduce the amounts of plastic waste generated, improve separate collection in households, and reduce unwanted impurities in the plastic products and packaging produced and offered on the market, which make their subsequent recycling difficult.

### *5.2 Construction and Demolition Waste*

The analyses show that the requirements of the Regulation on construction waste management and the use of recycled construction materials, which determines the

**Table 1** The place of the Republic of Bulgaria for the implementation of the circular economy according to Eurostat research

| Country | Year | Unit of measurement | |
|---|---|---|---|
| **2015** | | **2020** | |
| *Recycling of household waste* | | | |
| Bulgaria | 29,4 | 34,6 | % |
| Germany | 66,7 | 69,6 | % |
| Malta | 10,9 | 10,5 | % |
| EU (27) | 44,9 | 48,6 | % |
| *Recycling of all waste, except for basic mineral waste* | | | |
| Bulgaria | 17 (2016) | 23 (2018) | % |
| Belgium | 81 (2014) | 87 | % |
| Malta | 27 (2014) | 25 | % |
| EU (27) | 54 (2014) | 53 | % |
| *Recycling of packaging waste by packaging type* | | | |
| Bulgaria | 64,1 | 61,2 (2019) | % |
| Belgium | 81,5 | 79,7 | % |
| Hungary | 37,1 | 40,0 | % |
| EU (27) | 66,6 | 64,3 | % |
| *Recycling of electronic waste (e-waste)* | | | |
| Bulgaria | 96,5 | 66,7 (2018) | % |
| Liechtenstein | 127,1 | 89,8 (2018) | % |
| Iceland | 47,2 | 24,7 (2018) | % |
| EU (27) | 35,7 | 38,9 (2018) | % |
| *Biowaste recycling* | | | |
| Bulgaria | 43 | 34 | Kilogram |
| Luxembourg | 111 | 183 | Kilogram |
| Cyprus | 21 | 6 | Kilogram |
| EU (27) | 75 | 93 | Kilogram |
| *Utilization of construction and demolition waste* | | | |
| Bulgaria | 96 (2014) | 96 | % |
| Netherlands | 100 (2014) | 100 | % |
| France | 77 (2014) | 64 | % |
| EU (27) | 87 (2014) | 89 | % |
| *Use of recycled materials* | | | |
| Bulgaria | 3,0.1 | 4,9 | % |
| Netherlands | 25,8 | 33,8 | % |
| Romania | 1,7 | 1,4 (2021) | % |
| EU (27) | 11,3 | 11,7 (2021) | % |

(continued)

**Table 1** (continued)

| Country | Year | Unit of measurement | |
|---|---|---|---|
| *Trade in recyclable raw materials* | | | |
| Bulgaria | 141,606 | 200,602 | Tones |
| Netherlands | 7,384,951 | 8,928,527 | Tones |
| Malta | 2190 | 1247 | Tones |
| EU (27) | 42,766,932 | 46,838,227 | Tones |
| **Metal** | **Year** | | **Units of measurement** |
| | **2013** | **2019** | |
| **Contribution of recycled materials to raw material demand—End-of-life recycling rates (EOL-RIR)** | | | |
| Iron | 22 | 31,5 | % |
| Yttrium | 0 | 31,4 | % |
| Zinc | 0 | 31 | % |
| Aluminum | 35 | 12 | % |
| Nickel | 32 | 17 | % |
| Platinum | 35 | 25,3 | % |

*Source* Eurostat/ https://www.eufunds.bg/bg
***Note 1*** The EOL-RIR indicator measures for a given raw material what proportion of its input into the production system comes from the recycling of "old scrap" (or "end-of-life scrap"), i.e. scrap and waste resulting from the processing of products at their end-of-life (EOL). EOL-RIR does not take into account scrap that originates from production processes (i.e. "new scrap" or "process scrap"); 2. In second place is the country with the highest percentage, in third place is the country with the lowest percentage, and in fourth place is the average value for the countries of the European Union

percentage of recycled construction materials that should be used in certain construction activities, are partially implemented. The construction sector needs to take a leading role in the prevention and reuse of construction waste in partnership with local authorities. Public institutions should also play a much more serious role by using public procurement for the design of construction sites with the use of recycled construction materials. The lack of a platform for the exchange of information and good practices does not allow for a connection to be made between the sources of different types of construction and demolition waste, the availability of different types of recycled construction materials at regional recycling facilities, and the construction sector in the country.

### 5.3 *Food Waste*

The total amount of food waste generated in Bulgaria is significant—around 500 thousand tons—and improvements in production processes and their overall management could lead to increased resource efficiency and a significant reduction in carbon emissions. The country is in the process of preparing a National Program for the Prevention and Reduction of Food Losses, which will cover all stages of the food chain: primary production; processing and manufacturing; retail and other distribution; restaurants and catering services; and households. Concerning food waste from households, a very small part of it is currently diverted from landfills. Introducing mandatory separate collection of bio-waste from households, simultaneously with the completion of municipal bio-waste treatment systems, will allow the recovery of around 200 thousand tons of food waste per year.

### 5.4 *Biomass and Bio-Based Products*

Energy produced from biomass exceeds 1 million tons, which represents over 10% of the total energy consumption in the country. In Bulgaria, households consume 70% of the energy produced from biomass. Although the potential for using waste wood for energy production allows for an even higher share of biomass in the country's energy balance, the priority should be its processing, not energy recovery. The strategy confirms that there is a very large untapped potential in the recovery of solid agricultural waste from corn stalks for grain, sunflower, and others. The quantities of solid agricultural waste in the country are estimated at over 2 million tons per year. The current practice of not recovering a large part of this agricultural waste represents a waste of resources for processing and production.

### 5.5 *Major Raw Materials*

The mining, processing, and metallurgy industries in Bulgaria are very well developed and provide a major contribution to the national economy. Some of the raw materials of critical importance for the EU are present on the territory of the country as major minerals, while other components are found as isomorphic impurities in the minerals of other minerals. Exploring the potential for the extraction of critical raw materials in the exploitation of existing deposits and future exploration activities can lead to significant economic benefits.

# 6 Vision for the Development of the Circular Economy in Bulgaria

The transition to a circular economy will provide Bulgaria with economic growth, a clean environment, social well-being, and a society with a high level of environmental awareness that thinks about future generations. Bulgaria's policy for the transition to a circular economy will be implemented through a green and competitive economy, less waste and more resources, and an economy for the benefit of consumers. Individual productions will be connected so that they exist in symbiosis, with the state committed to supporting this process. The country will contribute to the provision of critical raw materials in the European Union. The amounts of landfilled waste will be reduced to a minimum, and the rest will be returned to the production cycle or recycled.

The preparation of the Strategy and Action Plan for the Transition to a Circular Economy of the Republic of Bulgaria for the period 2021–2027 is in implementation of measure 589 "Preparation of a National Strategy in Relation to the Circular Economy Package" of the Management Program of the Government of the Republic of Bulgaria for the period 2017–2021, as one of the set goals (Goal 147) to achieve resource efficiency by applying the hierarchy in waste management, preventing its formation, promoting its reuse and recovery through recycling, reducing landfilling and limiting its harmful impact on the environment and human health.

## 6.1 *Main Objectives of the Plan/program of Bulgaria*

The transition to a circular economy will provide Bulgaria with economic growth, a clean environment, social well-being, and a society with high ecological awareness that thinks about future generations. The country's policy for the transition to a circular economy will be implemented by setting the following strategic objectives in the draft Strategy for the Transition to a Circular Economy: green and competitive economy; less waste and more resources; an economy that benefits consumers. Specific objectives have been proposed for each of the strategic objectives.

As a result of the transformation of the economy, it is expected that resource efficiency will be increased and the added value of industrial production will increase. The consumption of some products is expected to be replaced by services, and others to become suitable for longer use. It is expected that individual productions will have the opportunity to be connected so that they exist in symbiosis. In this way, our country is expected to contribute to the provision of critical raw materials in the European Union. It is expected that the quantities of landfilled waste will be reduced to a minimum, and the rest will be returned to the production cycle or recycled.

## 6.2 Financing of the Plan/program (State, Municipal Budget or International Programs, Other Financial Institutions)

The preparation of the Strategy and Action Plan for the Transition to a Circular Economy of the Republic of Bulgaria is financed through funds provided within the budget line under priority axis 6 "Technical Assistance" of the Operational Program "Environment" 2014–2020. The financing of the implementation of the measures included in the Action Plan for the Transition to a Circular Economy is planned to come from various sources, including owned and attracted funds of enterprises, from programs financed by the European Structural and Investment Funds for the period 2021–2027, state and municipal budgets, etc.

Due to its nature, the Action Strategy for the Transition to a Circular Economy is a cross-sectoral strategy. It builds on and builds on the measures set out in the drafts of new sectoral strategies and programs, the main ones being in the fields of economy, environmental protection, and regional development.

The Strategy and Action Plan for the Transition to a Circular Economy do not fall into hierarchical dependence with other plans and programs at the national and local level, but are consistent with and do not contradict existing plans, programs, and other strategic documents at the European and national level, such as the European Green Deal, the New Circular Economy Action Plan of 2020 of the European Commission, the Plastics Strategy in the Circular Economy of 2018, the National Development Program Bulgaria 2030.

The transition to a more circular economy, where the value of goods, materials, and resources is retained in the economy for as long as possible and waste generation is minimised, is a key contribution to the EU's efforts to build a sustainable, low-carbon, and competitive resource-efficient economy. The circular economy is expected to increase the competitiveness of the European Union economy, protect businesses from resource scarcity, and help create new business opportunities for innovative and efficient ways of production and consumption. At the same time, the circular economy will help save energy and prevent irreversible damage caused by the use of resources at a rate that exceeds the earth's capacity to renew them—this is about the climate, biodiversity, and air, soil, and water pollution. The transition is expected to have a positive impact on reducing carbon dioxide emissions. By promoting more sustainable activities in key sectors and new business opportunities, the preconditions will be created for harnessing the potential of the circular economy to create growth and jobs. Actions at the European Union level support the circular economy along the entire value chain—from production to consumption, repair and recycling, waste management, and secondary raw materials that are reused in the economy. These guidelines are reflected in the draft Strategy and Action Plan for the Transition to the Circular Economy of the Republic of Bulgaria for the period 2021–2027.

## 6.3 Environmental Issues Relevant to the Plan/programme

About half of total greenhouse gas emissions and over 90% of biodiversity loss and water scarcity are due to resource extraction and processing of materials, fuels, and food. The European Union industry has initiated certain activities to reduce the negative impact, but it still accounts for 20% of greenhouse gas emissions in the region. Change is possible if the current "linear" form of management in industry is replaced, in which raw materials are extracted, traded, and processed into goods, and finally disposed of as waste or emissions. The solution is to increase the use of recycled materials, increase the efficiency of the resources used, extend the life of products, etc.

The problem of difficulties for enterprises in adapting environmental legislation to the activities of the specific organisation arises due to two reasons. On the one hand, this may be due to a lack of knowledge of the individual processes in the production cycle, and on the other hand, to a lack of knowledge in the field of environmental legislation and environmentally friendly practices. This leads to a lack of environmental expertise. This, in turn, in addition to the lack of motivation and interest in implementing ecodesign practices due to low demand, can also be attributed to a lack of knowledge and human resources.

In 2008, the Council approved an Action Plan on Sustainable Consumption and Production and Sustainable Industrial Policy. It includes a number of proposals for sustainable consumption and production that will contribute to improving the environmental performance of products and support the demand for more sustainable goods and production technologies. At the same time, industry is encouraged to take advantage of innovation to meet the challenge. The main element of the Action Plan is the preparation of a dynamic framework for improving the energy and environmental performance of products and promoting their acceptance by consumers. The challenge is to create an environmentally friendly product cycle by: improving the overall environmental performance of products throughout their life cycle, encouraging and stimulating the demand for better products and production technologies, and supporting consumers to make better choices through consistent and simplified labelling.

The new European Union Circular Economy Action Plan, adopted in 2020, aims to create a well-functioning market for secondary raw materials. Secondary raw materials face a number of challenges arising from their competition with primary raw materials, with aspects of this competition covering both the safety of raw materials and their efficiency, availability, and price. Bulgaria is a relatively small market for secondary raw materials. Recycling and the use of secondary raw materials in production are significantly lower than the European Union average. There is significant potential for the use of secondary raw materials in production and the replacement of natural raw materials, which is a prerequisite for growth in the sector.

Water scarcity has increased in some parts of the European Union in recent decades, which is damaging to the environment and the economy. In addition to measures to use water more efficiently, the reuse of treated wastewater under safe and cost-effective conditions is a valuable but underused way to increase water supply and relieve pressure on overexploited water resources in the European Union. Water reuse in agriculture will also contribute to nutrient recycling by acting as a substitute for solid fertilisers. The decline in groundwater levels is one of the main threats to the water environment of the European Union, in particular for agricultural irrigation, but also for industrial use and urban development. Member States may also use reclaimed water for other purposes, such as industrial, public works, and environmental purposes.

Since joining the European Union, Bulgaria has increased its spending on environmental protection and restoration and has achieved improvements in its environmental performance, but some significant challenges remain. The generated municipal waste has been on a steady downward trend, with an increase in the share of treated waste and an increase in the coverage of the population with organized waste collection systems, but reducing the share of landfilled waste remains a challenge. Good waste management is a building block of the circular economy and contributes to preventing the negative impact of waste on the environment and health. Significant results have been achieved in recent years, but some challenges remain. In January 2018, a European Strategy for Plastics in a Circular Economy was published, which aims to protect the environment from plastic pollution while promoting innovation in the sector. As fundamental measures to achieve the above-mentioned goals, the European Union wants to reduce the consumption of single-use plastics, limit the use of plastic microparticles, and make all plastic packaging recyclable by 2030. Microplastic pollution is also a key target of the strategy, to reduce it through the application of REACH principles in products where microparticles are added, through regulations to minimise microparticles released from tyres, textiles, and paints, the provision of information on the condition of products, or a combination of both.

The volume of construction and demolition waste is one of the largest among the different types of waste in Europe. Although existing estimates show that the share of construction and demolition waste generated is lower than the average for the European Union, its quantity remains significant and comparable to that of municipal waste generated. Construction waste management is relatively well-regulated in Bulgaria, but there are insufficient measures to enforce legislation and encourage recycling.

## *6.4 Significance of the Plan/programme for the Implementation of Community Environmental Legislation*

The concept of a circular economy is embedded in a number of European Union documents, including the Commission Communication of 2 December 2015 on Closing the loop—an EU action plan for the circular economy. The action plan focuses on actions at the European Union level that have high added value. As part of the

European Green Deal, in March 2020 the European Commission published a new Circular Economy Action Plan with measures to accelerate the transition to a sustainable model and the recovery of natural resources, thereby reducing the impact on the environment and society. The new action plan aims to: establish a framework for sustainable product policy; key value chains in the field of products; a more effective waste policy aimed at their prevention, reuse, and recovery; create circular models for the benefit of people, regions, and cities.

## 7 Specific Objectives and Main Measures in Bulgaria

By using natural resources more efficiently, society can live healthier lives, save money, create jobs, boost its economy, and respect the planet's boundaries. Resource efficiency and the wise use of natural resources are key factors for the competitiveness of the economy in a new circular model of production and consumption. Businesses play a major role in the transition to a more resource-efficient economy by adapting their business models and value chains, but the pace and scale of these changes will depend on the extent to which an enabling policy framework can be put in place.

Resource efficiency makes economic sense. By using fewer resources more efficiently, Bulgaria can maintain a competitive advantage, create green growth, and sustainable jobs, and better protect the environment. The green economy is about managing the scarce resources of the environment. Therefore, the country aims to reduce resource consumption, promote the introduction of circular business models, ensure the conditions for connectivity between enterprises in the economy, and focus on the extraction of critical raw materials.

To achieve the goal, the Strategy proposes to achieve the following specific goals, along with the measures to them:

### *7.1 Specific Goal 1.1. Higher Resource Productivity*

In essence, the goal focuses on ways to effectively use natural resources and increase their productivity in industry, prioritizing key value chains in the production of electronics and ICT, food, textiles, batteries, and automotive parts, and in construction.

Main measures: Creation of a center to support businesses in the transition to a circular economy; Technologies for increased resource efficiency; Extension of the application deadline for projects that envisage the implementation of new ones; Analysis of the legislation on issuing production permits; Analysis of complex permits to unify the approach to similar waste from production and their possible subsequent use as resources in other production; Financing small and medium-sized enterprises for the introduction of resource-efficient technologies in the processing industry, with priority given to food, textile, production of plastics and plastic products, accumulators, and batteries, will have a direct effect on increasing resource productivity; Promoting water reuse; Supporting small and medium-sized enterprises in carrying out an analysis of the points of intensive water use in the production process and demonstrating the economic benefit of water reuse; Improving the awareness of small and medium-sized enterprises about the available technologies for closing the water use cycle; Analysis of the effect of water reuse on various agricultural crops; Analysis of the possibilities for applying price incentives for water reuse in enterprises.

## 7.2 Specific Objective 1.2. New Business Models

Circular business models serve to reduce the extraction and use of natural resources. As a result, the generation of industrial and household waste also decreases. Circular business models use already existing materials and products as input materials and therefore their environmental footprint is significantly smaller than that of traditional business models.

The strategy focuses on assessing and introducing incentives for more efficient use of natural resources, supporting enterprises in financing the costs of switching to new models and eco-design of products, as well as analyzing the possibilities for increasing the requirements for mandatory content of recycled materials in certain product groups.

Main measures: Financing small and medium-sized enterprises in the processing industry for the introduction of eco-design, with priority in key value chains—electronics, information and communication technologies, textiles, and plastics, as well as for the introduction of models that allow the use of services instead of products; Creation of a mechanism for subsequent financing of eco-design project proposals through a financial instrument Conducting information training campaigns for businesses; Preparation of specialized training programs for "green" management and "green" business in vocational schools or as an additional qualification in vocational training centers; Analysis of the possibilities for increasing the requirements for mandatory content of recycled materials in certain product groups and ways to strengthen control in construction.

## 7.3 Specific Objective 1.3. Connectivity in the Economy

The transition of heavy industry and manufacturing to a circular economy can be a challenging task. Creating partnerships, joint work and exchange of materials between enterprises can increase sustainability and economic profits while reducing environmental impact and costs. To achieve connectivity in the economy, resource market platforms will be created and promoted. Areas in industrial zones and parks will be provided with priority to groups of enterprises participating in industrial symbiosis. The utilization of biomass from agriculture in the economy will be stimulated.

The implementation of the Strategy for the transition to a circular economy seeks connectivity between all chains and participants in the transition process: Research centers to create resource efficiency technologies for business, institutions to initiate voluntary agreements with businesses for better management of different types of waste, waste from one production to be used as a resource by another (industrial symbiosis).

Main measures: Development and implementation of innovations related to the circular economy, namely recycling, energy and resource-efficient technologies; Development and implementation of innovations related to the circular economy, namely recycling, energy and resource-efficient technologies; Creation and maintenance of a Platform for the results of implemented technologies for resource efficiency, ecological products, and industrial symbiosis; Promotion of the application of environmental criteria, where possible, in public procurement documentation for the award of activities under projects for which it is appropriate, financed by programs; Adoption of a decision on the priority provision of areas in industrial zones and parks to groups of enterprises participating in industrial symbiosis; Structuring of a financial instrument for industrial symbiosis for large processing enterprises in the sectors identified in the previous analysis; Study of the applicability of technologies for processing agricultural waste into resources for other productions; The construction of installations for recycling and utilization of biomass of agricultural origin.

## 7.4 *Specific Objective 1.4. Bulgaria Contributes to the Supply of Critical Raw Materials*

The Strategy specifies that the country has the potential to extract and recycle many of the raw materials of critical importance for the European Union, as a result of which economic benefits can be realized. In addition, additional mineral raw materials such as copper, lead and zinc—from metallic minerals, as well as zeolites, kaolin-quartz sands and bentonites—from industrial minerals and rocks, are of particular importance for the economy of Bulgaria.

The Strategy gives priority attention to the raw materials of critical importance and to strengthening the role of the country in the extraction, production and recycling of certain raw materials of critical importance within the European Union, taking into account the leading role of the extractive and processing industries and metallurgy in the country's economy, the presence of a significant number of deposits, proven reserves, including in closed deposits and the potential for finding new mineralizations of economic interest for a significant number of metallic minerals and industrial minerals and rocks, the presence of experience in the extraction and processing of raw materials, the presence of scientific and technical potential for their exploration, development, processing and recycling.

The extraction and recycling of critical raw materials will be considered a matter of paramount national interest and will receive the necessary political and administrative support from state institutions. Future legislation should ensure acceleration and facilitation of the implementation of investment projects, removing existing obstacles to their implementation and at the same time ensuring compliance with high environmental protection standards.

Bulgaria has good potential for the extraction of some of the raw materials of critical importance for the European Union. On the territory of the country there are deposits of graphite, barite, fluorite, manganese, tungsten, etc. Another part of these materials appear as isomorphic impurities of the main minerals and can be extracted along the way, such as molybdenum, titanium, antimony, germanium, strontium, cerium, etc. In addition, additional mineral raw materials such as copper, lead and zinc—from metallic minerals, as well as zeolites, kaolin-quartz sands and bentonites.

Main measures: Preparation of a national list of critical raw materials; Amendments to national legislation to give priority to the exploration, extraction, processing, and recycling of raw materials included in the national list of critical raw materials; Ensuring appropriate institutional organization and capacity of institutions ensuring the availability of geological information and the management of exploration activities, the extraction and processing industry in the country; Studying the potential for extraction of critical raw materials in the exploitation of existing deposits and future exploration activities; Reviewing the available information and conducting additional studies on the content of critical raw materials in accumulated waste in closed sites; Identifying projects and investment needs in the field of extraction and processing of critical raw materials; Implementing solutions to increase circularity and reduce the amount of mining waste.

The aim is to generate less and less waste by promoting activities for the reuse, repair, refurbishment, and processing of products. Incentives will be provided for the construction of reuse centers in cities. New liability schemes are planned to be introduced to encourage the environmental behavior of enterprises. High-quality recycling will be promoted by encouraging the separate collection and providing more and better quality secondary raw materials for processing enterprises. Landfill waste will be minimized by expanding separate collection systems and new technologies and installations for processing remaining waste. Landfill waste will also be reduced by changing the way municipal waste is collected and introducing fair payment for services depending on the amount of waste.

## 7.5 *Specific Objective 2.1. Less Waste*

The specific objective mainly focuses on waste prevention as the most preferred option in the national waste management policy. Since municipalities will play a key role in the transition to a circular economy, they will need to develop municipal programs for the transition to a circular economy as stand-alone documents or as part of their development programs. More rational use of material resources and improvement of supply and distribution logistics should contribute to reducing waste generation in production processes.

Main measures: Implementation of new business models aimed at extending the useful life of the product, providing more repair and maintenance options, and using sharing platforms by consumers; Reduction of food waste and some categories of single-use plastic products and packaging; Introduction of measures at the municipal level for waste prevention and transition to a circular economy; Possibilities for introducing additional higher deductions for generated waste for municipalities reporting levels of municipal waste generation significantly exceeding the national average; National legislation in the field of single-use plastics; State institutions, public organizations, and local authorities should be leaders in implementing sustainable consumption models, being the first to limit the use of single-use plastic products in public administration and introducing a ban on the use of single-use plastic products and packaging no later than 2026; Measures to increase the quantities and improve the quality of separately collected plastic packaging waste; Measures adopted to provide information to consumers aimed at preventing and sustainable consumption of plastics and promoting opportunities for reuse and recycling through coordinated and continuous actions to increase public awareness and engagement; Funding and support for scientific organizations and businesses in the development of technologies and products leading to the prevention of plastic waste, the development of alternative replacement materials, new methods for recycling plastics, the development of new types of plastics with better environmental performance, including bio-based plastics and biodegradable plastics.

## 7.6 Specific Objective 2.2. More Opportunities for Sustainable Use

The strategy emphasizes the need for more products to be designed for longer life, ready to be disassembled, repaired, and eventually recycled.

Systems for separate collection of bulky waste, especially in large municipalities, should take into account the possibilities for the reuse of old furniture and electrical equipment, as well as other categories of products used in the household. The will of construction entrepreneurs will be relied on to reach a voluntary agreement with the central and local administrations concerning their commitments to the use of sustainable materials.

Main measures: Municipalities should set measures in their municipal waste management programs for the next programming period for sustainable use and reuse of products; Municipalities should provide sites for free delivery of separately collected waste from households; Strengthening the role of the construction sector in the management of construction waste; Pilot project for a regional installation for the treatment and recycling of construction and demolition waste.

## 7.7 *Specific Objective 2.3. More Recycled Waste, Better Quality Raw Materials*

The specific objective pays special attention to the separate collection of waste, which should become a common and conscious practice and way of behaving. In addition, the separate collection of recyclable fractions, bio-waste, and hazardous waste from households will become mandatory. A complete change in the way of collecting mixed household waste and a transition to paying for services proportional to the quantity and/or volume of waste is also necessary.

Main measures: New, more ambitious recycling targets should be considered for the different waste categories, as well as the introduction of new innovative management models; Necessary and comprehensive change in the way mixed household waste is collected and the gradual introduction of collection with individual containers; Transition to payment for services proportional to the quantity and/or volume of waste; Construction of waste transfer sites from households; The scope of requirements for documentation and reporting of packaging placed on the market and the quantities of separately collected, reused, recycled, and recovered packaging waste will be expanded; Additional measures to increase the quantities and improve the quality of separately collected plastic packaging waste; Additional, stricter requirements for mandatory separate collection of plastic waste generated in industry, commercial, and administrative facilities. The requirements for extended producer responsibility will also include plastic films used in agriculture; The trade and service sector should play a greater role in implementing measures for plastic waste prevention, reuse, take-back, separate collection and recycling; Analysis of the existing system for separate collection of packaging waste and assessment of the potential for achieving the new higher recycling targets; Introduction of a system for mandatory deposit and take-back of plastic containers made of PET and possibly other plastics is one of the possible measures for achieving the new higher recycling targets; Support for the construction of additional recycling capacities for certain categories of plastics for which there is no recycling option in the country; Inclusion of new product and waste categories in the requirements for extended producer responsibility, such measures will be considered initially for textile clothing and footwear, as well as some plastic products and single-use items; Improving the interaction between recovery organizations and municipalities with the aim of expanding separate collection systems, especially in small settlements; Access to financing and support for the introduction of new technologies for the treatment, recycling, and recovery of widespread waste in order to provide technological solutions and markets; Requirements for the use of recycled construction materials in public procurement for the design, construction, and maintenance of roads, as well as the design and construction of buildings; Coverage and proper reporting of products and packaging placed on the market as a result of internet sales and coverage of the relevant quantities by extended producer responsibility systems; Stricter requirements for auditing and control of the quantities of packaging placed on the market, declared by the obligated parties.

## *7.8 Specific Objective 2.4. No Landfilled Waste*

To achieve the specific objective, actions will be taken towards achieving high levels of reuse and recycling and the construction of new installations for the treatment of municipal waste will require additional and higher costs. The construction of new ones implies an increase in the quantities of sludge formed and requires the implementation of measures to limit the hazardous substances contained in them and to ensure appropriate recovery options. Increasing the quantities of secondary raw materials used in the production of products and articles, expanding the market for secondary raw materials by including new materials and production, effective implementation of national end-of-waste criteria for certain categories of materials and other measures to improve the quality of secondary raw materials are also part of the planned actions.

Main measures: Mandatory separate collection of bio-waste, predetermine the leading role of composting and anaerobic digestion in future regional waste management systems; Expansion of the capacity for mechanical–biological treatment of mixed waste and waste incineration with energy recovery will contribute to minimizing the amount of landfilled waste; Assessment of the necessary capacity for municipal waste treatment; Involving the private sector in achieving municipal waste recycling targets and the construction of municipal waste treatment facilities should provide additional investments; New sectoral Sludge Management Plan. The plan should consider the potential for using sludge in the reclamation of sites in the extractive industry, exploring the possibilities for applying phosphorus recycling techniques or biogas production in a joint anaerobic digestion process with other suitable biodegradable waste, providing opportunities for the separate and joint incineration of sludge for the purpose of energy production. Water and Sewerage Companies should provide for sludge treatment measures; The use of sludge in agriculture will be limited only to compost produced from sludge, meeting the end-of-waste requirements; Financing for projects related to the development and implementation of new technologies for separation, processing, and treatment of waste ensuring a higher degree of separation of secondary raw materials; Platforms for the exchange of information related to the demand and supply of secondary raw materials, recycled construction materials, and biomass; Additional financing for the completion of regional waste management systems; Additional indicators introduced for monitoring the costs of municipal waste; A ban on spending the revenues from the "household waste" fee for other activities in municipalities and control over the spending of funds will be strengthened, including through inspections by the National Audit

Office; Clear rules and criteria for spending the funds from deductions for waste disposal.

The spending of the funds from deductions should be prioritized for financing investment costs related to prevention, preparation for reuse, recycling, and recovery, as well as for measures and incentives related to prevention and separate collection. Financing of operating costs for pre-treatment, recycling, and recovery should be eligible only for municipalities where the amount of the household waste fee exceeds 80% of the eligibility threshold, calculated in relation to the income of households in the municipality. Financing disposal activities, as well as the purchase of containers and equipment for the collection of mixed household waste, will not be eligible. Municipalities where the amount of accumulated deductions exceeds the deductions due for a period of two years should submit a plan for spending the collected funds.

The strategy foresees a periodic assessment of the results achieved and the infrastructure built for separate collection, preparation for reuse, recycling, and recovery by municipalities. In case the first such assessment provided shows unsatisfactory results, the possibility of partial or full replacement of the waste landfill levy with a landfill fee, payable and spent at a national level, will be considered.

The strategic objective foresees the application of more "green" criteria in public procurement; consumption models based on renting and sharing of goods and services, or the choice of reusable products that are subject to repair, to be supported, as long as they are truly sustainable and affordable; the circular economy will stimulate consumers to take advantage of the opportunity to make informed choices and to take an active part in the ecological transition.

## *7.9 Specific Objective 3.1. Better Informed Consumers*

The transition to a circular economy requires that citizens make informed choices about products that are safe and have the lowest possible environmental impact. In this regard, the focus along the supply chains of such products will be on ecodesign.

Manufacturers will be encouraged to design, produce, and advertise products that have been assessed for their environmental impact at all stages of the life cycle. Retailers will offer these products in an appropriate manner, providing reliable and useful information. Consumers will be informed about the rules for the use of eco-labels and the certification schemes applied.

Key measures: Additional information and training on eco-labels, certification schemes, and their role in building a circular economy; Training should be targeted at different age and social groups; All actors along the supply chain should take action to appropriately present eco-friendly products in retail outlets; Financing of small and medium-sized enterprises in the processing industry, which receive funding for eco-design, as well as for the promotion of ecological products and new consumption models, with priority in key value chains—electronics, ICT, textiles and plastics; Provides for a study of the mechanism and applicability of voluntary agreements as a tool to stimulate retail chains to provide ecological products in an appropriate manner.

## 7.10 *Specific Objective 3.2. Sustainable Consumption Patterns*

The circular economy means sustainable patterns of behavior of the administration, business, and consumers. The promotion of opportunities for repair, rental, and sharing of products or purchasing a service instead of a product through electronic platforms and information campaigns, the creation of reuse centers in key locations in cities in order to facilitate access for citizens, familiarization with the requirements for "green" criteria, will stimulate the production of environmentally friendly products.

Changing behavior patterns towards sustainability is possible and necessary by raising consumer awareness of the market for second-hand products, rental/leasing, and repair through communication campaigns, as well as activities focused on the education of young people, such as including appropriate topics in the curricula of schools and universities. This measure can be divided into communication campaigns and activities focused on education.

The practice of mostly small private companies carrying out repairs of a number of goods—electrical appliances, clothes, shoes, furniture, bicycles, etc. is popular in our country. The creation of electronic registers on the website of the respective municipality/district municipal administration would provide additional opportunities for improving citizen awareness.

Main measures: Creation of reuse centers in key locations in cities. These centers can even be designed as temporary markets. State and especially municipal authorities can support the logistical and/or organizational activities of reuse center s, for example by providing municipal premises/areas specifically for such centers and promoting the positive effects of reuse; Public engagement should be increased in terms of preventing food waste and participating in separate bio-waste collection systems. The participation of the population in separate bio-waste collection systems will be of key importance for reducing their amount; Research on behavior patterns can be the subject of joint projects between scientific organizations and consumer protection organizations; Introducing the concept of "green" public procurement in the additional provisions to the Public Procurement Act will provide a clearer opportunity for contracting authorities to use this tool; Periodic updating of the Practical Guide for Awarding "Green" Public Procurement and training of municipalities is planned.

### *7.11 Specific Objective 3.3. Social Green Economy*

The transition to a circular economy will have a positive effect on job creation. There is potential in this area for companies engaged in repair and refurbishment activities, which will be supported to be present in key locations and in reuse centres in cities. The benefits of the circular economy will be expanded by promoting social entrepreneurship and strengthening social inclusion.

Main measures: Opportunities are envisaged to support companies engaged in repair activities to be present in reuse centres in cities. This can be implemented through public–private partnerships or implemented as municipal centres, financed from municipal budgets through landfill levies or other sources of funding; Repair activities can extend the life of products and are an example of sustainable behaviour models. Companies in the sector can operate as social enterprises; The activities of these enterprises can be further supported by financing qualification courses for people from vulnerable groups involved in repair activities.

## 8 Analysis of the State and Development of the Circular Economy in Bulgaria

High-tech economic development poses a challenge to humanity by reducing the resources available for its maintenance. The recent use of cheap resources, their availability, as well as relatively less care for the environment, have led to serious problems related, on the one hand, to general environmental pollution and the accumulation of an incredible amount of waste, and on the other hand, to the physical depletion of conventional resources. The new paradigm for sustainable development, defining the trinity between economy, environment, and social activities, requires a categorical and new attitude toward raw materials, technologies, and waste management. The transformation of thinking and actions towards reducing greenhouse gas emissions, saving fossil fuels, and increasingly using renewable energy sources is gaining momentum in the period 2010–2015. The chronology in the order Agenda for Sustainable Development, greenhouse gases, and renewable energy sources caused many changes in several directions: attention to climate change, the occurrence of which is still controversial among scientists, experts, and politicians; reduction of production and regulation of the use of fossil fuel products; closure of coal-fired power plants; debate on the nature of nuclear power plants; bans on the production and use of plastic products; emphasis on waste management and treatment technologies. In addition, the era of the circular economy began. The concept itself has been developed since the 1970s.

## *8.1 Evolution of Terminology, Revolution of Technologies. Zero-Waste Technologies of the Twentieth Century Versus the Circular Economy of the Twenty-First Century*

The industry of the twentieth century, designed to meet the needs of various sectors—mechanical engineering, transport, road construction, agriculture and forestry, maritime industries, etc., focuses on product lines with less attention to the environment, on the one hand, because nature is still able to absorb various types of pollution, science has not yet identified the connection between pollution and climate change, the discovery and widespread use of plastics is perceived as an extremely progressive achievement. To extract more oil and coal, to produce more metal, mineral fertilizers, and gasoline, to provide the necessary asphalt for roads and building materials with diverse applications is almost equivalent to a directive. In this hectic product age, it is still realized that efficiency requires saving materials and resources and this is a necessary condition for progress.

"Zero-waste technologies" is a term that precedes "Circular economy". It is known that there are no zero-waste technologies in the world—waste is generated at different stages of each technology—technological (when establishing or changing technological regimes), production (defective and poor-quality products), consumer (loss of purpose). Each technology seeks to reduce the volume of technological waste, introducing a requirement for their reuse in the technology itself or another related use. Here, science enters the role of creator of new technologies or products. Technologies are being born for the recovery of discarded rubber products, glass (glass crumbs), the return of paper to technological production lines, and the recovery of plastic products (technological and consumer waste). Special lines are being created for collecting paper, processing plastic waste (specially created technological lines), and production lines for grinding glass are being created. Plastic tomato pegs, bricks, tiles, and other irresponsible plastic products appear. Various inert materials are being tested as fillers in the production of plastic products on a massive scale (including phosphogypsum, fly ash, bentonite, and ground waste from thermosetting resins), and new plastic products based on polymer mixtures are being created. However, these technological solutions cannot cancel the linear nature of the economy and overcome the ever-growing consumer demand and mass disposal of obsolete goods and products, filling landfills (for which there is still enough land). A limit has been reached that causes. Accumulating waste causes pollution, metal waste is a pure loss for the economy, plastic does not decompose and enormously increases the volume of landfills, indiscriminate waste disposal and poor management of landfills or the presence of "illegal dumps" are becoming threatening. Soil, air, and water are polluted. In addition, the overproduction of goods, and the rapid change of household appliance models increase waste flows, even without the life cycle of the objects having expired, but also without being ready for their disposal. The focus of the economy continues to be production. Technologies focus mainly on the optimization of production processes in order to reduce waste. These technologies often work within the linear model of the economy, where resources are extracted, used, and then discarded. Technological innovations introducing elements for the recycling and reuse of materials have not been widely distributed or integrated into the overall economic system. "Zero-waste technologies" have failed to cope and thus the new paradigm of "circular economy" is born, which must bring order. The circular economy is a development of zero-waste technologies but with significantly higher political, economic, and social responsibility. The statement about the purely linear model of the economy of the twentieth century, i.e. a model in which raw materials are used, things are created from them, they are consumed and the remains are thrown away, is not entirely true. The change was caused by the technological revolution, which was a result of overconsumption and overproduction, the accelerated implementation of new technological and technical models, a change in the paradigm of quality of life, the digital transition, of the saturation of the environment with traces of all this. Progress, of course, is on the side of the circular economy, which must find solutions to the already global problems of inconsistencies between the over satisfaction of certain groups of the population with products and goods and the growing need for new resources (Kokalarov et al., 2021; Petrova et al., 2025d; Kokalarov, 2019).

## 8.2 *The Circular Economy Model*

The circular economy is an innovative model of production and consumption that aims to minimize waste and maximize the use of resources. This approach is important for sustainable development and environmental protection. The circular economy offers a comprehensive model that includes not only production but also consumption, with an emphasis on minimizing waste and maximizing resource efficiency. It seeks to create a closed resource cycle, where waste from one process becomes raw material for another. The circular economy is closely linked to sustainable development and social responsibility, encouraging innovations that are environmentally friendly and economically viable.

The main characteristics of the circular economy include: Waste minimization—the aim is to reduce the amount of waste generated during the production process and after the use of products; Resource maximization—the aim is to preserve the value of products, materials, and resources for as long as possible; Repair and recycling—products are designed so that they can be easily repaired and recycled, rather than thrown away; Innovation and new business models—it includes new approaches, such as the sharing economy and services instead of ownership, which reduce the need for new resources.

The key terms of the circular economy "reduce", "reuse", "repair", and "recycle" contain the philosophy, which is already becoming leading in terms of production and consumption and in part of the research work on sustainability, durability and specific qualities of materials and products.

The main steps to implement the circular economy paradigm in the European Union include: 2015 The European Commission presented an Action Plan for the

Circular Economy, which laid the foundations for a new strategy for sustainable development and resource management; 2018 New legislative proposals aimed at increasing recycling and reducing waste were adopted; 2020 The European Union continued to develop initiatives related to the circular economy, including new rules on packaging and e-waste.

The circular economy is important because it aims to minimize waste and extend the life cycle of products, which is essential for sustainable development. In practice, this means sharing, borrowing, reusing, repairing, and recycling existing materials and products for as long as possible. When a product reaches the end of its life, the materials it is made of can continue to be used in other ways. This can be done over and over again, minimizing waste disposal. This model relies on large quantities of cheap and accessible materials and energy sources.

## 8.3 Circular economy—A Revolutionary Model or an Upgrade

The zero-waste technologies of the 1980s are part of the linear model, while the circular economy offers an integrated approach. Zero-waste technologies focus primarily on production, while the circular economy covers the entire life cycle of products. The circular economy encourages broader innovations and new business models that not only reduce waste, but also create new economic opportunities. The circular economy provides a full production—consumption—waste cycle and provides a new serious raw material resource for the economy. The circular economy is simultaneously an accelerated development of the concept of zero-waste, an analytical model and a new approach to using the planet's raw material base while adhering to the principles of sustainable development and implementing high-tech solutions.

The not-so-perfect solutions in development and the accelerated movement towards the accumulation of waste problems and loss of resources is associated with the abandonment of traditional products and products, the replacement of a number of glass packaging with single-use plastic, the abandonment of the collection and processing of glass products, the production and use of single-use plastic products, the too rapid change of models, product designs and aggressive advertising to replace those that are already "out of fashion", the imposition of an aggressive design model in the packaging and sale of goods—designer packaging has an extremely short life, but on the other hand produces serious amounts of waste.

### 8.4 Education and Science and the Movement Towards a Circular Economy

The transition from waste-free technologies to a circular economy has its reflex to education and science, which change their course both conceptually and technologically. The problem of recycling materials is present in quite old textbooks in some technological disciplines (for example: paper recycling) in textbooks from the beginning of the plastic era (aging, stabilization, and recycling of polymers). The question is how quickly and adequately a problem "enters" education. Waste management is still not sufficiently represented (waste is different for different types of materials and needs specific consideration in different specialties). The circular economy is still only a concept in education. Even sustainable development. Here and there, topics related to sustainable development sneak in, and we need to seriously talk about "education for sustainable development". For science, this is also a permanently treated problem, it's just that the movement is not as fast as it is at the moment. Science is also guided by modernity and has always preferred the creation of new, rather than the treatment of old (discarded) materials. Science is the one that discovers and promotes, for example, the wonderful characteristics of plastics as a material and discovers the contradiction in the use of plastic products—between one of its most important qualities—sustainability and the need for degradation after use. Flooding the market with single-use plastic products, and replacing traditionally used glass and metal packaging with plastic polymers are part of the solution and are also a product of science. The question of the participation of education and science is related not only to their professional attitude. Policies, social dimensions, and the environment in which they develop play a significant role. Today, they need new tools and prerequisites to be adequate both thematically and as a dynamic of compliance with the problems.

Under the SMART Project, BG05SFOP001-2.025–0037-C01, funded by the Operational Program "Good Governance" 2014–2020, co-financed by the European Union through the European Social Fund, a survey was conducted on the applicability of the circular economy in Bulgaria.

The survey surveyed 161 respondents (0.03% of all 412,878 non-financial enterprises in the Republic of Bulgaria). The surveyed employees work in 143 enterprises of different sizes: micro (average number of employees less than 10 people, and annual turnover not exceeding 3,900,000 BGN, and/or value of assets not exceeding 3,900,000 BGN); small (average number of employees less than 50 people, and annual turnover not exceeding 19,500,000 BGN, and/or value of assets not exceeding 19,500,000 BGN); medium (average number of employees less than 250 people, and annual turnover not exceeding 97,500,000 BGN, and/or value of assets not exceeding 84,000,000 BGN); large (average number of employees exceeds 250 people annual turnover exceeds 97,500,000 BGN, and/or the value of assets exceeds 84,000,000 BGN).

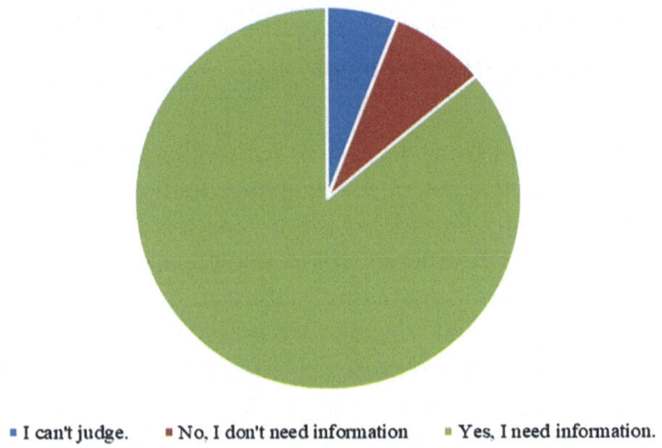


**Fig. 2** Need for more and regular information. *Source* https://www.eufunds.bg/bg

We will illustrate only some of the results through the shown Fig. 2 which provide a clear answer to the questions posed and do not even need a detailed comment, but they do require in-depth analysis.

Of the respondents, 124 (77%) declare that they have no implemented projects in the field of circular economy. This is a true reflection of the situation in this sector. The remaining 37 respondents have implemented 47 projects in the circular economy with the following distribution: in the field of industrial production, 24 projects have been implemented; in the field of household waste, 13 projects have been implemented; Micro - 45%; Small - 31% Medium - 15% Large - 9%.

Types of enterprises by size:

in the field of agriculture, 7 projects have been implemented;
in the field of construction, 3 projects have been implemented.

These results are far from even modest and the reasons for this are specified by the respondents as follows:

there are no incentives for its implementation according to 75% of the respondents;
the principles of the circular economy and the Strategy of Bulgaria 2021–2027 are not known, according to 72% of the respondents;
bureaucratic and regulatory obstacles, such as a lack of instructions from relevant departments and developed guidelines for the implementation of real projects according to 61% of respondents.

It is also interesting to note the number of negative responses: 8% of respondents do not believe that there is a lack of incentives for its implementation; according to 9%, there is sufficient awareness of the issue; and only 3% of respondents believe that there are no bureaucratic obstacles to the implementation of projects in the circular economy (Terziev et al., 2025e; 2025a).

Bulgaria is still in the initial phase of implementing the principles of the circular economy. The National Strategy and the Action Plan, adopted by Decision No. 832 of the Council of Ministers on 26.10.2022, are still a draft and have not been approved by the parliament (submitted for public discussion on the platform of the Council of Ministers on 4.11.2022), from which it follows that: the institutions responsible for its implementation cannot plan the relevant events and tasks, as well as that there is no funding provided for both the institutions and the other measures provided for in the Strategy for the Accelerated Implementation of the Principles of the Circular Economy. Without an approved Circular Economy Strategy and Plan, we cannot utilize the funds from the National Recovery and Sustainability Plan, is the conclusion of a study by the World Bank and the Circular Economy Council at the Bulgarian Chamber of Commerce and Industry in 2021. The current situation has not changed, respectively, the activities set out in the Action Plan for the Implementation of the Strategy for Transition to a Circular Economy 2022–2027 are no longer being implemented.

Bulgaria is significantly lagging behind in the implementation of the strategy and the planned activities, which is a result of the political crisis in the last few years and the inability to elect a regular government. All this and a complex of economic and other social factors further complicate the situation and create obstacles to the implementation of the national strategy (Circular economy: what it is and why it matters, publication in the EP, 2023d; Nozharov, 2022; Special Report 17/ 2023: Circular economy, 2023e; 2024e; UMG "St. Ivan Rilski", 2024f; Ministry of Environment and Water, 2024g).

## 9 Conclusion

The European Union produces over 2.1 billion tonnes of waste a year. It is updating its legislation to encourage the transition to a more sustainable model known as the circular economy (European Parliament. Climate and environment, 2024d). The circular economy is a model aimed at extending the life cycle of products. In practice, this means sharing, borrowing, reusing, repairingand recycling existing materials and products for as long as possible. When a product reaches the end of its life, the materials it is made of can continue to be used in other ways. This can be done over and over again, minimising waste disposal. The concept of a circular economy builds on the traditional linear model, where raw materials are used, made into things, consumed and the rest thrown away. This model relies on large quantities of cheap and accessible materials and energy sources. Part of the traditional model is to build defects into appliances so that they stop working on a certain date or after a certain number of cycles, forcing consumers to buy new ones. The European Parliament is against this wasteful practice (European Parliament. Climate and environment, 2024d). Reusing and recycling products reduces the need to extract natural resources and limits the associated damage to ecosystems that threaten biodiversity. Recycling raw materials reduces the risks associated with their supply—the EU becomes less dependent on global crises, price fluctuations and geopolitical pressures. This is particularly true for key raw materials needed to develop technologies to meet climate goals, such as batteries and electric motors (European Parliament. Climate and environment, 2024d).

Efficient use of resources is key to increasing the competitiveness of the economy, generating growth and creating new jobs. It is estimated that the transition to a circular economy could create 700,000 jobs in the EU by 2030. Designing products and materials in a way that allows for multiple uses and recycling will stimulate innovation in different sectors of the economy. Consumers will also have access to more sustainable and cost-effective products that will save them money in the long term and improve their quality of life (European Parliament. Climate and environment, 2024d). The European Union wants to build a circular and climate-neutral economy by 2050. To achieve this, the EU has adopted many new measures to reduce waste and improve products on the market. New or updated rules concern ecodesign, packaging, the right to repair products, waste management and misleading environmental claims (European Parliament. Climate and environment, 2024d; European Environment Agency (2025f).

## References

https://www.eca.europa.eu/en/publications?ref=SR-2023-17. (2024e). Retrieved December 15, 2024.

Circular economy: what it is and why it matters, publication in the EP. (2023d). Brussels, Publication date 24-05-2023.

European Environment Agency. (2025f). https://www.eea.europa.eu/en, Retrieved March 15, 2025.

European Parliament. Climate and environment. (2024d). https://www.europarl.europa.eu/topics/bg/article/20151201STO05603/krghovata-ikonomika-kakvo-predstavliava-tia-i-zashcho-e-vazhna, Retrieved 25 December 2024.

Georgiev, M., & Terziev, V. (2023c). The labor market policies and social development in Bulgaria. *Optimizing Energy Efficiency During a Global Energy Crisis*, edited by Meltem Okur Dinçsoy and Hamit Can, IGI Global, pp. 239–281. https://doi.org/10.4018/979-8-3693-0400-6.ch015.

Kokalarov, M., Sakakushev, B., & Parvanov, S. (2021). Generalized methodology for application of the photogrammetric method for measurement in heavy engineering. *2021 XXXI International Scientific Symposium Metrology and Metrology Assurance (MMA)*, (07-11 September 2021), Sozopol, Bulgaria, pp. 1–5, https://doi.org/10.1109/MMA52675.2021.9610966

Kokalarov, M. (2019). Predetermination of the uncertainty budget and error analysis of the photogrammetric method in heavy machinery. *Proceedings of the 30th International DAAAM Symposium "Intelligent Manufacturing & Automation"* (23–26th October 2019, Zadar, Croatia), Volume 30, No. 1, pp. 0914–0920, Ed. B. Katalinic/Published by DAAAM International, Vienna, Austria, EU, 2019, ISBN: 978-3-902734-22-8, ISSN: 1726-9679, Vienna, Austria. https://doi.org/10.2507/30th.daaam.proceedings.127.

Ministry of Environment and Water. (2024g). Strategy and Action Plan for the Transition to a Circular Economy of the Republic of Bulgaria for the Period 2021–2027. Retrieved December 18, 2024.

Nozharov, S. (2022). The social costs of the circular economy in the EU. Prof Marin Drinov Publishing House of BAS, Sofia, ISBN: 978-619-245-199-8.

Petrova, T., Petrov, Z., & Solovev, D. (2025d). Strategies for organizational development in modern societies. In M. Okur Dinçsoy & H. Can (Eds.), *Achieving Business Excellence Through Triple Transformation*, pp. 241–276. IGI Global Scientific Publishing. https://doi.org/10.4018/979-8-3693-9435-9.ch009.

Special Report 17/2023: Circular economy. (2023e). Slow transition by Member States despite EU action.
Terziev, V., & Georgiev, M. (2023b). Challenges to active social policies in Bulgaria, gabriell-e-lit, Sofia, Bulgaria, 2023, 356 p., ISBN: 978-619-7705-20-1.
Terziev, V., & Georgiev, M. (2024a). New social transformations and changes in societies. *112th International Scientific Conference on Economic and Social Development—"Creating a unified foundation for Sustainable Development: Interdisciplinarity in Research and Education"*—Varazdin, 4–5 June, 2024, Economic and Social Development (Book of Abstracts), Cakovec, Croatia, 2024, pp. 24–25, ISSN: 1849-7543.
Terziev, V., & Georgiev, M. (2024b). Social development and the contribution of science. *112th International Scientific Conference on Economic and Social Development—"Creating a unified foundation for Sustainable Development: Interdisciplinarity in Research and Education"*—Varazdin, 4–5 June, 2024, Economic and Social Development (Book of Proceedings), Cakovec, Croatia, 2024, pp. 209–231, ISSN: 1849-7535.
Terziev, V., & Georgiev, M. (2024c). Social transformations and social programming. *112th International Scientific Conference on Economic and Social Development—"Creating a unified foundation for Sustainable Development: Interdisciplinarity in Research and Education"*—Varazdin, 4–5 June, 2024, Economic and Social Development (Book of Proceedings), Cakovec, Croatia, 2024, pp. 251–256, ISSN: 1849-7535.
Terziev, V., & Georgiev, M. (2024d). New social transformations and changes in societies. *112th International Scientific Conference on Economic and Social Development—"Creating a unified foundation for Sustainable Development: Interdisciplinarity in Research and Education"*—Varazdin, 4–5 June, 2024, Economic and Social Development (Book of Proceedings), Cakovec, Croatia, 2024, pp. 311–329, ISSN: 1849-7535.
Terziev, V., Georgiev, M., & Simeonov, S. (2025a). Transformation and change of the social and educational system. gabriell-e-lit, Sofia, 2025, 446 p. ISBN: 978-619-7705-65-2.
Terziev, V., Georgiev, M., & Simeonov, S., (2025e). The labour market in bulgaria in a period of social transformations. *ISRG Journal of Arts Humanities & Social Sciences (ISRGJAHSS)*, Volume—III Issue-I (January-February) 2025, pp. 33–45, ISSN: 2583-7672 (Online). https://doi.org/10.5281/zenodo.14615015.
Terziev, V. (2022a). COVID-19 pandemics and its impact on Bulgarian universities in the context of the new challenges to social system and labour market policies—Internauka Publishing House CUL, Kyiv, 2022, 138 p., ISBN: 978-611-01-2647-2.
Terziev, V. (2022b). Higher education system and labour market policies in Bulgaria in COVID-19 crisis and post-crisis periods—Internauka Publishing House CUL, Kyiv, 2022, 168 p., ISBN: 978-611-01-2645-8.
Terziev, V. (2023a). Innovation development of industrial business organizations. *Optimizing Energy Efficiency During a Global Energy Crisis*, edited by Meltem Okur Dinçsoy and Hamit Can, IGI Global, pp. 282–306. https://doi.org/10.4018/979-8-3693-0400-6.ch016.
Terziev, V. et al. (2025b). Social enterprise is often a part of the strategic development of a country. *Achieving Business Excellence Through Triple Transformation*, edited by Meltem Okur Dinçsoy and Hamit Can, IGI Global, pp. 141–166. https://doi.org/10.4018/979-8-3693-9435-9.ch006.
Terziev, V. et al. (2025c). Exploring social inequalities and social entrepreneurship opportunities. *Achieving Business Excellence Through Triple Transformation*, edited by Meltem Okur Dinçsoy and Hamit Can, IGI Global, pp. 277–302. https://doi.org/10.4018/979-8-3693-9435-9.ch010.
UMG "St. Ivan Rilski". (2024f). Project "Rethinking sustainable development in European regions through the use of circular economy business models". Retrieved December 18, 2024.

# Macroeconomic Implications for the Symbiotic Relationship Between Climate Change, Economic Growth and Agricultural Growth in Africa

**Prosper E. Edoja, Rosemary N. Okoh, Goodness C. Aye, and Rangan Gupta**

**Abstract** The symbiotic relationship between agricultural growth, economic growth, and climate change in Africa is explored in this research, with emphasis on the macroeconomic implications of these relationships. Using panel data for five African countries (South Africa, Kenya, Egypt, Democratic Republic of Congo, and Nigeria) from 1961 to 2021, the research employs sound econometric techniques, including Fully Modified Ordinary Least Squares (FMOLS) and Dynamic Ordinary Least Squares (DOLS), to investigate the long-run effects of climate change on income and agricultural income. The study also tests the Environmental Kuznets Curve (EKC) hypothesis, which proposes an inverted U-shaped relationship between economic growth and environmental degradation. The results indicate a long-run cointegrating relationship among $CO_2$ emissions, income, and agricultural income, with climate variables like temperature and $CO_2$ emissions having significant impacts on economic and agricultural outcomes. The evidence is found to show bidirectional causality between GDP and $CO_2$ emissions, indicating that economic growth determines environmental degradation and vice versa. The FMOLS model has mixed results to validate the EKC hypothesis, but the DOLS model supports it, which indicates that after a certain level of income, greater economic growth could reduce $CO_2$ emissions. The research highlights the necessity for African economies to harmonize economic growth with environmental protection, stressing climate-smart agriculture and adherence to international environmental treaties such as COP21. Policymakers are called upon to look into the long-term consequences of climate change on economic and agricultural expansion, especially for risk-prone areas such as Africa, whose impacts of climate change are disproportionately high.

---

P. E. Edoja (✉) · R. N. Okoh
Faculty of Agriculture, Department of Agricultural Economics, Delta State University, Abraka, Delta State, Nigeria
e-mail: edojaprosper@delsu.edu.ng

G. C. Aye
Department of Agricultural Economics, Joseph Sarwarn Tarka University, Makurdi, Nigeria

G. C. Aye · R. Gupta
Department of Economics, University of Pretoria, Hatfield, South Africa

 
P. Singh et al. (eds.), *Circular Economy and Environmental Resilience*,
https://doi.org/10.1007/978-3-031-93091-1_13

**Keywords** Climate change · Economic growth · Agricultural growth · Environmental Kuznet Curve (EKC) · Africa · FMOLS · DOLS

## 1 Introduction

Climate change, as defined by the United Nations Framework Convention on Climate Change (UNFCCC), refers to "a change in climate attributed directly or indirectly to human activity that alters the composition of the global atmosphere, in addition to natural climate variability observed over comparable time periods" (UN, 1992). Climate is inherently dynamic, characterized by long-term warming and cooling cycles (Hu & Guan, 2017). However, recent rapid and extensive changes are too extreme to be considered "normal," and these changes are closely linked to increased atmospheric carbon due to human activities (Brown et al., 2016; Edoja et al., 2016; IPCC, 2007). As one of the defining challenges of the twenty-first century, climate change is a major driver of environmental degradation and a variety of socioeconomic issues, including poverty, economic stagnation, and unsustainable development in health and safety (Dumrul & Kilicarslan, 2017; Swart et al., 2003). It has directly impacted state expenditures on flooding, forest conservation, and pollution control, while also affecting terms of trade, agricultural yields, labor productivity, natural resource depletion, and social welfare through price increases, floods, pollution, and droughts (Edoja et al., 2016; Koop & Van Leeuwen, 2016). The Intergovernmental Panel on Climate Change (IPCC) further defines climate change as "a change in the state of the climate identified by changes in the mean and/or variability of its properties that persist for decades or longer" (Jhariya et al., 2019).

The severity of climate change's impact on food security, hunger, and nutrition is influenced by factors such as population growth, income levels, and economic development. Socioeconomically disadvantaged communities are likely to face more pronounced consequences. Stige (2006), Wolfe et al. (2005), and Orindi et al. (2006) argue that rising global food demand could exacerbate food shortages and worsen hunger and malnutrition, particularly in impoverished nations. In contrast, industrialized countries, with their advanced technology, automated agriculture systems, and financial resources, are better equipped to mitigate the effects of climate change (Dwivedi et al., 2021). Developing nations like Nigeria, Egypt, and Kenya, however, face critical challenges due to high temperatures, weak adaptive capacities, and lack of early warning systems.

Nigeria, like many African countries, is already experiencing the adverse effects of climate change. Vulnerability is exacerbated by extreme environmental degradation, particularly from petroleum sector activities, heavy reliance on rain-fed agriculture, and limited access to agricultural technology. These factors have reduced the supply of ecosystem goods and services, diminishing the coping capacity of rural communities (Okoh, 2013). Agriculture, being highly sensitive to climate variability, is expected to bear the brunt of climate change impacts, as temperature and precipitation are direct inputs into agricultural production (Aye & Edoja, 2017; Barnwall & Kotani, 2013; Deschenes & Greenstone, 2007; Deressa et al., 2005).

Changes in agricultural output due to climate variability can negatively impact inflation, unemployment, current account deficits, and government budgets. Reduced agricultural production can drive up food prices, increase unemployment in rural areas, and widen current account deficits due to increased food imports (Chang & Vo, 2020; Monamodi, 2024). Additionally, government compensation for climate-related damages to producers can strain public finances (Bayraç and Doğan, 2016; Maharjan & Joshi, 2012).

Interestingly, agriculture both affects and is affected by climate change. While climate change impacts agricultural productivity, agriculture contributes to global warming through practices such as enteric fermentation, improper manure management, rice cultivation, biomass burning, livestock production, and fossil fuel use, all of which release significant amounts of greenhouse gases like $CO_2$, methane, and nitrous oxide (Omotoso & Omotayo, 2024). Other contributing factors include soil management practices, land-use changes, water use, and food waste.

The competition among nations to expand their economies often leads to adverse environmental consequences, including pollution, overexploitation of natural resources, deforestation, habitat degradation, and climate change. This is particularly evident in African countries like Nigeria, Egypt, and South Africa (Phimphanthavong, 2013).

This paper explores the bidirectional relationship between climate change and macroeconomic performance, focusing on the economic and agricultural sectors in Africa. The research addresses the following questions:

1. Is there a short-run relationship between climate change and agricultural income in Africa?
2. Is there a long-run relationship between climate change and agricultural income in Africa?
3. Is there a short-run relationship between climate change and income in Africa?
4. Is there a long-run relationship between climate change and income in Africa?
5. Is there a causal relationship between agricultural income and $CO_2$ emissions?
6. Is there a causal relationship between GDP and $CO_2$ emissions?
7. Can the Environmental Kuznets Curve (EKC) hypothesis be validated in Africa?

The study is structured as follows: a literature review, theoretical framework, model specification, methodology, findings, and policy implications. To better understand the impact of economic expansion on the environment, this paper focuses on Africa, where environmental concerns such as deforestation, soil erosion, water pollution, and $CO_2$ emissions are prevalent. Due to data limitations, the study uses $CO_2$ emissions per capita, temperature, relative humidity, and rainfall as indicators of environmental degradation. $CO_2$ emissions, which account for approximately 75% of greenhouse gas emissions (Abbasi & Riaz, 2016), are a key focus, particularly in the context of the Environmental Kuznets Curve (EKC) hypothesis, which posits that environmental degradation increases with economic activity before eventually declining (Tutulmaz, 2015). $CO_2$ emissions primarily result from burning fossil fuels, industrial processes, and waste management, making them a critical area of study in understanding the interplay between economic growth and environmental sustainability.

## 2 Literature Review

Numerous empirical investigations have been conducted to examine the effects of environmental change on agricultural income in Africa. Additionally, there are research that investigate the macroeconomic consequences of climate change on agriculture, the economy, and income. However, considering the recent upsurge in notoriety of the subject of global warming, there exists a growing effort to accurately measure the effect of climate change consequences regarding the world's economy. Innovative methods either seek to capture the prior effects caused by environmental change throughout the past 100 years (Acevedo et al., 2018; Cashin et al., 2017; Fischer et al., 2005; Gebreegziabher et al., 2011; Hsiang, 2016; Kahn et al., 2019; Zhai et al., 2009) or estimate the potential consequences of climate change liable to varying suppression approaches (Dietz & Stern, 2014; Katsouyanniet et al., 2001; Nordhaus & Yang, 1996; Nordhaus, 2013; Parry et al., 2007; Rotmans & Dowlatabadi, 1998; Wade & Jennings, 2016; Weitzman, 2012). Some relevant literature is summarized in Tables 1 and 2.

## 3 Empirical Model

In this study, African continents (proxied by South Africa, Kenya, Egypt, Democratic Republic of Congo and Nigeria) are observed for the years 1961–2021. Macroeconomic data on income (GDP, CPI) and agricultural income hereafter called (AGDP),), and trade openness (TO) were retracted from World Development Indicator (WDI, 2022), as well as Climate change data for GHG $CO_2$ emissions, Rainfall, Temperature (TEMP), and Humidity). The present study adheres to the established rules for time series modeling, including of two methodological building blocks: unit root testing as a need for making the right modeling method choice (VAR/VECM—standard abbreviations of the Vector Auto regression and Vector Error-Correction frameworks for modelling time series, respectively), and proper model specification and estimation within the selected modelling approach. In time series modeling, however, we employ a panel dataset rather than a single nation, and its analysis is different from a univariate scenario in terms of unit root tests and estimate techniques. The vast majority of the recent papers dealing with the empirical analysis of the intertwining effects between $CO_2$ emission and size of the economy used panel datasets and panel data estimation methods (Al-Mulali, 2011, 2012 in Mitic et al., 2017). According to Al-Mulali (2012) and Mitic et al. (2017), utilising panel

**Table 1** Potential microeconomic implications of climate change on African economies

| Researcher and year | Focus of study | Methodology | Key findings | Policy implications |
|---|---|---|---|---|
| Sandalli (2021) | Macroeconomic implications of climate change in Sub-saharan Africa | Panel data analysis for 84 OECD and Sub-saharan African countries 1970–2018 | Increased temperatures significantly reduce income growth, particularly in hotter and poorer sub-saharan Africa countries | Underscores the importance of addressing climate changes disequilibrium impact on global sustainable development |
| Kahn et al. (2019) | Long-term effects of climate change on economic activity | Stochastic growth model using panel data for 174 countries from 1960 to 2014 | Persistent temperature deviation from historical norms negatively affects per-capita real output growth. Changes in rainfall do not yield significant effects. Projections show a substantial reduction in world real GDP per capita by 2100 without mitigation measures | Highlights the importance of adhering to agreements like Paris Agreement to combat climate change adverse economic impacts |
| Asafu-Adjaye et al. (2021) | Macroeconomic consequences of climate change in Africa | Panel vector auto-regression model and computable general equilibrium model | Temperature shocks are inflationary and lead to a significant decline in real output in African countries. Climate-induced reduction in agricultural production negatively affect income, food supply, prices and carbon emission | It calls for technical advancement in climate smart agriculture as for the reduction of climate change effects in Africa |
| Kiley (2021) | Relationship between temperature and economic growth risk | Quartile regressions | Consistently demonstrate a strong impact of temperature on the likelihood of severe economic recessions across various dimensions | Calls for proactive climate mitigation strategies to reduce economic risks associated with temperature fluctuations |

*Source* Author's compadium 2025

**Table 2** Literature on previous research on the macroeconomic implications of climate change

| Study and author | Main findings |
|---|---|
| Asafu-Adjaye (2021) | Predicts detrimental effects of climate change on African economies, with the greatest impact on Southern Africa (−6% growth by 2050), followed by North Africa (−1.4%) and East Africa (−0.6%) |
| Fischer et al. (2005) | Anticipates a significant reduction in suitable rainfed land for cereals by the 2080s, potentially leading to the cessation of wheat production in Africa |
| Gebreegziabher et al. (2011) | Shows that climate change's economic impact on Ethiopian agricultural productivity will worsen after 2030, resulting in a potential 30% reduction in average income over 50 years |
| Cline (2007) | Projects agricultural output declines for all African countries, with greater losses in non-irrigated regions (e.g., Sudan, Senegal, Niger, Mali) and milder losses where irrigation is prevalent (e.g., South Africa, Zambia) |
| Stige et al. (2006) | Foresees significant maize production reductions in southern Africa under increased El Niño-Southern Oscillation (ENSO) conditions, assuming no adaptation |
| Abou-Hadid (2006) | By 2050, climate change may lead to a decline in Egypt's national crop production, with estimates ranging from a loss of 11% for rice to 28% for soybeans, compared to the current climate conditions |
| Seo and Mendelson (2027) | Examination of environmental variability impact on livestock farming in Africa. predictions should include A decrease of 32% in anticipated net gain with a 2.5 °C warming and a 70% reduction with a 5 °C warming. Rainfall effects were found to be relatively similar |
| Rojas-Downing et al. (2017) | The impact of the changing climate on Africa's livestock farming is anticipated to be constrained by heightened animal water consumption and the need for more crop land to meet the demand for livestock feed |
| Hertel et al. (2010) | Estimates a decline of 10–22% in the production of selected food crops in African countries under various climate change scenarios, with South Africa experiencing a potential 25–42% decline in coarse grain production |

*Source* Author's compadium 2024

datasets offers several advantages. In econometric modeling as opposed to individual time series data. The paper emphasises the benefits of employing VECM and co-integration analysis methodologies in conducting unit root testing, highlighting their capacity to control heterogeneity, enhance the level of independence, and provide stable parameter estimates.

## *3.1 Panel Unit Root Tests—Methodology*

Hossfeld, 2016 classified panel unit root tests into the First and Second generations. The two significant tests mentioned here are the Levin-Lin-Chu, 2002 (LLC) and the ImPesaran-Shin, 2003 (IPS). The tests are extensions of the traditional unit root test for modelling univariate time series using the augmented Dickey-Fuller (ADF) method, assuming individual independence strictly. The ADF unit root test for the univariate case of the random process yt is determined by approximating the test equation.

$$\Delta y_{it} = \rho y_{t-1} + \sum_{p=1}^{p} \varphi_p \Delta y_{t-p} + \gamma_l' D_l + \varepsilon_t, t = 1, \ldots, T \quad (1)$$

in which case the deterministic terms' vector, $D_l$, $l = \{1, 2, 3\}$, indicates whether the method consists both a perpectual term and a time pattern (empty set), merely a static phrase devoid of any temporal progression, or both. $D_l = \left\{ \begin{array}{c} emptyset \\ 1 \\ 1, t \end{array} \right\},$

While the error term with identical independence and normal distribution is denoted by εt, $\varepsilon_t \sim N(0,\sigma^2)$. The ADF test statistically compares the alternative—that process yt is stationary—with the null hypothesis, which states that $y_t$ is stationary. When writing the Eq. (1) in mathematical syntax, it is similar to testing $H_0$: $\rho = 0$ whichcontains the unit root. It is analogous to checking $H_0$: $\rho = 0$ opposed to the substitute $H_1$: $\rho < 0$ in mathematical notation of Eq. (1).

The error term with identical independence and normal distribution is represented by εt, where εt follows a normal distribution with mean 0 and variance σ2. The ADF test compares the alternative hypothesis, which assumes that process yt is stationary, with the null hypothesis, which assumes that yt is also stationary. When expressing Eq. (1) in mathematical notation, it resembles the hypothesis test H0: $\rho =$ 0, which includes the presence of a unit root. In mathematical notation of Eq. (1), it is comparable to testing the null hypothesis H0: $\rho = 0$ against the alternative hypothesis H1: $\rho < 0$.

The ADF test statistics are used for stationarity test of the null hypothesis that process $y_t$ has the unit root, comparing the alternative H0: $\rho = 0$ against H1: $\rho <$ 0. Estimating the following equation serves as the foundation for the ADF test in a panel case:

$$\Delta y_{i,t} = \rho_i y_{i,t-1} + \sum_{p=1}^{p_i} \varphi_{ip} \Delta y_{i,t-p} + \gamma_{li}' D_{li} + \varepsilon_{it}, t = 1, \ldots, T, i = 1, \ldots, N \quad (2)$$

The subscript i, is essentially an individual application of Eq. (1). It is assumed that the mistakes εi,t $\sim N(0, \sigma^2)$ are independent for each person. In addition to the perpectual term (individual fixed effect) and temporal trend, time dummies $\theta_t$ may also be included in vectors with predictable elements.

Another constraint is that the Ordinary Least Squares (OLS) estimator, the most widely used estimator in regression research) is deployed to estimate regression using ADF on combined (pooled) panel data, based on the assumption that each person follows the same auto-regressive method. The LLC test compares null values with alternative values, while the IPS test allows unique autoregressive processes among individuals, allowing statistical inference based on individual t-statistics.

$$\bar{t}_{NT} = N^{-1} \sum_{i=1}^{N} t_{iT}(P_i, \varphi_{i1}, \ldots, \varphi_i p_i) \tag{3}$$

where the t-statistic for testing the unit root in the ith individual process is indicated by$t_{iT}(P_i, \varphi_{i1}, \ldots, \varphi_i p_i)$; lag order $P_i$ is usually chosen based on some information requirement. To test the null hypothesis $H_0$: $\rho i = 0 \forall i$ against the alternative hypothesis $H_1$: $\exists_i \in \{1,\ldots,N\}$, $\rho_i < 0$, $\bar{t}_{NT}$ is utilized.

First-generation tests can hinder cross-sectional dependency by considering temporal fixed effects in ADF specification. Second-generation unit root tests, such as the Cross-sectional IPS test (CIPS), reduce cross-sectional independence. CIPS uses Cross-sectional ADF regression (CADF) with delayed cross-sectional averages to account for common factors. The IPS process is followed in test statistics and inference.

$$\Delta y_{i,t} = \rho_i y_{i,t-1} + \varphi_i \bar{y}_{t-1} + \varphi_i \Delta \bar{y}_t + \gamma_{li}' D_{li} + \varepsilon_{it}, t = 1, \ldots, T, i = 1, \ldots, N \tag{4}$$

The CIPS statistic is calculated using the group-mean of t-statistics derived from specific CADF equations, as explained in (3).

## 3.2 Co-integration Analysis—Methodology

This section evaluates the econometric modeling strategy proposed by Al-mulali (2011, 2012) and Mitic et al. (2017) for the link between economic expansion and carbon dioxide emissions. The modified technique consists of Granger causality testing and co-integration testing, derived from a panel data sets analysis framework. In the final stage, a more explicit modeling approach based on OLS-wise estimation methodologies for Dynamic OLS (DOLS) and Fully Modified OLS (FMOLS) is employed.

Pedroni (2004) developed the most extensive panel co-integration test framework based on the EG process, which filters out short-run influences and individual-specific deterministic patterns. Pedroni split his test statistics into two categories: tests that assume distinct processes (grouped) and tests that assume a shared process (pooled). Pedroni's (2004) panel co-integration testing is based on the co-integrating equation that is supplied.

$$y_{i,t} = \beta_i' x_{i,t} + \gamma_{li}' D_{li} + \varepsilon_{i,t} \tag{5}$$

where $x_{i,t}$ is an independent variable, or in more general case, m-dimensional vector of independent variables

When an independent variable is denoted by $x_{i,t}$, or more broadly, an m-dimensional vector of independent variables

$$x_{i,t} = {}' x_{i,t-1} + \varepsilon_{i,t} \tag{6}$$

Co-integration testing uses auxiliary regressions of residuals, derived from (5) using the EG technique, with two specifications: semi-parametric and parametric, depending on the test's parameterization.

$$\hat{\varepsilon}_{i,t} = \rho \hat{\varepsilon}_{i,t-1} + \mu_{i,t} \tag{7}$$

$$\hat{\varepsilon}_{i,t} = \rho \hat{\varepsilon}_{i,t-1} + \sum_{p=1}^{p_i} \phi_{ip} \Delta \hat{\varepsilon}_{i,t-p} + \mu^{*}_{i,t} \tag{8}$$

Pedroni (2004) used the partitioned vector $z_{i,t} = \left[ y_{i,t}, x_{i,t} \right]$ to describe the true data generating process $z_{i,t} = z_{i,t-1} + \xi_{i,t}$, where the error process is stationary and has an asymptotic covariance matrix Ω_i. This representation helps derive properties of co-integration testing.

$$\Omega_i = \begin{bmatrix} \Omega_{11i} \ \Omega_{12i} \\ \Omega_{21i} \ \Omega_{22i} \end{bmatrix}, \tag{9}$$

The long-run variance and covariance of a matrix are represented by $\Omega_{11i}$, $\Omega_{22i}$, and $\Omega_{21i} = \Omega_{12i}$, respectively. Combining convergent and dynamic covariance can decompose $\Omega_i$, using the invariance principle.

$$\Omega_i = \Omega_i^0 + \Gamma_i \Gamma_i', \tag{10}$$

Pedroni triangulated the $\Omega_i$ matrix to simplify co-integration test statistics, resulting in a lower triangular matrix $L_i$ with a partition shown as

$$L_i = \begin{bmatrix} L_{11i} \ L_{12i} \\ L_{21i} \ L_{22i} \end{bmatrix}, \tag{11}$$

in which $L_{11i} = (\Omega_{11i} - \Omega_{21i}/\Omega_{21i}^{-1}\Omega_{21i}, L_{12i} = 0, L_{21i} = \Omega_{21i}\Omega_{21i}^{-1/2}, L_{22i} = \Omega_{21i}^{1/2}$.

The following statistics for the co-integration test are computed under the assumption of cross-sectional independence or errors, $E[\xi_{i,t}, \xi_{k,l}] = 0 \ \forall$ i,t,k,l, v-statistic panel

$$Z_{\hat{v}NT} = \left( \sum_{i=1}^{N} \sum_{t=1}^{T} \hat{L}_{11i}^{-2} \hat{\varepsilon}_{i,t-1}^{2} \right) \tag{12}$$

ρ-statistic panel

$$Z_{\widehat{\rho}NT-1} = \left( \sum_{i=1}^{N} \sum_{t=1}^{T} \widehat{\varepsilon}_{i,t-1}^{2} \right) - 1 \sum_{i=t}^{N} \sum_{t=1}^{N} \left( \widehat{\varepsilon}_{i,t-1} \Delta \widehat{\varepsilon}_{i,t} - \widehat{\lambda}_{i} \right); \tag{13}$$

Semi-parametric (panel t-statistic)

$$Z_{tNT} = \left( \overset{\sim}{\sigma}{}^{2}_{NT} \sum_{i=t}^{N} \sum_{t=1}^{T} \widehat{\varepsilon}_{i,t-1}^{2} \right)^{-1/2} \sum_{i=1}^{N} \sum_{t=1}^{T} \widehat{\varepsilon}_{i,t-1} \Delta \widehat{\varepsilon}_{i,t} - \widehat{\lambda}_{i} \tag{14}$$

parametric (Panel t-statistic)

$$Z_{tNT}^{*} = \left( \tilde{S}_{NT}^{*2} \sum_{i=1}^{N} \sum_{t=1}^{T} \widehat{L}_{11i}^{-2} \widehat{\varepsilon}_{i,t-1}^{2} \right)^{-1/2} \sum_{i=1}^{N} \sum_{t=1}^{T} \widehat{L}_{11i}^{-2} \widehat{\varepsilon}_{i,t-1} \Delta \widehat{\varepsilon}_{i,t}; \tag{15}$$

Group ρ-statistic

$$\tilde{Z}_{\widehat{\rho}NT-1} = \sum_{i=1}^{N} \left[ \left( \sum_{t=1}^{T} \widehat{\varepsilon}_{i,t-1}^{2} \right)^{-1} \sum_{i=1}^{N} \left( \widehat{\varepsilon}_{i,t-1} \Delta \widehat{\varepsilon}_{i,t} - \widehat{\lambda}_{i} \right) \right] \tag{16}$$

semi-parametric (Group t-statistic)

$$\tilde{Z}_{tNT} = \sum_{i=1}^{N} \left[ \left( \widehat{\sigma}_{i}^{2} \sum_{t=1}^{T} \widehat{\varepsilon}_{i,t-1}^{2} \right)^{-1/2} \sum_{i=1}^{N} \left( \widehat{\varepsilon}_{i,t-1} \Delta \widehat{\varepsilon}_{i,t} - \widehat{\lambda}_{i} \right) \right]; \tag{17}$$

parametric (Group t-statistic)

$$\tilde{Z}_{tNT} = \sum_{i=1}^{N} \left[ \left( \widehat{S}_{i}^{*2} \sum_{t=1}^{T} \widehat{\varepsilon}_{i,t-1}^{2} \right)^{-1/2} \sum_{i=1}^{T} \left( \widehat{\varepsilon}_{i,t-1} \Delta \widehat{\varepsilon}_{i,t} \right) \right], \tag{18}$$

The set of equations, $\widehat{\lambda}_i$, $\widehat{S}_i^2$, $\widehat{\sigma}_i^2$, $\overset{\sim}{\sigma}{}^2_{NT}$, $\widehat{S}_i^{*2}$, $\tilde{S}_{NT}^{*2}$, represent the dynamic covariance, contemporaneous variance, and long-run variances of $\widehat{\mu}_{i,t}$, and $\widehat{\mu}_{i,t}^2$. The estimates of nuisance parameters are derived from residuals from (7) and (8).

$$\widehat{\lambda}_i = T^{-1}\sum_{s=1}^{K_i}\omega_{sKi}\sum_{t=s+1}^{T}\widehat{\mu}_{i,t}\widehat{\mu}_{i,t-s};\ \widehat{S}_i^2 = T^{-1}\sum_{t=1}^{T}\widehat{\mu}_{i,t};\ \widehat{\sigma}_i^2 = \widehat{S}_i^2 + 2\widehat{\lambda}_i; \tag{19}$$

$$\widetilde{\sigma}^2_{NT} = N^{-1}\sum_{i=1}^{N}\widehat{L}_{11i}^{-2}\widehat{\sigma}_i^2;\ \widehat{S}_i^{*2} = T^{-1}\sum_{t=1}^{T}\widehat{\mu}^*_{i,t};\ \widetilde{S}^{*2}_{NT} = N^{-1}\sum_{i=1}^{N}\widehat{S}_i^{*2},$$

where $\omega_s k_i = 1 - k\ /(K_i + 1)$ is a Newey-West kernel-based estimator with lag window $K_i$ (weights for all lags beyond $K_i$ are 0), and $\omega_s k_i$ stands for linearly decreasing weights. Newey-West weights and the long-run covariance estimate are used to generate nuisance estimates $L_{11i}^{-2}$.

Pedroni (2004) demonstrates that cointegration tests converge to the normal distribution standard, allowing for within-dimension and between-dimension tests. However, it struggles with managing structural discontinuities, a common issue with GDP and energy consumption. The Westerlund (2007) panel cointegration test resolves this issue by discovering structural cracks endogenously and using bootstrapping errors to account for cross-sectional dependence. When cointegration is present, FMOLS and DOLS OLS-based estimators are commonly used for panel regressions.

$$\widehat{\beta}_{NT} = \left(\sum_{i=1}^{N}\sum_{t=1}^{T}(x_{i,t} - \overline{x}_i)^2\right)^{-1}\sum_{i=1}^{N}\sum_{t=1}^{T}(x_{i,t} - \overline{x}_i) \tag{20}$$

Pedroni and Mitic et al. recommend using FMOLS for cointegration analysis on panels, with the regularized OLS modified by pooled FMOLS estimate.

$$\widehat{\beta}_{FM} = \left(\sum_{i=1}^{N}\widehat{L}_{22i}^{-1}\sum_{t=1}^{T}(x_{i,t} - \overline{x}_i)^2\right)^{-1}\sum_{i=1}^{N}\widehat{L}_{11i}^{-1}\widehat{L}_{22i}^{-1}\left(\sum_{t=1}^{T}(x_{i,t} - \overline{x}_i)y^*_{i,t} - T\delta_i\right) \tag{21}$$

In this case, $y^*_{i,t} = (y_{i,t} - \overline{y}_i) - \left(\frac{\widehat{L}_{21i}}{\widehat{L}_{22i}}\right)\Delta x_{i,t} + \left(\frac{\widehat{L}_{21i} - \widehat{L}_{22i}}{\widehat{L}_{22i}}\right)\beta(x_{i,t} - \overline{x}_i)\ and\ \widehat{\delta}_i \equiv \widehat{L}_{21i} + \widehat{\Omega}^0_{21i} - \left(\frac{\widehat{L}_{21i}}{\widehat{L}_{22i}}\right)\left(\widehat{\Gamma}_{22i} + \widehat{\Omega}^0_{22i}\right)$, and the definitions of additional variables are found in (9)–(11). When estimating dynamic co-integrated panels, Pedroni (2001) developed the FMOLS to address heterogeneity difficulties and changes in individual reactions to short-term departures from co-integrating equilibrium in dynamic co-integrated panels. By adding individual-specific intercepts and allowing error processes' serial correlation features to differ across panelists individually, the FMOLS effectively addresses these concerns and improves the accuracy of the regression.

Kao and Chiang (2001) expanded the DOLS estimator for panel analysis by developing finite sample characteristics for OLS, DOLS, and Pedroni's FMOLS, using a regression for DOLS estimation.

$$y_{i,t} = \beta_i' x_{i,t} + \sum_{j=-q}^{q} \zeta_{ij} \Delta x_{i,t+j} + \gamma_{li}' D_{li} + \varepsilon_{i,t} \tag{22}$$

where q stands for the leads and lags that are often chosen based on an informational criterion. The DOLS estimator, based on Monte Carlo simulations, outperforms other estimators for unbiased estimates in infinite samples, including FMOLS and OLS. It also reduces endogeneity in the model by reducing lead and lag differences in regressor augmentation, thus robustly correcting endogeneity in explanation-provider factors (Lean & Smyth, 2010; Afonso & Jalles, 2012).

## 3.3 *Panel Causality Test*

The study examined the short-term effects of economic development and control variables on $CO_2$ emission and the causal relationship between them. Causality measures the antecedent and informative value of one variable, like GDP, for another variable, such as $CO_2$. The panel causality test, suggested by Dumitrescu and Hurlin (2012), was applied in this investigation, accounting for heterogeneity in data and cross-section dependence using critical values derived from the block bootstrap approach. The VAR model was used to assess Granger non-causality in a panel data situation.

$$\mathrm{y}_{\mathrm{i,t}} = \lambda_{2\mathrm{I}} + \sum_{\mathrm{k}=1}^{\mathrm{k}} \alpha_{2\mathrm{I}}^{(\mathrm{k})} \mathrm{y}_{1,\mathrm{t-k}} \sum_{\mathrm{k}=1}^{\mathrm{k}} \beta_{2\mathrm{I}}^{(\mathrm{k})} \mathrm{X}_{\mathrm{I,t-k}} + \varepsilon_{2I,t}, \ \mathrm{I} = 1, \ \ldots, \ \mathrm{N}, \mathrm{t} = 1, \ldots, \ \mathrm{T} \tag{23}$$

The study uses a panel with fixed-in-time individual impacts, a constant lag order, and balanced parameters. The parameters $\alpha_{\mathrm{I}}^{(\mathrm{k})}$ and $\beta_{\mathrm{I}}^{(\mathrm{k})}$ allow for group variation. The error terms, $\varepsilon_{I,t}$, are independently and identically distributed among persons, with $\mathrm{E}(\varepsilon_{1,t} = 0)$ and $\mathrm{E}(\varepsilon_{I,t}) = \sigma_{t,i}^2$,

The Homogenous Non causation (HNC) hypothesis null hypothesis states that all cross-units in the panel do not show causation from x to y.

$$H_0 : \beta_i = 0 \forall_i = 1, \ldots, N \tag{24}$$

As per the alternative, a causal relationship exists between x and y for a minimum of one cross-unit.

$$H_0 : \beta_i = 0 \forall_i = 1, \ldots, N \tag{25}$$

$$\beta_i \neq 0 \forall_i = N_i + 1, N_i + 2, \ldots, N \tag{26}$$

The W bar and Z bar test statistics for the Homogenous Non Causality (HNC) hypothesis are calculated using a cross-sectional average of the N standard individual Wald statistics of Granger non-cause tests, also known as the W bar statistic, according to Mitic et al., 2017.

$$W_{N,T}^{Hnc} = \frac{1}{N}\sum_{i=1}^{N} W_{i,T} \tag{27}$$

The Z bar statistic is a measure of the standard deviation of a sample, equivalent to the standardized average statistic for a fixed T sample.

$$Z_{N}^{Hnc} = \sqrt{\frac{N}{2XK} X \frac{(T-2K-5)}{(T-K-3)} X \left[\frac{(T-2K-3)}{(T-2K-1)} W_{N,T}^{Hnc} - K\right]_{N\to\infty}} \xrightarrow{d} N(0,1) \tag{28}$$

The Granger causality test is conducted, excluding data from one cross section from the lagged values of data from the subsequent cross section, to test the resilience of the first method, assuming all cross sections are identical.

# 4 Empirical Results

## 4.1 *Descriptive Statistics*

The data analysis used mean, median, and standard deviation to understand the properties of the data. The mean represents the average value of the sample, while the standard deviation measures the extent of the deviation from the mean. The summary statistics for $CO_2$ and energy consumption are presented in Table 3. The mean and standard deviation for agricultural imports, exports, food production index, real exchange rate, foreign direct investment, and consumer price index are also presented. The data showed that variables were asymmetrically distributed, with kurtosis generally smaller or larger than 3. The null hypothesis that variables are normal was rejected.

## 4.2 *Cross-Sectional Dependency Test*

The study used a balanced panel dataset and cross-sectional dependency test approaches to identify cross-sectional dependency. Variables were converted to logarithmic terms for seamless unit root tests. Existing literature provides tests like Breusch-Pagan LM (1980), Pesaran scaled LM (2004), and Baltagi, Feng, and Kao bias-corrected scaled LM (2012), with the Pesaran CIPS used for this research. The

**Table 3** Descriptive statistics

| | LAGDP | LGDP | LCO2 | LCPI | LEXCH | LHUM | LRAIN | LTEMP | LTO | FDI |
|---|---|---|---|---|---|---|---|---|---|---|
| Mean | 21.909 | 24.172 | 11.658 | −0.460 | −0.718 | 4.014 | 5.978 | 3.138 | 24.336 | 1.62E + 09 |
| Median | 22.228 | 24.291 | 11.727 | 2.343 | 1.781 | 4.038 | 6.606 | 3.186 | 23.994 | 4.23E + 08 |
| Maximum | 25.466 | 27.076 | 13.234 | 9.513 | 7.596 | 4.359 | 7.771 | 3.310 | 31.727 | 4.13E + 10 |
| Minimum | 17.252 | 20.491 | 10.035 | −30.970 | −29.374 | 3.636 | 1.850 | 2.793 | 16.571 | −7.4E + 08 |
| Std. Dev | 1.875 | 1.586 | 1.036 | 9.083 | 8.685 | 0.228 | 1.774 | 0.147 | 2.643 | 3.45E + 09 |
| Skewness | −0.417 | 0.005 | 0.034 | −2.440 | −2.344 | −0.218 | −1.349 | −0.986 | 0.418 | 6.482627 |
| Kurtosis | 2.653 | 2.129 | 1.502 | 7.682 | 7.293 | 2.343 | 3.175 | 2.687 | 2.878 | 68.70369 |
| Jarque- Bera | 10.529 | 9.790 | 29.062 | 577.305 | 520.186 | 8.033 | 94.450 | 51.483 | 8.105 | 48,962.05 |
| Probability | 0.005 | 0.007 | 0.000 | 0.000 | 0.000 | 0.018 | 0.000 | 0.000 | 0.017 | 0 |

*Source* Author's findings 2025

cross-sectional test results presented in the Table 4, shows results from two statistical tests: the CIPS and the Truncated CIPS tests. Both tests are employed to determine if there is cross-sectional dependence among the variables in the data set. The CIPS and Truncated CIPS tests produce test statistics for each variable. These statistics are used to test the null hypothesis that there is cross-sectional dependence. From the Table 4 AGDP, $CO_2$, FDI, HUM, TEMP, TO, and RAIN showed significant cross-sectional dependence as their test statistics falls on any of $p < 0.1$ (*), $p < 0.05$ (**), or $p < 0.01$ (***) indicating the rejection of the null hypothesis which states that there is no Cross-sectional dependence. The CPI and EXCH on the other hand showed significant cross-sectional dependence at the 10% level, while GDP has significant cross-sectional dependence at the 5% level. The outcomes are consistent between the CIPS and Truncated CIPS tests, which suggest robustness in the test results. Based on the results observed from the table, it could be implied that Variables like FDI, TEMP, and RAIN, which show high significance at 1%, are strongly cross-sectionally dependent, indicating that their behaviors are interrelated across different sections of the data. Whereas, variables with lower significance levels, such as CPI and EXCH, the cross-sectional dependence is weaker but still present. In all, the test indicates that most variables in the dataset are significantly cross-sectionally dependent, which has important implications for modeling and inference in the study.

Based on the stationarity data presented in Table 5, you will observed that the Levin, Lin & Chu (LLC) Test, a panel unit root test that assumes a common unit root process; Im, Pesaran and Shin (IPS) Test, which allows for individual unit root processes; ADF—Fisher Chi-square Test, this Combines individual Augmented Dickey-Fuller tests to provide a test statistic for the panel; and the PP—Fisher Chi-square Test, that Combines individual Phillips-Perron tests to provide a test statistic for the panel, were all carried out to summarise the numerous unit root tests which are statistical tests used to determine whether a time series is stationary.

The results from the table above having a p-value of 0.05 indicates that the series is stationary. Therefore, we reject the null hypothesis that the series has a unit root (i.e., it is non-stationary). As per the Levin, Lin & Chu (LLC) Test, the result at level for variables such as $CO_2$, GDP, TO, RAIN are significant at various levels ($p < 0.01$ or $p < 0.05$), implying stationarity. However, at First Difference, almost all variables

**Table 4** Cross-sectional dependency tests

| Test | AGDP | CO2 | CPI | EXCH | GDP |
|---|---|---|---|---|---|
| CIPS | −2.520** | −1.676* | 1.617* | −1.695* | −1.863* |
| Truncated CIPS | −2.520** | −1.676* | 1.617* | −1.695* | −1.863* |
| Test | FDI | HUM | TEMP | TO | RAIN |
| CIPS | −4.069*** | −2.540** | −2.669** | −1.989* | −4.878*** |
| Truncated CIPS | −4.069*** | −2.540** | −2.669** | −1.989* | −4.878*** |

*Note* Null hypothesis: cross-section dependence. Levels of significance: * $p < 0.1$, ** $p < 0.05$, *** $p < 0.01$
*Source* Author's findings 2025

**Table 5** Unit root tests

| Tests | | Variables | | | | | | | | | |
|---|---|---|---|---|---|---|---|---|---|---|---|
| | | AGDP | $CO_2$ | CPI | RXR | GDP | FDI | HUM | TEMP | TO | RAIN |
| Levin, Lin & Chu | Level | −0.470 (0.319) | −5.086 (0.000) | −1.246 (0.106) | 0.340 (0.633) | −1.408 (0.080) | −0.175 (0.430) | −0.874 (0.191) | 4.012 (1.000) | −1.546 (0.061) | −4.172 (0.000) |
| | First Diff | −2.990 (0.001) | −3.089 (0.001) | −1.484 (0.069) | −3.604 (0.000) | −6.471 (0.000) | 2.370 (0.991) | −3.953 (0.000) | −4.590 (0.000) | −8.339 (0.000) | – |
| Im, Pesaran and Shin | Level | 1.709 (0.956) | −1.160 (0.123) | 2.229 (0.987) | 3.172 (0.999) | 1.399 (0.919) | 0.473 (0.682) | −2.399 (0.008) | 2.000 (0.977) | −0.115 (0.454) | −4.939 (0.000) |
| | First Diff | −6.493 (0.000) | −6.731 (0.000) | −2.565 (0.005) | −5.299 (0.000) | −6.986 (0.000) | −7.850 (0.000) | −7.185 (0.000) | −6.968 (0.000) | −8.560 (0.000) | |
| ADF—Fisher Chi-square | Level | 2.984 (0.982) | 29.910 (0.001) | 2.278 (0.994) | 1.011 (1.000) | 5.132 (0.882) | 8.063 (0.623) | 20.917 (0.022) | 5.969 (0.818) | 11.415 (0.326) | 46.260 (0.000) |
| | First Diff | 61.684 (0.000) | 65.499 (0.000) | 22.250 (0.014) | 49.243 (0.000) | 67.624 (0.000) | 79.310 (0.000) | 71.919 (0.000) | 101.881 (0.000) | 88.413 (0.000) | |
| PP—Fisher Chi-square | Level | 2.641 (0.989) | 50.397 (0.000) | 0.694 (1.000) | 0.613 (1.000) | 4.600 (0.916) | 36.129 (0.000) | 18.064 (0.054) | 14.216 (0.163) | 26.752 (0.003) | 126.678 (0.000) |
| | First Diff | 152.077 (0.000) | 202.453 (0.000) | 38.846 (0.000) | 100.990 (0.000) | 155.375 (0.000) | 192.663 (0.000) | 142.580 (0.000) | 113.071 (0.000) | 135.922 (0.000) | |

*Notes* LLC, IPS, ADF-Fisher and PP-Fisher examine the null hypothesis of non-stationarity, and ** indicates statistical significance at the 5% level. Probabilities for Fisher-type tests were computed by using an asymptotic $\chi 2$ distribution. All other tests assume asymptotic normality. The lag length is selected using the Modified Schwarz Information Criteria. All variables are in natural logarithms (LN)
*Source* Author's findings 2025

(take for instance, AGDP, $CO_2$, RXR, GDP, HUM, TEMP, TO) become stationary, meaning that they were integrated of order 1 (I(1)). For the Im, Pesaran and Shin (IPS) Test at Level, only HUM and RAIN are stationary with significant p-values ($p < 0.05$). Whereas, at First Difference, all variables became stationary, indicating that differencing the data makes them stationary. As regards the ADF - Fisher Chi-square Test, the result at Level establishes that $CO_2$, HUM, and RAIN are stationary with significant p-values while at First Difference, all variables become stationary. While, the PP - Fisher Chi-square Test shows that at Level, variables such as $CO_2$, GDP, FDI, TEMP, TO, and RAIN are stationary. While at First Difference, all variables are stationary.

In all, Stationarity at Level, showed that a few variables ($CO_2$, RAIN, and sometimes HUM) were stationarity at their levels, it can therefore be inferred that their statistical properties do not change over time without transformation. However, after first differencing, a large number of variables became stationary, indicating they are originally non-stationary and follow a unit root process. These series are integrated of order 1 (I(1)). Owing to the fact that a majority of variables are stationary after first differencing, the onus for the transformation (e.g., differenced) of the data before being used in any regression analysis to avoid spurious results became imperative. Whereas, the stationary nature of variables like $CO_2$, TEMP, and RAIN at level or after first difference implies they can be used for further analysis, such as forecasting or examining relationships with other variables.

## *4.3 Results of Cointegration Analysis*

To determine whether the regressions are spurious, the results of the panel cointegration tests must be examined. Given the results, it is appropriate to test the cointegrating relationship between the three variables. This paper employs two kinds of panel cointegration tests, i.e. Pedroni's (2004) and Kao's (1999) panel cointegration tests. For the Within-Dimension Results, the result in the Table 6 showed that Panel v-Statistic with a test statistic of (−0.576) and Panel rho-Statistic with a test statistic of (−0.704) failed being significant at the p-values of ($p = 0.718$) and ($p = 0.241$) respectively. Meaning that, there is no cointegration. Whereas, the Panel PP-Statistic and Panel ADF-Statistic test statistics of (−1.645) and (−1.821) turned out to be significant at the 5% level with the p-values of ($p = 0.050$) and ($p = 0.034$) respectively, indicating evidences of cointegration. More so, the Between-Dimension Results for Group rho-Statistic and Group PP-Statistic showed the test statistic of (0.238) and (−0.984) with the p-values of ($p = 0.594$) and ($p = 0.162$), indicative of no cointegration for the two different statistics respectively. However, the Group ADF-Statistic which has the test statistic of (−1.565) is significant at the 10% level ($p = 0.059$), providing weak evidence of cointegration. From the given observations, it has become evidently clears that the Panel PP-Statistic and Panel ADF-Statistic provide significant evidence of cointegration within the panel, suggesting a long-run relationship between income and climate change variables,

**Table 6** Cointegration test between income and climate change

| | Test statistic | Prob | | Test statistic | Prob |
|---|---|---|---|---|---|
| Within-dimension | | | Between-dimension | | |
| Panel v-Statistic | −0.576 | 0.718 | | | |
| Panel rho-Statistic | −0.704 | 0.241 | Group rho-Statistic | 0.238 | 0.594 |
| Panel PP-Statistic | −1.645** | 0.050 | Group PP-Statistic | −0.984 | 0.162 |
| Panel ADF-Statistic | −1.821** | 0.034 | Group ADF-Statistic | −1.565* | 0.059 |

*Note* Alternative: common AR coefficients for within-dimension, individual AR coefficients for between dimension. Levels of significance: * $p < 0.1$, ** $p < 0.05$, *** $p < 0.01$
*Source* Author's findings 2025

with the Group ADF-Statistic also showing a weak evidence of cointegration. While the ones with no cointegrating relationships indicates that these tests do not find evidence of a long-run equilibrium relationship.

The significant results from the Panel PP-Statistic and Panel ADF-Statistic imply that there is likely a long-run equilibrium relationship between income and climate change across the panel. This suggests that, despite short-term fluctuations, these variables are tied together over the long run. Whereas, the lack of significance in some other statistics, especially in the between-dimension tests, indicates that the evidence of cointegration is not uniformly strong across all dimensions of the data.

### 4.3.1 Cointegration Test Between Agricultural Income and Climate Change

The cointegration results in the Table 7 assess the long-run relationship between Agricultural Income and Climate Change. The test is divided into within-dimension and between-dimension tests, with several statistics provided for each. **For Within-dimension Tests, the Panel v-Statistic showed (−0.648, p = 0.741) while the Panel rho-Statistic showed (−0.734, p = 0.231). The outputs indicate** no cointegration due to the non-significant p-value. Whereas, the **Panel PP-Statistic (−1.776, p = 0.038)**, evidencing the presence of cointegration at the 5% significance level, implying a long-run relationship between the variables. The **Panel ADF-Statistic (−1.666, p = 0.048)**, also indicated the presence of cointegration at the 5% significance level, further supporting the existence of a long-term relationship. **However, the Between-dimension Tests, the Group rho-Statistic (−0.194, p = 0.423)**, shows no evidence of cointegration, with the p-value being non-significant. The **Group PP-Statistic (−1.617, p = 0.053)**, showed marginal significant at the 10% level, suggesting weak evidence of cointegration. **Group ADF-Statistic (−1.669, p = 0.048)**, indicates cointegration at the 5% significance level, confirming a long-run relationship between the variables.

The foregoing observations show mixed evidences of cointegration, with some statistics (especially within the dimension) strongly indicating a long-run relationship

**Table 7** Cointegration test between agricultural income and climate change

| | Test statistic | Prob | | Test statistic | Prob |
|---|---|---|---|---|---|
| Within-dimension | | | Between-dimension | | |
| Panel v-statistic | −0.648 | 0.741 | | | |
| Panel rho-statistic | −0.734 | 0.231 | Group rho-Statistic | −0.194 | 0.423 |
| Panel PP-statistic | −1.776** | 0.038 | Group PP-Statistic | −1.617 | 0.053 |
| Panel ADF-statistic | −1.666** | 0.048 | Group ADF-Statistic | −1.669 | 0.048** |

*Note* Alternative: common AR coefficients for within-dimension, individual AR coefficients for between dimension. Levels of significance: * $p < 0.1$, ** $p < 0.05$, *** $p < 0.01$
*Source* Author's findings 2025

between Agricultural Income and Climate Change, while others do not. However, the significant Panel PP-Statistic, Panel ADF-Statistic, and Group ADF-Statistic suggest that, on balance, there is evidence of cointegration, implying a long-term relationship between Agricultural Income and Climate Change.

## *4.4 Cointegration Test Between Agricultural Income and Climate Change*

The Kao (1999) residual analysis test showed that ADF t-statistic value is −3.761 and The p-value is 0.000. The ADF t-statistic is highly significant with a p-value of 0.000. This result strongly indicates the presence of cointegration between Income and Climate Change. In other words, there is a long-run equilibrium relationship between income and climate change. Therefore, we accept the null hypothesis which state that there is a longrun equilibrium relationship between income and climate change (Table 8).

**Table 8** Kao (1999)'s residual cointegration test results income and climate change

| | t-statistic Prob | t-statistic Prob |
|---|---|---|
| ADF | −3.761 | 0.000 |

*Note* Alternative: common AR coefficients for within-dimension, individual AR coefficients for between dimension. Levels of significance: * $p < 0.1$, ** $p < 0.05$, *** $p < 0.01$
*Source* Author's findings 2025

**Table 9** Kao (1999)'s residual cointegration test results Agricultural Income and Climate Change

| | t-statistic Prob | t-statistic Prob |
|---|---|---|
| **ADF** | −3.383 | 0.000 |

*Note* Alternative: common AR coefficients for within-dimension, individual AR coefficients for between dimension. Levels of significance: * $p < 0.1$, ** $p < 0.05$, *** $p < 0.01$
*Source* Author's findings 2025

## *4.5 Cointegration Test Between Agricultural Income and Climate Change*

**ADF t-statistic showed t**he value of −3.383 and a p-value of 0.000. The ADF t-statistic is significant, with a p-value of 0.000, indicating strong evidence of cointegration between Agricultural Income and Climate Change. This suggests that there is a long-term equilibrium relationship between these two variables, meaning that they are likely to move together over time despite short-term deviations. We therefore accept the null hypothesis of there being a long-run relationship between Agricultural income and climate change (Table 9).

## *4.6 FMOLS and DOLS Estimates of the Long Run Effect of Climate Change on Income (LGDP as Dependent Variable) and on Agricultural Income*

The Fully Modified Ordinary Least Squares (FMOLS) and the Dynamic Ordinary Least Squares (DOLS) are econometric techniques used to estimate the long-run relationship between variables in the presence of cointegration, addressing issues of endogeneity and serial correlation.

**FMOLS Estimates**

The coefficient of carbon dioxide emissions ($LCO_2$) is 1.009 for LGDP, indicating a positive and significant 1% effect on income. While for agricultural income (LAGDP), the coefficient is 0.640, also positive and significant at 1%. Temperature (LTEMP) has a positive effect on both LGDP (2.430) and LAGDP (9.678), with statistical significance of 5% noted for LGDP. The results presented in the Table 10 and further reveals that Consumer Price Index (LCPI) has a strong positive and significant effect on both LGDP (0.852) and LAGDP (0.674). With exchange rate (LEXCH) having a negative and significant effect on both LGDP (−0.766) and LAGDP (−0.645). Same also is Trade openness (LTO), which has a positive and significant effect on LGDP (0.080) and also a significant effect on LAGDP (0.640). However, humidity (LHUM) showed a coefficient of 2.969 for LGDP, suggesting a positive effect, though it is not statistically significant. For LAGDP, the coefficient is 1.819, similarly positive but not significant. Likewise, Rainfall (LRAIN) which has

**Table 10** FMOLS and DOLS Estimates of the long run effect of Climate change on Income (LGDP as dependent variable) and on Agricultural Income

| Variable | LGDP | | LAGDP | |
|---|---|---|---|---|
| | FMOLS | DOLS | FMOLS | DOLS |
| LCO2 | 1.009 (6.879***) | C (2.728**) | 0.640 (3.292***) | |
| LHUM | 2.969 (0.883) | 3.432 (0.858) | 1.819 (0.337) | 0.894 (3.409**) |
| LRAIN | 0.009 (0.065) | 0.118 (0.598) | 0.068 (0.351) | −1.911 (−0.273) |
| LTEMP | 2.430 (1.150**) | 2.261 (0.790**) | 9.678 (3.049***) | 0.064 (0.226) |
| LCPI | 0.852 (10.768***) | 0.564 (5.667***) | 0.674 (6.513***) | 9.859 (2.399*) |
| LEXCH | −0.766 (−10.702***) | −0.779 (−9.150***) | −0.645 (−6.361***) | 0.627 (6.415***) |
| LTO | 0.080 (2.348***) | 0.390 (6.379***) | 0.640 (3.292***) | −0.603 (−6.363***) |
| Adj. R-squared | 0.969 | 0.986 | 0.948 | 0.971 |

*Notes* Asymptotic distribution of t-statistic is standard normal as T and N go to infinity. ** Indicates that the parameter is significant at the 5% level
*Source* Author's findings 2025

a negligible effect on both LGDP and LAGDP, with coefficients of 0.009 and 0.068, respectively, both of which are statistically insignificant.

**DOLS Estimates**

The coefficient for $LCO_2$ under DOLS is statistically significant and positive for both LGDP and LAGDP. The coefficient for LHUM is higher for LGDP (3.432) than for LAGDP (0.894), but only significant in the latter at the 5% level. LRAIN shows a small and insignificant effect on LGDP and a negative but insignificant effect on LAGDP. Whereas, temperature has a significant positive effect on LGDP (2.261), though its impact on LAGDP is minimal and insignificant. The LCPI remains significant at 1% and 5% levels respectively with positive coefficients for both LGDP and LAGDP. Exchange rate has a consistently negative impact (−0.779 with a t-statistic of −9.150) on LGDP and positive impact (0.627 with a t-statistic of 6.415) on LAGDP under DOLS, with 1% significant level for both variables. While, Trade openness is significantly positive for LGDP (0.390) but has a significant negative impact on LAGDP (−0.603).

In all, FMOLS and DOLS provide estimates that adjust for endogeneity and serial correlation, making them robust for long-run estimations. The findings suggest that climate variables such as $CO_2$ and temperature have significant long-run effects on income and agricultural income. Economic factors like CPI, exchange rate, and trade openness also show significant impacts, although their direction and magnitude vary between GDP and agricultural GDP. The adjusted R-squared values indicate a high level of explanatory power for both methods, implying that the models are well-fitted to the data. The outcome of this investigation validates the stand of Amaefule et al., 2023, who posits that through its direct effects on unexpected rainfall, drought, and flood, $CO_2$ emissions disrupt agricultural production, which thwarts efforts to reduce poverty through employment channels and increases hunger by causing food spoilage and famine.

## *4.7 Dumitrescu and Hurlin (2012) Panel Causality Test*

The null hypothesis that changes in GDP do not cause changes in $CO_2$ emissions is strongly rejected, as indicated by the significant W-Stat (9.167) and Zbar-Stat (7.346), with a p-value of 0.000, indicating significance at 1% level. This result suggests that GDP changes have a significant effect on $CO_2$ emissions, indicating that economic growth leads to increased $CO_2$ emissions in the panel data. Similarly, the null hypothesis that changes in $CO_2$ emissions do not cause changes in GDP is also strongly rejected. The significant W-Stat (5.706) and Zbar-Stat (3.760), with a p-value of 0.000 imply that $CO_2$ emissions also have a significant impact on GDP. This suggests that environmental factors like $CO_2$ emissions can influence economic growth in the panel.

The findings indicate **bidirectional causality** between GDP and $CO_2$ emissions. This means that not only does economic growth influence environmental degradation (increased $CO_2$ emissions), but environmental factors also play a role in shaping economic outcomes. These results highlight the interdependent relationship between economic activity and environmental impact, suggesting that policies aimed at economic growth must also consider their environmental consequences, and vice versa. This contradicts Aye and Edoja's (2017) claim that GDP alone has predictive ability for $CO_2$ emission. The findings are significant deviations from previous research (Table 11).

**Table 11** Dumitrescu and Hurlin (2012) panel causality test

| | W-Stat | Zbar-Stat | Prob |
|---|---|---|---|
| D(GDP) does not homogeneously cause D($CO_2$) | 9.167 | 7.346 | 0.000 |
| D($CO_2$) does not homogeneously cause D(GDP) | 5.706 | 3.760 | 0.000 |

*Notes* The W bar test statistic corresponds to the cross sectional average of the N standard individual Wald statistics of Granger non causality tests. The Z bar statistic corresponds to the standardized statistic (for fixed T sample)
*Source* Author's findings 2025

## 4.8 *Environmental Kuznets Curve (EKC) Hypothesis*

The Environmental Kuznets Curve (EKC) hypothesis suggests that the relationship between environmental degradation (e.g., $CO_2$ emissions) and economic growth (GDP) follows an inverted U-shape. Initially, as an economy grows, environmental degradation increases, but after reaching a certain level of income, further economic growth leads to environmental improvement.

**FMOLS and DOLS Estimates**

The LGDP observed from the Table 12, for FMOLS shows the coefficient is −0.702, but it is not statistically significant, indicating no strong evidence of a linear relationship between GDP and $CO_2$ emissions. Likewise, the DOLS, with a coefficient of 0.861, also not significant, positing no clear linear effect of GDP on $CO_2$ emissions in the dynamic model. However, the $LGDP^2$ FMOLS, shows a coefficient of 0.020 which is statistically significant at the 1% level, indicating a positive relationship between the square of GDP and $CO_2$ emissions. This suggests that as GDP increases, $CO_2$ emissions increase at an increasing rate. *Whereas the DOLS for* $LGDP^2$ *shows the coefficient is −0.011 and statistically significant at the 1% level, indicating a negative relationship between the square of GDP and* $CO_2$ *emissions. This suggests a decreasing rate of increase in* $CO_2$ *emissions as GDP grows, supporting the EKC hypothesis.* LEXCH for both FMOLS (0.036) and DOLS (0.058) coefficients are positive and significant, indicating that an increase in the exchange rate is associated with higher $CO_2$ emissions. The LTO FMOLS coefficient is −0.025 but not significant, suggesting no clear effect of trade openness on $CO_2$ emissions. Whereas the DOLS coefficient is −0.072 and significant, indicating that greater trade openness is associated with lower $CO_2$ emissions, potentially supporting the idea that more open economies adopt cleaner technologies. For the LFDI, both FMOLS and DOLS coefficients are positive and significant, suggesting that foreign direct investment leads to an increase in $CO_2$ emissions, which could be due to the "pollution haven" effect, where investments are made in countries with less stringent environmental regulations.

It can therefore be surmised from the foregoing that the results present mixed evidence for the EKC hypothesis. The positive and significant coefficient for $LGDP^2$ in the FMOLS model suggests a potential violation of the EKC hypothesis, indicating that $CO_2$ emissions may increase with economic growth even at higher income levels. However, the DOLS model, with a significant negative coefficient for $LGDP^2$, supports the EKC hypothesis, implying that after reaching a certain income level, further growth leads to reduced $CO_2$ emissions. The results highlight the complexity of the relationship between economic growth and environmental degradation, suggesting that policy measures must be tailored to the specific economic and environmental context of a country.

**Table 12** Table

| | $LCO_2$ | |
|---|---|---|
| Variable | FMOLS | DOLS |
| LGDP | −0.702 (−0.388) | 0.861 (0.827) |
| $LGDP^2$ | 0.020 (0.008)*** | −0.011 (0.016)*** |
| LEXCH | 0.036 (0.015)** | 0.058 (0.017)*** |
| LTO | −0.025 (0.017) | −0.072 (0.034)** |
| LFDI | 0.000 (0.000)** | 0.000 (0.000)** |
| Adj. R-squared | 0.986 | 0.989 |

*Note* Standard errors in parentheses. Levels of significance: * $p < 0.1$, ** $p < 0.05$, *** $p < 0.01$. FMOLS, Fully Modified OLS; DOLS, Dynamic OLS; OLS, Ordinary Least Squares
*Source* Author's findings 2025

## 5 Discussion

The cointegration analysis for African economies involved the use of DOLS and FMOLS methodologies, as established by Pedroni (2004) and Kao and Chiang (2017) respectively. The results clearly demonstrate the presence of a statistically significant long-term cointegrating link between carbon emissions and income, as well as between $CO_2$ emissions and agricultural revenue. The modified $R^2$ is around 0.94, indicating that the specified model effectively explains the variation in the data. This conclusion aligns with the findings of Mitic et al., (2017) and Tamazian and Rao (2010), who examined the relationship between GDP and $CO_2$ emissions in a cohort of transitional states. The main difference between Tamazian and Rao (2010) and Mitic et al., (2017) and our research lies in the fact that they focused on estimating static and dynamic contemporaneous correlations between income and $CO_2$, while our study examines the long-term cointegration of Africa by selecting a representative country from each of the five African regions. Another notable distinction is our ability to validate the EKC hypothesis tests, which they neglected to accomplish.

Africa would have the greatest impact of climate change under all climatic scenarios with a temperature increase of over 1.5 degrees Celsius. Although Africa has the lowest emissions and the least amount of greenhouse gas emissions, it is experiencing significant collateral damage that presents systemic risks to its economies, infrastructure investments, water and food systems, public health, agriculture, and livelihoods. This damage has the potential to reverse the modest development progress made by the continent and push it back into higher levels of extreme poverty. In order to alter this narrative, it is imperative to devote significant focus on the African economy, agriculture, and the interdependent connection between macroeconomic factors and climate change. Therefore, the results of this study, which have confirmed the EKC hypothesis and shed more light on the causal connection between $CO_2$ emissions and GDP, as well as $CO_2$ emissions and AGDP, should be carefully considered. This will help Africa to improve its image and effectively manage its pursuit of a better quality of life and stronger economies. It is important to recognise that economic development is accompanied by $CO_2$ emissions, and industrialization is closely linked to $CO_2$ emissions.

Furthermore, African economies as a collective entity must also comply with global policy incentives such as COP 21 and make further endeavours to achieve the objectives established by this and other international agreements. Despite not being prominent contributors to global $CO_2$ emissions or other greenhouse gases, a considerable number of African countries have ratified the COP 21 agreement as a demonstration of their endorsement for a sustainable economy. The findings of this study are thus capturing the interest of policymakers, highlighting the pressing requirement to review and modernise existing policies and laws. Additionally, it emphasises the necessity to establish novel environmental policies and regulations aimed at reducing pollutant emissions, especially those caused by human activities. Carbon capture and storage (CCS) is widely recognised as a crucial strategy for attaining $CO_2$ emission reduction targets. It is considered one of the specific methods to mitigate $CO_2$ emissions (Leung et al., 2014). Efforts are underway to create technologies that can transport $CO_2$ emissions from major point sources of fossil fuel use to secure geological storage sites, instead of releasing them into the atmosphere (Gibbins et al., 2008; Herzog & Golomb, 2004).

## 6 Conclusion and Final Remarks

This study has explored the intricate relationship between climate change, economic growth, and agricultural income in Africa, employing advanced econometric techniques such as Fully Modified Ordinary Least Squares (FMOLS), Dynamic Ordinary Least Squares (DOLS), and panel cointegration tests. The findings provide critical insights into the long-run and causal relationships between these variables, offering valuable implications for policymakers and future research directions.

### *6.1 Key Findings*

1. **Long-Run Relationships:** The cointegration analysis revealed significant long-run relationships between climate change variables (such as $CO_2$ emissions and temperature) and both income (GDP) and agricultural income (AGDP). This suggests that climate change has a persistent impact on economic and agricultural performance in Africa, underscoring the need for sustainable policies to mitigate these effects.
2. **Bidirectional Causality:** The Dumitrescu and Hurlin (2012) panel causality test confirmed a bidirectional causal relationship between GDP and $CO_2$ emissions.

This implies that economic growth drives environmental degradation, while environmental factors also influence economic outcomes. This interdependence highlights the necessity for balanced policies that promote economic development while addressing environmental sustainability.

3. **Environmental Kuznets Curve (EKC) Hypothesis:** The study provided mixed evidence for the EKC hypothesis. While the FMOLS model suggested that $CO_2$ emissions increase with economic growth, the DOLS model supported the EKC hypothesis, indicating that after a certain income threshold, further growth could lead to reduced emissions. This duality underscores the complexity of the relationship between economic development and environmental degradation in Africa.
4. **Impact of Climate Variables:** Climate variables such as temperature and $CO_2$ emissions were found to have significant effects on both GDP and agricultural income. For instance, higher temperatures positively impacted agricultural income, likely due to the reliance on rain-fed agriculture in many African countries. However, the negative effects of $CO_2$ emissions on agricultural productivity highlight the vulnerability of the sector to climate change.
5. **Economic Factors:** Exchange rates, trade openness, and foreign direct investment (FDI) were also found to influence $CO_2$ emissions and economic growth. For example, trade openness was associated with lower $CO_2$ emissions in the DOLS model, suggesting that more open economies may adopt cleaner technologies. Conversely, FDI was linked to increased emissions, possibly due to the "pollution haven" effect.

## *6.2 Policy Implications*

The findings of this study have several important policy implications:

- **Sustainable Development:** African nations must prioritize sustainable development strategies that balance economic growth with environmental preservation. Policies should focus on reducing $CO_2$ emissions through cleaner technologies, renewable energy, and carbon capture and storage (CCS) initiatives.
- **Agricultural Resilience:** Given the vulnerability of the agricultural sector to climate change, investments in climate-resilient agricultural practices, such as drought-resistant crops, irrigation infrastructure, and early warning systems, are crucial. These measures can enhance food security and reduce the economic impacts of climate variability.
- **Economic Diversification:** To mitigate the adverse effects of climate change on economic growth, African countries should diversify their economies away from climate-sensitive sectors like agriculture. Encouraging industrialization and technological innovation can reduce dependence on agriculture and create more resilient economic structures.
- **Global Cooperation:** African nations should actively participate in global climate agreements, such as COP 21, to access funding and technology transfers that

support climate adaptation and mitigation efforts. International cooperation is essential for addressing the transboundary nature of climate change.

## 6.3 Future Research Directions

While this study provides valuable insights, several areas warrant further investigation:

- **Regional Disparities:** Future research could explore regional disparities within Africa to understand how different sub-regions are affected by climate change and economic growth. This could help tailor policies to specific regional needs.
- **Sectoral Analysis:** A more detailed sectoral analysis, focusing on industries such as manufacturing, energy, and services, could provide a clearer picture of how climate change impacts different sectors of the economy.
- **Technological Innovations:** Research into the role of technological innovations in mitigating climate change and promoting sustainable growth is essential. This could include studies on renewable energy adoption, carbon capture technologies, and climate-smart agriculture.
- **Policy Evaluation:** Future studies could evaluate the effectiveness of existing climate policies in Africa and identify best practices for reducing emissions while promoting economic growth. This could involve case studies of successful policy implementations in other developing regions.

## 6.4 Final Remarks

This study underscores the urgent need for African nations to address the dual challenges of climate change and economic development. The findings highlight the interconnectedness of economic growth, agricultural productivity, and environmental sustainability, emphasizing the importance of integrated policies that promote resilience and sustainability. By adopting a proactive approach to climate adaptation and mitigation, Africa can not only safeguard its economic future but also contribute to global efforts to combat climate change. The road ahead is challenging, but with targeted policies, international cooperation, and continued research, Africa can navigate the complexities of climate change and achieve sustainable development.

# References

Abbasi, F., & Riaz, K. (2016). CO2 emissions and financial development in an emerging economy: An augmented VAR approach. *Energy Policy, 90*(C), pages 102–114.

Abidoye, B. O., & Odusola, A. F. (2015). Climate change and economic growth in Africa: An econometric analysis. *Journal of African Economies, 24*(2), 277–301.

Abou-Hadid, A. F. (2006). Assessment of impacts, adaptation, and vulnerability to climate change in North Africa: Food production and water resources. *A Final Report Submitted to Assessments of Impacts and Adaptations to Climate Change (AIACC).*Washington, DC.
Acevedo, M. F., Harvey, D. R., & Palis, F. G. (2018). Food security and the environment: Interdisciplinary research to increase productivity while exercising environmental conservation. *Global food security, 16*, 127–132. https://doi.org/10.1016/j.gfs.2018.01.001
Adams, R., Brian H. Stephanie, L and Leary, N. 1998. Effects of global climate change on agriculture: an interpretative review. *Climate Research, 11*: 19–30, 19 98
Addison, T. (2015). Thirty years in Africa's development from structural adjustment to structural transformation? (WIDER Working Paper No. 2015/119 Thirty). Helsinki, Finland.
Al-Mulali, U. (2011). Oil consumption, CO2 emission and economic growth in MENA countries. *Energy, 36*, 6165–6171.
Al-mulali, U. (2012). Factors affecting CO2 emission in the Middle East: A panel data analysis. *Energy, 44*, 564–569.
Amponsah, L., Hoggar, G. K., & Asuamah, S. Y. (2015). Climate change and agriculture: Modeling the impact of carbon dioxide emission on cereal yield in Ghana. *Agriculture and Food Sciences Research, 2*(5), 32–38.
Asafu-Adjaye, J., Mahadevan, R., & Sterner, T. (2021). Macroeconomic consequences of climate change in Africa: A panel analysis. *Environmental Economics Journal, 12*(3), 45–67.
Aye, G. C., & Edoja, P. E. (2017). Effect of economic growth on CO2 emission in developing countries: Evidence from a dynamic panel threshold model. *Cogent Economics & Finance, 5*(1), 1379239.
Baig, I. A., Chandio, A. A., Ozturk, I., Kumar, P., Khan, Z. A., & Salam, M. (2022). Assessing the long and short-run asymmetrical effects of climate change on rice production: Empirical evidence from India. *Environmental Science and Pollution Research, 29*, 34209–34230. https://doi.org/10.1007/s11356-021-18014-z
Barnwal, P., & Kotani, K. (2013). Climatic impacts across agricultural crop yield distributions: An application of quantile regression on rice crops in Andhra Pradesh, India. *Ecological Economics, 87*, 95–109.
Bayraç, H. N., & Doğan, E. (2016). Türkiye'de İklim Değişikliğinin Tarım Sektörü Üzerine Etkileri. *Eskişehir Osmangazi Üniversitesi İİBF Dergisi, 11*(1), 23–48.
Boko, M. (2007). Climate change: Impacts, vulnerabilities, and adaptation in developing countries. Climate Change 2007: Impacts, Adaptation and Vulnerability. The Working Group II Contribution to the IPCC Fourth Assessment Report, (433–467).
Brown, M. A., Smith, T., & Johnson, R. (2016). Human-induced climate dynamics: A meta-analysis. *Climatic Change, 134*(2), 301–315.
Burke, M., Hsiang, S. M., & Miguel, E. (2015). Global non-linear effect of temperature on economic production. *Nature, 527*(7577), 235–239.
Cashin, P., Mohaddes, K., & Raissi, M. (2017). Fair weather or foul? The macroeconomic effects of El Niño. *Journal of International Economics, 106*, 37–54.
Chang, C., & Vo, D. H. (2020). *Contemporary issues in business and economics*. MDPI.
Cline, W. R. (2007). *Global warming and agriculture: Impact estimates by country*. Peterson Institute.
Cuervo, J. & Gandhi, V.P. (1998). Carbon taxes: Their macroeconomic effects and prospects for global adoption: a survey of the literature, Washington, IMF Working Paper, WP/98/73.
Dell, M., Jones, B. F., & Olken, B. A. (2012). Temperature shocks and economic growth: Evidence from the last halfcentury. *American Economic Journal: Macroeconomics, 4*(3), 66–95.
Deressa, T., Hassan, R., & Poonyth, D. (2005). Measuring the economic impact of climate change on South Africa's Agriculture: The case of sugarcane growing regions. *Agrekon, 44*(4), 524–542.
Deschenes, O., & Greenstone, M. (2007). The economic impacts of climate change: Evidence from agricultural output and random fluctuations in weather. *The American Economic Review, 97*(1), 354–385.

Dietz, S., & Stern, N. (2014). Endogenous growth, convexity of damages, and climate risk. *Journal of Environmental Economics, 68*(1), 133–153.
Dumrul, Y., Kilicarslan, Z., (2017). Economic impacts of climate change on agriculture: empirical evidence from ARDL approach for Turkey. *Journal of Business, Economics and Finance (JBEF), 6*(4), 336–347. Permemant link to this document: https://doi.org/10.17261/Pressacademia.2017.766
Dwivedi, Y. K., Hughes, L., Kar, A. K., Baabdullah, A. M., Grover, P., Abbas, R., Andreini, D., Abumoghli, I., Barlette, Y., Bunker, D., Kruse, L. C., Constantiou, I., Davison, R. M., De, R., Dubey, R., Fenby-Taylor, H., Gupta, B., He, W., Kodama, M., & Wade, M. (2021). Climate change and COP26: Are digital technologies and information management part of the problem or the solution? An editorial reflection and call to action. *International Journal of Information Management, 63*, Article 102456. https://doi.org/10.1016/j.ijinfomgt.2021.102456
Edoja, P. E., Aye, G. C., & Abu, O. (2016). Dynamic relationship among CO2 emission, agricultural productivity and food security in Nigeria. *Cogent Economics & Finance, 4*(1), 1204809.
Fankhauser, S., & McDermott, T. K. J. (2014). Understanding the adaptation deficit: Why are poor countries more vulnerable to climate events than rich countries? *Global Environmental Change, 27*, 9–18. https://doi.org/10.1016/j.gloenvcha.2014.04.014
Fischer, G., Shah, M., Tubiello, F. N., & Van Velhuizen, H. (2005). Socio-economic and climate change impacts on agriculture: an integrated assessment, 1990–2080. *Philosophical Transactions of the Royal Society B: Biological Sciences, 360*(1463), 2067–2083.
Gebreegziabher, Z., A. Mekonnen, A., & Kassahun M. M. (2011). Crop-Livestock inter-linkages and climate change implications for Ethiopia's agriculture: a Ricardian approach. Photocopy. Unpublished research, Environmental Economics Policy Forum for Ethiopia, Addis Ababa.
Gibbins, J. & Chalmers, H. (2008). Carbon capture and storage. *Energy Policy, 36*, 4317–4322.
Hanjra, M. A., & Qureshi, M. E. (2010). Global water crisis and future food security in an era of climate change. *Food Policy, 35*, 365–377.
Hertel, T. W., Burke, M. B., & Lobell, D. B. (2010). The poverty implications of climate-induced crop yield changes by 2030. *Global Environmental Change, 20*(4), 577–585.
Herzog, H., & Golomb, D. (2004). Carbon capture and storage from fossil fuel use. *Encyclopedia Energy, 1*, 1–11.
Hsiang, S. (2016). Climate econometrics. *Annual Review of Resource Economics, 8*(1), 43–75.
Hu, A., & Guan, Q. (2017). *China: Tackle the challenge of global climate change*. Taylor & Francis.
IPCC (Intergovernmental Panel on Climate Change) (2007a). Summary for policymakers. Climate change 2007: The physical science basis, Working Group I Contribution to IPCC Fourth Assessment Report: Climate Change 2007. Geneva.
Jhariya, M. K., Banerjee, A., Meena, R. S., & Yadav, D. K. (2019). *Sustainable agriculture, forest and environmental management*. Springer.
Kahn, M. E., Mohaddes, K., Ng, R. N. C., Pesaran, M. H., Raissi, M. and Yang, J. (2019). Long-term macroeconomic effects of climate change: A cross-country analysis. IMF Working Paper Fiscal Affairs Department. WP/19/215.
Kao, C., Chiang, M. (2001). On the estimation and inference of a cointegrated regression in panel data. In B. H. Baltagi, T. B. Fomby, R. C. Hill, (Eds.), *Nonstationary Panels, Panel Cointegration, and Dynamic Panels*. Advances in Econometrics; Emerald Group Publishing Limited, 2001; Volume 15, pp. 179–222.
Katsouyanni, K., et al. (2001). Confounding and effect modification in the short-term effects of ambient particles on total mortality: Results from 29 European cities within the APHEA2 project. *Epidemiology, 12*(5), 521–531.
Khurshid, N., Fiaz, A., Khurshid, J., & Ali, K. (2022). Impact of climate change shocks on economic growth: A new insight from non-linear analysis. *Frontiers in Environmental Science, 10*, 1039128. https://doi.org/10.3389/fenvs.2022.1039128
Kiley, M. T. (2021). Growh at risk from climate change. FEDS Working Paper No. 2021–54.

Koop, S. H. A., & Van Leeuwen, C. J. (2016). The challenges of water, waste and climate change in cities. *Environment Development and Sustainability, 19*(2), 385–418. https://doi.org/10.1007/s10668-016-9760-4

Lagarde, C. (2014). Africa rising—Building to the future. Maputo, Mozambique: IMF. Retrieved from http://www.imf.org/external/np/speeches/2014/052914.htm

Leung, D. Y., Caramanna, G., & Maroto-Valer, M. M. (2014). An overview of current status of carbon dioxide capture and storage technologies. *Renewable and Sustainable Energy Reviews, 39*, 426–443.

Maharjan, K. L., & Joshi, N. P. (2012). *Climate change, agriculture, and rural livelihoods in Nepal.* Springer.

Mason-D'Croz, D., Bogard, J. R., Herrero, M., Robinson, S., Sulser, T. B., Wiebe, K., Willenbockel, D., & Godfray, H. C. J. (2020). Modelling the global economic consequences of a major African swine fever outbreak in China. *Nature Food, 1*(4), 221–228. https://doi.org/10.1038/s43016-020-0057-2

McMillan, M., Rodrik, D., & Sepúlveda, C. (2017). *Structural change, fundamentals, and growth: a framework and case studies.* Intl Food Policy Res Inst.

Mishra, D., & Sahu, N. C. (2014). Economic impact of climate change on agriculture sector of coastal Odisha. *APCBEE Procedia, 10*, 241–245.

Mitic, P., Fedajev, A., & Radulescu, M. (2017). Panel cointegration analysis of energy consumption and GDP in transition economies. *Energy Policy, 105*, 254–263.

Monamodi, N. E. (2024). The impact of current account balance on economic growth in South Africa. *Economies, 12*(2), 39. https://doi.org/10.3390/economies12020039

Nordhaus, W. D. (2013). *The climate casino: Risk, uncertainty, and economics for a warming world.* Yale University Press.

Nordhaus, W. D., & Yang, Z. (1996). A regional dynamic general-equilibrium model of alternative climate-change strategies. *American Economic Review, 86*(4), 741–765.

Okoh, R. N. (2013). Petroleum exploitation and environmental degradation in Nigeria. *Journal of Sustainable Development, 6*(4), 1–12.

O'Neill, B. C., Kriegler, E., Ebi, K. L., Kemp-Benedict, E., Riahi, K., Rothman, D. S., Van Ruijven, B. J., Van Vuuren, D. P., Birkmann, J., Kok, K., Levy, M., & Solecki, W. (2015). The roads ahead: Narratives for shared socioeconomic pathways describing world futures in the 21st century. *Global Environmental Change, 42*, 169–180. https://doi.org/10.1016/j.gloenvcha.2015.01.004

Omotoso, A. B., & Omotayo, A. O. (2024). The interplay between agriculture, greenhouse gases, and climate change in Sub-Saharan Africa. *Regional Environmental Change, 24*, 1. https://doi.org/10.1007/s10113-023-02159-3

Orindi, V. A., Nyong, A., & Herrero, M. (2006). Pastoral livelihood adaptation to drought in Kenya. *International Institute for Environment and Development.*

Parry, M. et al. (2001). Millions at risk. Defining critical climate change targets and threats. *Global Environmental Change, 11*: 1–3

Pedroni, P. (2004). Panel cointegration: Asymptotic and finite sample properties of pooled time series tests with an application to the PPP hypothesis. *Journal of Economic Theory, 20*, 597–625.

Phimphanthavong, H. (2013). Economic growth and environmental degradation in Laos. *ASEAN Economic Bulletin, 30*(1), 64–76.

Rojas-Downing, M. M., Nejadhashemi, A. P., & Harrigan, T. (2017). Climate change and livestock production: A review. *Global Food Security, 12*, 84–98.

Rotmans, J. & Dowlatabadi, H. (1998). Integrated assessment modeling. In S. Rayner & E. Malone (Eds.), *Human Choice and Climate Change*, Vol. 3: The tools for policy analysis. Columbus, Ohio, USA, Batelle Press, 15.

Sandalli, A. (2021). The Macroeconomic implications of climate change on sub-Saharan Africa: A case for sustainable development. *Review of Business and Economics Studies., 9*(1), 8–36. https://doi.org/10.26794/2308-944X-2021-9-1-8-362021

Seo, S.N. & Mendelsohn, R. (2008). *A structural ricardian analysis of climate change impacts and adaptations in African agriculture*. Policy Research Working Paper Series 4603, The World Bank. https://openknowledge.worldbank.org/bitstream/.../WPS4603.txt

Stige, L. C., Stave, J., & Chan, K. S. (2006). ENSO and maize production in Southern Africa. *PNAS, 103*(50), 19690–19695.

Swart, R., Robinson J. & Cohen, S. (2003). Climate change and sustainable development: Expanding the options. *Climate Policy, 3*(Suppl. 1), 19–40.

Tol, R. S. (2009). The economic effects of climate change. *Journal of Economic Perspectives, 23*(2), 29–51.

Tutulmaz, O. (2015). Revisiting the Environmental Kuznets Curve hypothesis. *Ecological Economics, 119*, 219–230.

UN (1992). United nations framework convention on climate change. http://unfccc.int/resource/docs/convkp/conveng.pdf (26.04.2014).

United Nations Framework Convention on Climate Change. (1992). United Nations framework convention on climate change. UNFCCC.

Wade, R., & Jennings, M. (2016). Climate adaptation in low-income nations. *World Development, 78*, 1–15.

Weitzman, M. L. (2012). Fat-tailed uncertainty in climate economics. *Review of Environmental Economics, 6*(2), 275–292.

Wolfe, D. W., Schwartz, M. D., Lakso, A. N., Otsuki, Y., Pool, R. M., Shaulis, N. J. (2005). Climate change and shifts in spring phenology of three horticultural woody perennials in northeastern USA. Internat J Biometeorol 49:303–309. Meteorological Organization, Geneva.

Zhai, F., Lin, T., & Byambadorj, E. (2009). A general equilibrium analysis of the impact of climate change on agriculture in the People's Republic of China. *Asian Development Review, 26*(1), 206–225.

Zhai, F., & Zhuang, J. (2009). Agricultural impact of climate change: A general equilibrium analysis with special reference to Southeast Asia. ADBI Working Paper 131. Tokyo: Asian Development Bank Institute.

# Corporate Social Responsibility: A Pathway to Sustainability in the Fashion Industry

**Deepak Khagokpam, Avi Sharma, Radhey Shyam Sharma, and Vandana Mishra**

**Abstract** Corporate Social Responsibility (CSR) is crucial for promoting sustainability in the fashion industry, heavily impacted by fast fashion and overconsumption. This study investigates CSR's role in addressing these challenges through two main objectives: (i) providing a systematic theoretical analysis of the responsibilities of firms and consumers in developing effective CSR strategies, and (ii) evaluating the incorporation of sustainability aspects and performance in CSR reports from leading fashion companies. Using a benchmark scoring technique based on Global Reporting Initiative (GRI) standards, the study assesses the accountability and performance indicators in these reports for 2021 and 2022. The analysis reveals that both firms and consumers are vital in advancing sustainable practices. Firms must integrate comprehensive CSR strategies addressing social, environmental, and economic impacts, while consumers drive demand for sustainable products through informed choices and political consumerism. The evaluation of CSR reports indicates an increase in sustainability disclosures over the years, highlighting a trend towards greater transparency. However, a significant gap remains in providing comprehensive quantitative data for accurate performance assessment. The study identifies challenges in effectively integrating CSR within the fashion industry, including complex supply chains and inherent overconsumption. It underscores the critical role of consumer behavior

D. Khagokpam · R. S. Sharma · V. Mishra (✉)
Department of Environmental Studies, Bioresources and Environmental Biotechnology Laboratory, University of Delhi, Delhi, India
e-mail: vandanamishra77@gmail.com

A. Sharma
Department of Psychology, Aryabhatta College, University of Delhi South Campus, New Delhi, India

R. S. Sharma
Delhi School of Climate Change and Sustainability, Institute of Eminence, University of Delhi, Delhi, India

V. Mishra
Centre for Interdisciplinary Studies On Mountain and Hill Environment (CISMHE), University of Delhi, Delhi, India

CEMDE-DDA Biodiversity Parks Programme, University of Delhi, Delhi, India



P. Singh et al. (eds.), *Circular Economy and Environmental Resilience*,
https://doi.org/10.1007/978-3-031-93091-1_14

in driving sustainable practices, noting a gap between awareness and action due to limited accessible information and feasible alternatives. To address these challenges, the study suggests promoting slow fashion, leveraging economic incentives, and integrating cultural values into product design. Additionally, adopting standardized reporting frameworks and technological innovations can enhance supply chain transparency and traceability, fostering consumer trust and responsible consumption. This research has significant implications for improving CSR practices, encouraging stakeholder engagement, and fostering an integrated approach to sustainability in the fashion industry. Aligning CSR strategies with global sustainability goals can mitigate negative impacts and enhance corporate reputation. Future studies should focus on longitudinal analyses of CSR initiatives and the role of policy and regulation in promoting corporate sustainability, contributing to the ongoing discourse on sustainable fashion.

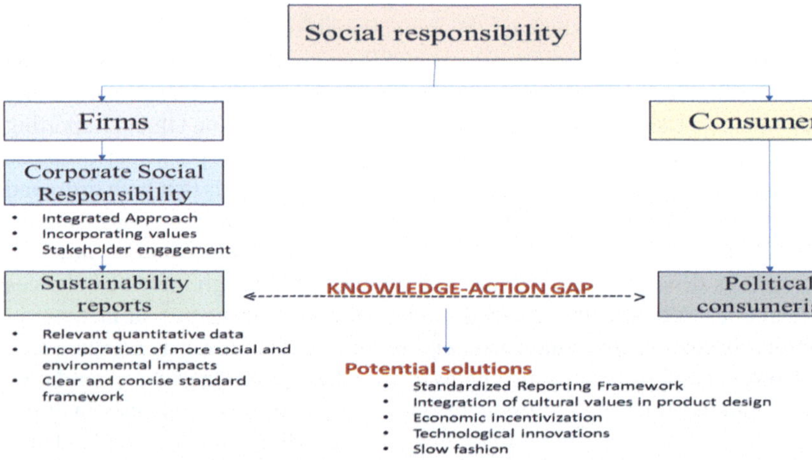


**Keywords** Corporate social responsibility · Sustainable fashion · Global reporting initiative · Consumer behavior · Supply chain transparency

## 1 Introduction

The fashion industry, a cornerstone of global culture and economy, wields considerable influence on environmental and social landscapes. Traditionally, clothing has served as a medium for personal expression and societal identity. However, the advent of fast fashion—characterized by rapid production cycles and inexpensive garments—has exacerbated environmental degradation and social inequities. The industry's extensive supply chain, from fiber production to consumer use and disposal, significantly contributes to pollution, resource depletion, and exploitative labor practices (Niinimäki et al., 2020; Rawat et al., 2016, 2018). These impacts underscore the urgent need for the industry to adopt sustainable practices that mitigate negative consequences while fostering economic and social well-being.

Corporate Social Responsibility (CSR) emerges as a promising pathway toward sustainability in the fashion industry. CSR involves self-regulating business practices that address social, environmental, and economic impacts, thereby enhancing corporate accountability and stakeholder trust (Carroll, 1999. Effective CSR communication, coupled with transparent and quantitative reporting, empowers consumers to make informed choices and promotes responsible consumption (Meseguer-Sánchez et al., 2021). Transparency and accountability in CSR practices not only build consumer trust but also drive industry-wide shifts towards more sustainable practices (Eccles et al., 2014; Freeman, 1984).

Current CSR practices in the fashion industry often lack comprehensive quantitative data, hindering meaningful performance analysis and comparison. Detailed quantitative reporting is essential for benchmarking progress and fostering continuous improvement in sustainability practices (GRI, 2021; Nikolaou & Tsalis, 2013). This study aims to evaluate the incorporation of sustainability aspects in CSR reports of selected fashion companies and assess their performance. The significance of this study lies in its potential to inform better CSR practices, encourage stakeholder engagement, and foster an integrated approach to sustainability in the fashion industry. The findings can guide firms in developing robust CSR frameworks that align with global sustainability goals, ultimately contributing to a more responsible and sustainable fashion sector. The strategic implementation of CSR can not only mitigate negative impacts but also enhance competitive advantage and corporate reputation (Berg et al., 2020; Kramer & Porter, 2011).

The fashion industry faces numerous challenges in integrating CSR effectively. The extensive supply chain, involving multiple stakeholders from farmers to consumers, complicates the implementation of uniform sustainability standards. Furthermore, the fast fashion model thrives on consumer demand for new trends and low-cost apparel, which often leads to overconsumption and waste (Fletcher, 2013). Addressing these issues requires a multifaceted approach that includes not only improving supply chain transparency but also changing consumer behavior towards more sustainable consumption patterns.

Several studies highlight the critical role of consumer behavior in driving the shift towards sustainable fashion (Beckmann, 2007). For instance, Joung (2014) found that consumers with higher environmental awareness are more likely to engage in sustainable purchasing behaviors. However, there remains a significant gap between consumer knowledge and action. This disparity is often due to a lack of accessible information and feasible alternatives to conventional fashion products (Niinimäki & Hassi, 2011). Therefore, enhancing consumer education and providing viable sustainable options are crucial for fostering responsible consumption.

The fashion industry has a significant opportunity to lead the way in corporate sustainability through effective CSR practices. By addressing the environmental and social impacts of their operations, fashion companies can enhance their reputation, build consumer trust, and contribute to global sustainability goals. This study aims to provide: (i) a systematic theoretical analysis of responsibilities of firms and consumers for developing more effective CSR strategies, and (ii) a comprehensive analysis of CSR practices in eight major fashion industry companies, identifying best practices and areas for improvement. Our findings will help facilitate the industry's transition towards a more sustainable future.

# 2 Methodology

## *2.1 Systematic Theoretical Analysis of CSR in the Fashion Industry*

### 2.1.1 Research Design and Literature Review

This study employs a systematic theoretical analysis to explore the responsibilities of firms and consumers in developing effective CSR strategies for a sustainable fashion industry. The research design includes a comprehensive literature review, drawing on academic journals, industry reports, and books selected based on relevance, credibility, and publication date. Key databases such as Google Scholar, JSTOR, PubMed, and Scopus are utilized with keywords including "Corporate Social Responsibility," "CSR evolution," "Creating Shared Value," "political consumerism," "consumer behavior," and "sustainable fashion."

### 2.1.2 Theoretical Framework

The theoretical framework examines the evolution of CSR from its inception in the 1950s to its current strategic form, incorporating key concepts like Carroll's four components of CSR (1999) and Kramer's Creating Shared Value (CSV) (2011). The review also addresses barriers to effective CSR implementation, such as stakeholder awareness and regulatory challenges. Consumer behavior theories, including the Value-Belief-Norm (VBN) theory (Cheng et al., 2014) and the Theory of Reasoned Action (Vallerand et al., 1992), are analyzed to understand how consumer values influence sustainable consumption.

### 2.1.3 Data Analysis and Synthesis

A thematic analysis identifies and synthesizes key themes from the literature, while a comparative analysis evaluates different CSR frameworks and consumer behavior models. This integrative approach highlights the interplay between corporate actions and consumer behavior, providing a holistic view of sustainability efforts. Strategic recommendations for firms and practical guidelines for consumers are developed to bridge the gap between CSR activities and consumer expectations.

## 2.2 *Sustainability Assessment in the Eight Selected Fashion Industry*

### 2.2.1 Scoring Framework for Sustainability Analysis

For this analysis, a scoring framework based on the balanced scorecard's thematic perspective, Global Reporting Initiative (GRI) standards, and scoring-benchmarking technique was used. This method was originally developed by Nikolaou and Tsalis (2013) to score sustainability reports of Greek firms. The main advantage of this method is its reliable and transparent scoring technique to evaluate corporate sustainability performance based on the sustainability reports published by the firms themselves. By incorporating the GRI standards, this method links CSR activities with the targets of the Sustainable Development Goals (SDG).

### 2.2.2 Integration of GRI Standards and Sustainable Development Goals (SDGs) in the Fashion Industry Analysis

The Global Reporting Initiative (GRI) is a global autonomous body offering the most commonly used reporting standards worldwide. These standards cover various topics, from environmental concerns to human rights and taxation. Additionally, the GRI provides a way to align CSR strategies and reporting practices with sustainable development goals (Dissanayake, 2021a, 2021b). Sector-specific disclosures offer flexibility for firms to choose the most relevant ones, thus enhancing legitimacy and providing room for continuous growth. Aligning with SDGs can be advantageous for companies by better representing consumer values and differentiating themselves with sustainable products (UNGC, 2018). The GRI standards provide a way to integrate SDGs into reporting practices (Fonseca et al., 2023). Therefore, only those companies were chosen for the analyses that (i) belong to the fashion industry, (ii) follow GRI standards, and (iii) have published sustainability reports for the years 2021 and 2022.

### 2.2.3 Application of Balanced Scorecard Framework to GRI Indicator

The balanced scorecard, developed by Kaplan and Norton, (1992), provides a comprehensive view of a firm's operations by addressing four key perspectives: financial, customer (or stakeholder), internal business, and innovation and learning. In this study, specific GRI indicators were categorized into these four themes. The method utilized the "Consolidated Set of GRI Standards," including all updated disclosures as of 2022. Eight GRI disclosures were used for the financial perspective, 33 for the stakeholder perspective, 17 for the internal perspective, and 4 for the learning and growth perspective, totalling 62 GRI disclosures (Fonseca et al., 2023).

Scoring Method

A benchmark scoring technique was used to evaluate the reports, employing two indicators: the accountability indicator and the performance indicator (Nikolaou & Tsalis, 2013). The accountability indicator measures the completeness of the report by scoring whether it includes the required disclosures for the four themes on a three-point scale: 0 indicates no information published for a specific Global Reporting Initiative (GRI) disclosure; 1 signifies that only qualitative information is published without related data; and 2 means quantitative data is published along with qualitative information.

The performance indicator, which applies only to disclosures with quantitative data, measures how well a firm has performed compared to the previous year. It also uses a threepoint scale: 1 indicates that the performance level decreased from the preceding year; 1.5 signifies that the performance level remained the same as the prior year; and 2 means that the performance level increased compared to the prior year.

Scoring Calculations and Aggregation

After scoring all disclosures across the four themes, the total accountability indicator score is calculated using the formula:

**Accountability indicator = AF + AS + AI + AL** where, AF is the accountability financial indicator, AS is the accountability stakeholder indicator, AI is the accountability internal indicator, AL is the accountability learning and growth indicator. The total performance indicator is calculated using the formula:

$$\textbf{Performance indicator} = \textbf{PF} + \textbf{PS} + \textbf{PI} + \textbf{PL}$$

where, PF is the performance financial indicator; PS is the performance stakeholder indicator; PI is the performance internal indicator; PL is the performance learning and growth indicator.

Finally, the sustainability score is calculated as:

$$\textbf{Sustainability score} = \textbf{Total accountability score} + \textbf{Total performance indicator}$$

# 3 Theorical Analyses of Responsibility of Firms and Consumers for Developing Effective CSR Strategies

A systematic theoretical analysis identified effective CSR strategies and assessed their implementation. Key findings indicate that fashion companies must integrate comprehensive CSR strategies addressing social, environmental, and economic impacts, emphasizing transparency and stakeholder engagement. Consumers drive sustainability through informed purchasing and political consumerism, highlighting the need for enhanced education and feasible sustainable alternatives to bridge the gap between awareness and action.

## *3.1 Corporate Social Responsibility: Evolution, Strategic Integration, and Firms' Responsibilities*

### 3.1.1 Evolution of Corporate Social Responsibility (CSR)

Corporate Social Responsibility (CSR) has evolved significantly since the 1950s. Initially seen as a societal obligation, by the 1960s, CSR incorporated social values and economic returns (Agudelo et al., 2019). By the mid-1970s, CSR aligned with social norms and values. Carroll (1999) conceptualized CSR into four components: economic, legal, ethical, and philanthropic responsibilities, emphasizing the balance between profit generation, legal compliance, ethical conduct, and community engagement (Fig. 1).

*Economic Responsibilities*: Economic responsibility is the foundation of Carroll's CSR pyramid, highlighting the obligation of businesses to be profitable. Profitability ensures survival, growth, and contribution to the economy through sustainability, innovation, and financial health. Key aspects include (i) ensuring long-term profitability through sustainable practices, (ii) investing in innovative processes and products to enhance efficiency, and (iii) maintaining strong financial performance to attract investment and provide stability for employees and stakeholders.

*Legal Responsibilities*: Legal responsibility requires businesses to operate within government laws and regulations, essential for maintaining legitimacy and avoiding penalties. This includes adherence to labor laws, environmental regulations, and consumer protection statutes (Locke et al., 2009). Key aspects include (i) strict compliance with laws and regulations, (ii) clear and honest reporting of business practices and financial performance, and (iii) robust governance structures to meet legal obligations and uphold ethical standards.

*Ethical Responsibilities*: Ethical responsibility extends beyond legal compliance, involving fairness, justice, and respect for individuals and communities affected by business operations. It requires companies to consider the broader impact of

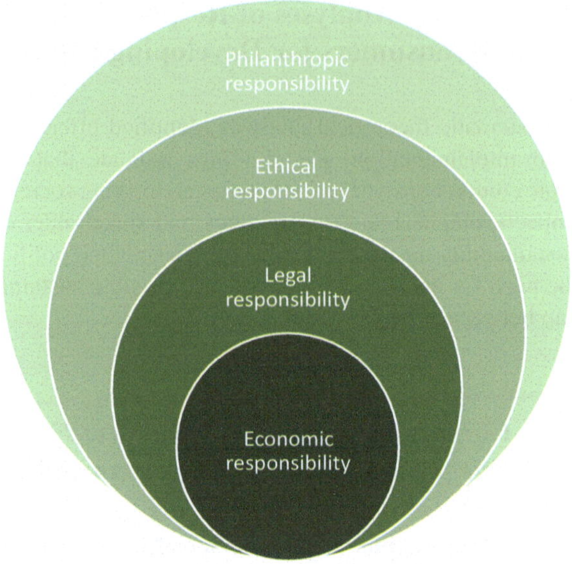


**Fig. 1** Four components of corporate social responsibility

their decisions and actions. Key aspects include (i) conducting business honestly and transparently, (ii) ensuring fair treatment of all stakeholders, and (iii) valuing individual rights and avoiding harm to people and the environment.

*Philanthropic Responsibilities*: Philanthropic responsibility involves voluntary activities that contribute to societal well-being, such as charitable donations, community service, and initiatives that improve quality of life. While not mandatory, these activities enhance a company's reputation and demonstrate its commitment to social good. Key aspects include (i) active participation in community development, (ii) donating resources to charitable causes, and (iii) implementing programs addressing social issues like education, health, and poverty.

Carroll's CSR framework highlights the need for businesses to balance profit generation with legal compliance, ethical conduct, and community engagement, ensuring a holistic approach to corporate responsibility.

### 3.1.2 Strategic Integration of CSR in the 21st Century

*CSR as a Core Strategic Business Component*: In the 2000s, Corporate Social Responsibility (CSR) evolved from peripheral activities into a core component of strategic business management, driven by the recognition that integrating CSR into corporate strategy could lead to sustainable competitive advantages and long-term profitability. Companies began to see CSR as a strategic tool to align business objectives with societal goals. This involves embedding social and environmental considerations into core operations and decision-making, linking CSR activities to the company's mission and strategic goals. This synergy enhances operational efficiencies, reduces costs, and mitigates risks while contributing positively to society and the environment.

*Creating Shared Value (CSV) for Strategic CSR Integration*: A pivotal concept that encapsulates the strategic integration of CSR is Michael Porter's Creating Shared Value (CSV), introduced in 2011. CSV goes beyond traditional CSR by integrating social and economic progress in a way that benefits both the company and the broader community. Porter argues that companies can create shared value by (i) Developing products and services that meet societal needs and open up new markets. For example, fashion companies can create sustainable clothing lines that cater to the growing demand for eco-friendly products; (ii) Enhancing the efficiency and sustainability of operations by improving resource utilization, reducing environmental impact, and fostering better working conditions. For instance, implementing energy-efficient processes and waste reduction practices in production; (iii) Strengthening the business environment in the locations where companies operate by supporting local suppliers, fostering innovation, and investing in community development.

This can involve collaborations with local artisans and promoting fair trade practices.

*CSR in the Fashion Industry*: The fashion industry has embraced strategic CSR to enhance brand image and credibility. With growing awareness of environmental and social impacts, consumers demand greater transparency and responsibility from brands. CSR practices in fashion include sustainable sourcing, fair labor practices, environmental stewardship, and consumer engagement. Sustainable sourcing uses environmentally friendly and ethically sourced materials. Fair labor practices ensure fair wages, safe working conditions, and ethical treatment of workers. Environmental stewardship reduces the environmental footprint through measures like reducing water and energy consumption and using eco-friendly dyes. Consumer engagement educates consumers about sustainable fashion and promotes responsible consumption patterns, such as clothing swaps and recycling programs.

Studies, such as those by Thorisdottir et al. (2020), have shown that effective CSR practices in the fashion industry not only improve the company's reputation but also build stronger relationships with consumers. Brands perceived as socially responsible are more likely to gain consumer trust and loyalty, which can lead to increased market share and financial performance. The strategic integration of CSR in the twenty-first century marks a significant shift in how companies approach social responsibility. By aligning CSR with business strategy, companies can achieve a balance between economic success and societal good. The concept of Creating Shared Value provides a robust framework for this integration, particularly in industries like fashion, where ethical practices and sustainability are increasingly paramount. As companies continue to innovate and adapt their CSR strategies, they contribute to a more sustainable and equitable global economy.

### 3.1.3 Barriers to Effective CSR Implementation

Despite the benefits of Corporate Social Responsibility (CSR), several barriers hinder its effective implementation in the fashion industry. Overcoming these barriers is crucial for achieving meaningful social and environmental accountability.

*Lack of Stakeholder Awareness*: Many stakeholders, including consumers, employees, suppliers, and investors, lack awareness and understanding of CSR, leading to apathy or resistance. For example, consumers may not prioritize sustainable fashion if they are unaware of its environmental and social impacts. To address this, firms should educate stakeholders through campaigns, workshops, and transparent communication, providing clear, accessible information about CSR efforts and their impacts to build trust and support.

*Managerial Commitment*: Some managers view CSR as secondary or a cost center, leading to inadequate resources and lack of integration into core operations. Senior leaders must champion CSR, integrating it into the company's mission and strategic goals. Demonstrating the long-term benefits of CSR, such as enhanced reputation and risk management, can garner managerial support. Training managers on CSR's value and offering incentives for achieving CSR goals can foster greater commitment and accountability.

*Financial Constraints*: Implementing CSR initiatives often requires significant financial investment, which can be a major barrier, especially for SMEs. Strategies to overcome this include conducting a thorough cost–benefit analysis to highlight CSR's long-term financial benefits, such as cost savings from energy efficiency and waste reduction, and exploring external funding sources like grants, subsidies, and partnerships with NGOs to alleviate financial burdens.

*Inconsistent Regulations*: The fashion industry operates in a global market with varying regulations, complicating uniform CSR practices. To address this, firms should adhere to global standards set by the GRI and SDGs, providing a consistent framework for CSR practices. Engaging with policymakers to advocate for more consistent and stringent regulations can level the playing field and encourage widespread CSR adoption.

*Proactive Participation and Transparent Reporting*: Firms must move beyond treating CSR as an obligation and adopt a proactive approach emphasizing genuine social and environmental accountability. This involves integrating CSR into the core business strategy, engaging stakeholders, and providing transparent reporting on CSR activities and outcomes. Key actions include embedding CSR into strategic planning, involving stakeholders in CSR initiatives and decision-making processes, and regularly publishing detailed and honest reports on CSR activities, progress, and challenges. By addressing these barriers through strategic actions, fashion firms can enhance their CSR implementation, contributing to a more sustainable and socially responsible business environment.

### 3.1.4 The Role of Managers and Corporate Decision-Making: Real World Examples

Managers are crucial in driving Corporate Social Responsibility (CSR) in the fashion industry, where ethical concerns about labor practices, environmental impact, and sustainability are significant. Managers who consider multiple stakeholders—employees, customers, suppliers, and the community—along with long-term business credibility, tend to outperform those focusing solely on short-term financial gains (Orlitzky, 2011).

*Ethical Leadership and Environmental Stewardship*: Patagonia exemplifies strong CSR commitment, driven by management's focus on sustainability and ethical practices. Their mission, "We're in business to save our home planet," is reflected in initiatives like the "Worn Wear" program, which promotes clothing repair, reuse, and recycling, enhancing brand reputation and customer loyalty.

*Balancing Fast Fashion with Sustainability*: H&M, a major fast fashion player, has faced scrutiny over its environmental and labor practices. In response, H&M's management introduced the "Conscious" collection with sustainable materials and committed to 100% sustainable or recycled materials by 2030. These CSR initiatives demonstrate how managerial decisions address stakeholder concerns and build long-term credibility. Similarly, The North Face, launched the "Clothes the Loop" program to recycle old clothing, focusing on environmental sustainability, promoting a circular economy and enhancing its eco-friendly reputation.

*Water Conservation and Worker Well-being*: Levi Strauss & Co. integrates CSR with a focus on environmental sustainability and worker well-being. The "Water < Less" initiative reduces water consumption, and the "Worker Well-being" program improves health and financial security for supply chain workers, balancing profitability with social and environmental responsibilities.

*Cruelty-free fashion*: Stella McCartney's brand, committed to cruelty-free fashion, appeals to consumers prioritizing animal rights and sustainability. Managerial decisions to avoid leather, fur, and animal products align with the brand's values and consumer expectations, reinforcing long-term credibility and financial success.

*Influence of Media and Public Perception*: Larger fashion firms often face intense media scrutiny, pressuring them to engage in robust CSR activities. Nike, after facing backlash in the 1990s over sweatshop labor, implemented comprehensive changes, including stricter labor standards and transparency reports, demonstrating the importance of responsive managerial decision-making.

*Consumer Perception and Socio-Political Contexts*: Consumer perception of CSR varies based on interests and socio-political contexts. In regions with high environmental awareness, sustainability-focused brands are prioritized, while in areas with labor rights concerns, ethical practices are key. Managers must navigate these priorities and tailor CSR strategies accordingly.

Managers in the fashion industry play a vital role in CSR success. By addressing multiple stakeholder concerns, focusing on long-term credibility, and responding to media scrutiny, they can drive positive social and environmental impacts while achieving better financial performance. Real-world examples from companies like Patagonia, H&M, Levi Strauss & Co., Stella McCartney, and The North Face illustrate how effective managerial decision-making in CSR leads to sustainable business practices and enhanced brand reputation.

## 3.2 *Influencing Corporate Behavior: The Consumer's Role in CSR*

Consumer behavior, shaped by cognitive biases, social influences, emotional responses, and socio-economic factors, plays a pivotal role in driving sustainable practices. Understanding these elements is crucial for promoting sustainability and addressing issues like overconsumption in the fashion industry.

### 3.2.1 Consumer Behaviour and Cognitive Influences

*Cognitive biases*: Cognitive biases lead to irrational decision-making that prioritizes immediate gratification over long-term benefits. For instance, the status quo bias makes consumers prefer familiar options, making sustainable products harder to adopt. A consumer might continue buying fast fashion because they are used to it, despite knowing its environmental impact. Confirmation bias leads consumers to seek information that confirms their pre-existing beliefs, ignoring evidence of affordable, sustainable options. For example, a consumer who believes sustainable fashion is too expensive might overlook budget-friendly sustainable brands like H&M Conscious. The anchoring effect causes initial information, such as price, to heavily influence decisions, deterring exploration of sustainable products. If a consumer sees a highpriced sustainable item first, they might assume all sustainable products are similarly priced and not consider other options.

*Social influences*: Social influences impact consumer behavior through societal and cultural norms. Peer pressure can drive the adoption of sustainable fashion if it becomes a social trend (Halder et al., 2020; Blasé et al., 2024). For example, when influencers and celebrities endorse eco-friendly brands like Stella McCartney or Patagonia, it encourages their followers to do the same. Cultural norms that emphasize collectivism and environmentalism promote sustainable choices. In countries like Sweden, where environmental consciousness is high, consumers are more likely to support sustainable fashion. Additionally, endorsements by celebrities or influencers provide social proof that boosts the appeal of sustainable products. Campaigns like Emma Watson's "The Press Tour," where she exclusively wore sustainable fashion, significantly raised awareness and interest in eco-friendly clothing.

*Emotional responses*: Emotional responses often drive consumer behavior, with emotions like guilt, pride, and fear influencing decisions. Awareness of the negative impacts of their choices can make consumers feel guilty, motivating them to choose more sustainable options (Hur & Kim, 2017). Brands like Everlane capitalize on this by promoting transparency about their production processes, which appeals to consumers' desire to make ethical choices. Conversely, positive emotions, such as pride from contributing to sustainability, reinforce eco-friendly behavior. Wearing brands like Reformation, which are known for their sustainable practices, can evoke pride in consumers. However, fear-based messaging must be balanced to avoid causing disengagement. Fear of environmental degradation can be a powerful motivator, but it needs to be paired with actionable steps consumers can take.

*Socio-economic factors*: Socio-economic factors like income, education, and cultural background significantly influence consumer choices (Niinimäki, 2010). Higher income levels generally provide more flexibility to choose sustainable products, though affordable options need better visibility. Brands like Uniqlo have introduced affordable, eco-friendly lines to reach a broader audience. Educated consumers are more aware of the impacts of their consumption and more inclined towards sustainable products. Cultural attitudes towards sustainability vary, requiring tailored marketing strategies to effectively promote sustainable choices. In Japan, for instance, the concept of "mottainai" (regret over waste) supports the adoption of sustainable practices.

Addressing these factors involves education and awareness campaigns to increase consumer knowledge about the benefits of sustainable fashion and debunk cost myths. Leveraging social influences through collaborations with influencers and encouraging consumer engagement on social media can promote sustainable fashion. Providing clear information about the benefits of sustainable fashion helps counteract cognitive biases, while emotional marketing that evokes pride and responsibility can resonate with consumers. Finally, making sustainable options accessible and competitively priced ensures consumers can make responsible choices, contributing to a more sustainable future. For instance, Zara's Join Life collection offers fashionable yet affordable sustainable options, demonstrating that eco-friendly choices do not have to be prohibitively expensive.

### 3.2.2 Political Consumerism as a Driver for Change

Political consumerism emerges from societal concerns about production and consumption systems (Austgulen, 2016). It is a form of consumer activism where individuals make purchasing decisions based on political, ethical, or environmental values. Consumers leverage their buying power through actions like boycotting, buycotting, lifestyle changes, and discursive strategies (Bostrom & Micheletti, 2019).

*Boycotting*: Boycotting involves avoiding products from companies with unethical practices, such as environmental degradation or labor exploitation. For example, numerous campaigns have targeted companies like H&M and Zara for their alleged involvement in labor exploitation. These boycotts raise awareness and put pressure on companies to improve their practices to regain consumer trust and avoid financial losses.

*Buycotting*: Conversely, buycotting involves purchasing from companies with ethical practices. By supporting brands that align with their values, consumers can encourage positive corporate behavior. In the fashion industry, brands like Patagonia and Everlane have benefited from buycotting. Patagonia's commitment to environmental sustainability and fair labor practices has garnered a loyal customer base. Similarly, Everlane's transparency about its production processes and costs has attracted consumers who prioritize ethical shopping.

*Lifestyle Changes*: Altering consumption habits to reduce environmental impact, such as buying less, choosing high-quality items, and opting for second-hand or recycled products, exemplifies slow fashion (Domingos et al., 2022). Brands like Eileen Fisher and Stella McCartney support slow fashion by offering durable, timeless pieces and repair and recycling programs.

*Discursive Strategies*: Using communication to raise awareness and influence public opinion on sustainability issues involves social media campaigns, public endorsements, and educational initiatives. Campaigns like Fashion Revolution's #WhoMadeMyClothes encourage consumers to question the origins of their clothing and demand greater transparency from brands. Influencers and celebrities amplify these messages, reaching broader audiences and promoting sustainable consumption.

*Impact of Political Consumerism*: The collective actions of political consumers compel companies to adopt more sustainable practices and align with consumer values. When a significant number of consumers participate in boycotts, buycotts, or lifestyle changes, the economic pressure on companies can lead to substantial changes in corporate behavior. For instance, in response to consumer demand for more sustainable products, many fashion brands have introduced eco-friendly lines and committed to reducing their environmental footprint. For example, Brands like Adidas have launched sustainable product lines, such as the Parley collection made from recycled ocean plastic, to appeal to environmentally conscious consumers. Similarly, Nike has introduced the Move to Zero campaign, aiming for zero carbon and zero waste across its operations.

*Challenges and Opportunities*: Political consumerism faces challenges like requiring widespread participation and sustained effort. Additionally, companies may engage in "greenwashing," where they portray themselves as more environmentally friendly than they are, to capitalize on the trend without making meaningful changes (Delmas & Burbano, 2011). However, genuine commitment to sustainability presents opportunities for innovation and market differentiation. Engaging transparently with consumers about their sustainability journey can further enhance trust and loyalty.

Political consumerism empowers consumers to influence corporate practices through their purchasing decisions. By engaging in boycotting, buycotting, lifestyle changes, and discursive strategies, consumers drive companies towards ethical and environmentally friendly practices, fostering a more sustainable fashion industry.

### 3.2.3 Theoretical Perspectives on Consumer Responsibility

Understanding theoretical frameworks like the Value-Belief-Norm (VBN) theory, Cognitive Dissonance theory, and the Theory of Reasoned Action is essential for promoting sustainable consumer behavior. These theories provide insights into how values, beliefs, and norms drive actions towards sustainability and underscore the importance of aligning consumer values with sustainable practices.

*Value-Belief-Norm (VBN) Theory*: VBN theory posits that individuals' core values, beliefs about environmental issues, and personal norms significantly influence their proenvironmental behavior (Cheng, 2014). Consumers with strong environmental and altruistic values are more likely to engage in sustainable fashion practices, such as purchasing ecofriendly products or supporting ethical brands. Educational campaigns that enhance awareness of environmental impacts can strengthen these beliefs and norms, driving responsible consumer behavior.

*Cognitive Dissonance Theory*: This theory suggests that individuals experience psychological discomfort when their actions contradict their beliefs or values. To reduce this discomfort, they are motivated to change their behavior or beliefs to achieve consistency. For instance, consumers who value environmental protection but buy fast fashion may shift to purchasing sustainable brands. Retailers can highlight the environmental benefits of their products to leverage this motivation.

*Theory of Reasoned Action*: This theory explains that an individual's intention to perform a behavior is influenced by attitudes towards the behavior and subjective norms. Positive attitudes towards sustainable fashion and social pressures from friends and influencers can increase the intention to buy eco-friendly products. Brands can enhance these attitudes and norms through targeted marketing and social proof.

*Aligning Consumer Values with Sustainable Practices*: Educational campaigns can increase awareness about the environmental and social impacts of fashion, enhancing values and beliefs aligned with sustainability. Initiatives like the "Fashion Revolution" campaign encourage consumers to question the origins of their clothing, fostering a deeper understanding of ethical implications.

*Behavioral Nudges and Social Influence*: Behavioral nudges, such as providing information on the negative impacts of fast fashion at the point of sale or offering incentives for sustainable purchases, can encourage sustainable choices (Rawat et al., 2022, 2023). Social influence through endorsements by celebrities and influencers can shape subjective norms and drive wider adoption of sustainable practices.

*Accessibility and Appeal of Sustainable Fashion*: Making sustainable fashion accessible and appealing is crucial. Ensuring that eco-friendly options are competitively priced and aesthetically pleasing can align consumer behavior with sustainable values without significant sacrifices. Brands like Patagonia and Reformation successfully combine style with sustainability, making it easier for consumers to choose eco-friendly options. By applying these theoretical perspectives, companies can promote sustainable consumption and reduce the environmental impact of fashion.

### 3.2.4 Educating and Empowering Consumers

Educating consumers about the social and environmental impacts of their purchases is essential for promoting responsible consumption. Knowledgeable consumers are more likely to support eco-friendly products and boycott firms with negative impacts (Kozar, 2013). Effective educational campaigns and transparent communication can bridge the gap between awareness and action.

*Importance of Consumer Education*: Consumer education involves providing information about the lifecycle of products, from raw material extraction to disposal (Kaur et al., 1922). By understanding these environmental and social impacts, consumers can make informed decisions that align with their values. For example, awareness of the environment protecting benefits of organic cotton compared to conventional cotton can influence consumers to choose more sustainable options.

*Engaging and Accessible Educational Campaigns*: Educational campaigns must be engaging and accessible to effectively influence behavior. Storytelling is a powerful tool; sharing the stories behind products can create an emotional connection and highlight the importance of ethical practices. Brands like Patagonia and TOMS use storytelling to convey their commitment to sustainability and social responsibility.

*Transparency in Information*: Providing clear information about sourcing, manufacturing processes, and sustainability goals builds trust and credibility. For instance, Everlane's "Radical Transparency" initiative discloses the true cost of its products, including labor, materials, and transportation, allowing consumers to verify claims and make informed choices.

*Certifications and Digital Platforms*: Certifications and eco-labels also help consumers identify and trust sustainable products. Recognized certifications such as Fair Trade, Global Organic Textile Standard (GOTS), and B Corporation assure consumers of a product's ethical and environmental standards. Leveraging digital platforms and social media can further enhance the reach and impact of educational campaigns. Interactive content like videos and infographics can make learning about sustainability more engaging and memorable.

*Bridging the Gap Between Awareness and Action*: While increasing awareness is crucial, it is equally important to bridge the gap between awareness and action. Consumers often face barriers such as perceived higher costs and limited availability of sustainable products. Effective educational campaigns can address these barriers by highlighting the long-term cost savings of sustainable products and promoting their accessibility. Retailers can improve accessibility by stocking eco-friendly options in stores and online, and by providing clear labelling.

*Encouraging Behavioral Change*: Encouraging behavioral change through practical tips and incentives, such as recycling programs and clothing swaps, can motivate consumers to act on their knowledge. Engaging in two-way communication with consumers, through surveys and social media interactions, can also enhance trust and commitment to sustainable practices. By fostering a more informed and engaged consumer base, brands can drive responsible consumption and contribute to a more sustainable future.

### 3.2.5 Challenges and Strategies for Sustainable Consumption

*Challenges*: Sustainable consumption faces significant hurdles, despite the potential of political consumerism and lifestyle changes. One challenge is the temporary nature of consumer activism, which often relies on short-term campaigns that lose momentum, failing to create lasting change. Without sustained engagement and continuous advocacy, these movements' impact diminishes, leading to a return to unsustainable practices. Another challenge is the lack of feasible alternatives. Consumers may want to make environmentally conscious choices, but viable options are often limited due to higher costs, limited availability, or inadequate performance compared to conventional products (Mishra et al., 2020). Additionally, systemic barriers, such as insufficient recycling infrastructure and limited access to sustainable resources, hinder responsible consumption habits. However, below given strategies can help address these challenges.

*Economic Incentives and Community Engagement*: Economic incentives can motivate sustainable practices. Initiatives offering financial rewards for recycling or buying eco-friendly products can drive consumer behavior towards sustainability. Additionally, engaging communities through education and participatory activities fosters collective responsibility. Community-based sustainability projects enhance social cohesion and create supportive environments for sustainable consumption (Joung, 2013).

*Promotion of Slow Fashion*: The fast fashion industry contributes to overconsumption and environmental degradation. Promoting slow fashion, emphasizing quality over quantity, can mitigate these impacts. Slow fashion encourages consumers to buy fewer, higher-quality items designed to last longer, reducing waste and promoting thoughtful purchasing. Highlighting durability and timeless style can shift consumer attitudes away from disposable culture.

*Integration of Cultural Values into Product Design*: Products reflecting cultural values foster long-term emotional connections, making consumers less likely to discard them (Brown & Vacca, 2022). Designing items that showcase cultural heritage, local craftsmanship, and personal significance enhances perceived value and longevity. Emotional bonds with possessions lead to maintenance and repair rather than replacement, reducing overconsumption and supporting cultural preservation and local economies (Chakraborty, 2023).

*Sustainable Supply Chain Management*: Ensuring products are sustainably sourced, manufactured, and distributed is crucial for promoting sustainable consumption (Garg et al., 2020; Mishra et al., 2023). Companies can use renewable materials, reduce energy consumption, and minimize waste throughout the supply chain. Transparency in the supply chain empowers consumers to make informed choices, fostering trust and loyalty toward sustainable brands.

*Legislative and Policy Support*: Governments can drive sustainable consumption by enacting policies for product labeling, setting environmental performance standards, and incentivizing sustainable business practices. Public awareness campaigns and educational initiatives further inform consumers about the benefits of sustainability. Targeted strategies addressing these challenges can create a balanced, sustainable consumption pattern across economic, environmental, and social dimensions.

## 3.3 Bridging the Gap Between Firms and Consumers

The transition towards a sustainable fashion industry necessitates a collaborative effort between firms and consumers. Firms must integrate comprehensive CSR strategies addressing social, environmental, and economic impacts, supported by proactive management and transparent reporting. *For example,* Patagonia's "Worn Wear" program encourages customers to repair and recycle clothing, enhancing sustainability and customer loyalty. H&M's "Conscious Collection" utilizes sustainable materials like organic cotton and recycled polyester, showing the potential for integrating eco-friendly practices into mainstream fashion.

*Similarly,* consumers, empowered through education and political consumerism, can drive demand for sustainable products and practices. For example, the Fashion Revolution's #WhoMadeMyClothes campaign has raised global awareness, prompting consumers to demand greater transparency from brands. Additionally, certifications such as Fair Trade and the Global Organic Textile Standard (GOTS) help consumers identify ethically produced garments, fostering informed purchasing decisions.

Standardized regulations and frameworks are crucial to prevent ambiguity and ensure credibility. The Global Reporting Initiative (GRI) standards offer a consistent framework for CSR reporting, aiding firms in disclosing their sustainability practices transparently. Governments can support this transition through legislation mandating environmental performance standards and incentivizing sustainable business practices. For instance, the European Union's Circular Economy Action Plan aims to make sustainable products the norm, encouraging recycling and reducing waste.

Aligning CSR strategies with consumer education and regulatory support bridges the gap between corporate actions and expectations, fostering a sustainable, socially responsible fashion industry.

# 4 Evaluating CSR Practices and Performance in the Fashion Industry

To evaluate the incorporation of sustainability aspects in the CSR reports of selected fashion companies and assess their performance, eight companies that were chosen include: Asics, Adidas Group, Aditya Birla Fashion and Retail Limited, Puma, H&M Group, Nike Inc., Prada, and PVH Corp (Fig. 2). The assessment was based on the accountability and performance indicators of their sustainability reports for the years 2021 and 2022.

## *4.1 Accountability Indicator Scores*

The accountability indicator measures the completeness of sustainability reports. In 2021, PVH Corp. and H&M Group achieved the highest scores, each earning 60 out of 124. Scores were generally lower in the financial perspective and learning and growth indicators across most companies. In 2022, H&M Group retained the highest score, reflecting its dedication to enhancing its sustainability practices. The trend of increasing disclosure efforts among topperforming companies indicates a commitment to improving transparency (GRI, 2021). Nike Inc. and other companies that scored lower are presented with opportunities to enhance their sustainability reporting practices. Puma's position in 2022 suggests a need for continued focus

**Fig. 2** Radar plots showing comparison of selected fashion brands based on accountability and performance scores for 2021 and 2022

on reporting practices. These findings emphasize the importance of comprehensive and transparent reporting in building a credible and sustainable fashion industry. The varying performance across companies underscores the need for industry-wide standards and consistent efforts to integrate CSR into core business strategies.

## *4.2 Performance Indicator Scores*

Performance indicators measure the quantitative progress of companies on various sustainability metrics. In 2021, PVH Corp. achieved the highest performance score, while other companies were at varying levels on the spectrum. By 2022, H&M Group attained the highest performance score, reflecting its proactive approach to integrating sustainability into its core operations. The overall improvement in scores from 2021 to 2022 highlights ongoing efforts to enhance sustainability practices across the industry (Nikolaou et al., 2019). Companies that initially had lower scores, like Puma, have shown a positive trajectory towards better sustainability metrics, indicating their commitment to continuous improvement. These results underscore the critical importance of ongoing enhancement and adaptation in sustainability practices within the fashion industry. Enhanced quantitative reporting and performance tracking are essential for companies to identify areas of improvement, set measurable goals, and achieve meaningful progress in their sustainability journeys.

## *4.3 Sustainability Scores*

The sustainability score combines both accountability and performance indicators. In 2021, PVH Corp. had the highest sustainability score of 96 out of 248, while Puma had the lowest with 36. The mean score was 61.19, with a standard deviation of 22.29. In 2022, H&M Group achieved the highest score with 103, and Puma remained the lowest with 47, although showing improvement from the previous year. The mean score for 2022 was 75.44, with a standard deviation of 16.49. These figures suggest an overall trend towards better sustainability reporting and performance across the industry. The increased mean score and decreased standard deviation from 2021 to 2022 indicate that more companies are enhancing their sustainability efforts and achieving more consistent results (Fig. 3).

H&M Group's leading position highlights its effective strategies and commitment to sustainability, setting a benchmark for the industry. Puma's progress, despite being the lowest, reflects positive steps towards improving their sustainability practices. These results emphasize the collective movement within the fashion industry towards integrating more robust and transparent sustainability measures, fostering a more responsible and sustainable future.

**Fig. 3** Radar plots showing comparison of selected fashion brands based on their sustainability scores for the years 2021 and 2022

## 4.4 *Analysis of Reporting Practices*

The study revealed that companies generally improved their accountability indicators by disclosing more information. However, a significant gap remains in providing comprehensive quantitative data. Although commitment to reporting is evident, companies must focus on including quantitative data alongside qualitative disclosures to facilitate accurate performance assessment. Future CSR strategies should emphasize measurable outcomes to demonstrate real progress in sustainability efforts (Eccles et al., 2014).

### 4.4.1 Environmental Disclosures

Environmental impact disclosures were primarily focused on water usage, waste production, and scope emissions. However, reports on biodiversity impacts and ecosystem services were sparse. For instance, none of the analyzed companies provided information on habitat protection or restoration progress. Only two companies disclosed energy consumption outside their operations, crucial for understanding the full environmental footprint (Niinimäki et al., 2020). Future disclosures need to address these gaps by recognizing the interconnectedness of ecosystems and providing holistic environmental impact assessments (Fonseca et al., 2023).

### 4.4.2 Social Disclosures

The analysis highlighted insufficient reporting on social issues such as retirement plans, government subsidies, and indirect economic impacts. Paternity and maternity leave disclosures were among the least reported, despite their importance in bridging the gender gap. For example, research indicates that paternity leave increases fathers' involvement in childcare and boosts maternal employment. Most companies reported on gender diversity but lacked depth in other social metrics. Enhancing social disclosures can improve corporate accountability and social impact (Dissanayake, 2021a, 2021b). The study suggests that while companies are increasing their disclosure efforts, there is a need for more detailed and quantitative data to fully evaluate their sustainability performance.

## 5 Actionable Strategies and Scope for Future Research Recommendations and Scope for Future Research

Roadmap for improving CSR practices and their scopes are discussed below as actionable strategies focussing on the practical and immediate aspects that need to be covered, and research opportunities which include theoretical and forward-looking aspects.

### *5.1 Enhanced Quantitative Reporting and Holistic Environmental Impact Assessments*

Companies should integrate more quantitative data into their sustainability reports to provide clear performance metrics, including measurable targets for carbon emissions, water usage, and waste reduction. Aligning with international standards such as the GRI ensures transparency and comparability across the industry, enhancing accountability and aiding stakeholders in tracking progress and making informed decisions (GRI, 2021). Firms should also conduct comprehensive environmental impact assessments beyond carbon footprints to include biodiversity impacts and ecosystem services. Detailed disclosures on these aspects offer a more accurate representation of a company's environmental footprint, helping to identify and mitigate negative ecosystem impacts, thereby preserving biodiversity and promoting sustainable resource use (Niinimäki et al., 2020).

### 5.2 Social Metrics Enhancement

Expanding the scope of social metrics in CSR reporting is crucial. Companies should report on social issues such as paternity leave, retirement plans, and indirect economic impacts on communities. This includes metrics on employee well-being, diversity and inclusion, and community engagement. By providing a more comprehensive view of their social responsibilities, companies can demonstrate their commitment to creating a positive social impact and fostering a more equitable society.

### 5.3 Consumer Behaviour, Education, and Stakeholder Engagement

Educating and engaging consumers and stakeholders is vital for driving sustainable fashion adoption and enhancing CSR initiatives. Companies should develop educational campaigns that inform consumers about the environmental and social impacts of their purchases.

Strategies like storytelling, eco-labels, and sustainability certifications can raise awareness and encourage responsible consumption (Connell, 2019). Engaging consumers as active participants in sustainability initiatives can foster loyalty and brand advocacy.

Active stakeholder engagement is crucial for identifying key sustainability issues and improving CSR initiatives' relevance and impact. Companies should regularly engage with employees, customers, suppliers, and community representatives (Freeman, 2010). This collaborative approach helps companies understand expectations, address concerns, and cocreate solutions for sustainable business practices. For example, Unilever's extensive stakeholder engagement framework involves regular consultations with community groups and environmental organizations, shaping their sustainability strategies and addressing critical social and environmental issues. Integrating these efforts ensures a comprehensive approach to sustainability, driving both consumer and corporate progress.

### 5.4 Integration of Circular Economy Principles

To promote sustainability, companies should adopt circular economy principles more rigorously. This includes implementing product take-back schemes, recycling, and upcycling initiatives. By designing products for durability, repairability, and recyclability, companies can reduce waste and resource consumption (Ellen MacArthur Foundation, 2017). Integrating these principles into business models can help companies transition towards more sustainable production and consumption patterns.

## 5.5 Technological Integration

The role of technology in enhancing CSR efforts presents another promising area for research. The adoption of blockchain technology can provide greater transparency and traceability in supply chains, ensuring that sustainability claims are verifiable (Saberi et al., 2019). Additionally, the Internet of Things (IoT) can be used to monitor and optimize resource use in real-time, reducing environmental impacts. Future studies could examine the adoption barriers and facilitators for these technologies within the fashion industry, as well as their impact on CSR performance.

## 5.6 Policy, Regulation, and Standardized Framework Adoption

Examining the role of policy and regulation in promoting CSR practices is crucial for future research. Comparative studies across different regulatory environments can reveal effective policy frameworks for encouraging corporate sustainability. Additionally, exploring the potential impact of new regulations, such as extended producer responsibility (EPR), on the fashion industry's sustainability efforts is essential (Desore, 2018).

The adoption of standardized reporting frameworks, like the GRI and the Sustainability Accounting Standards Board (SASB), provides a consistent and transparent method for companies to report their CSR activities and impacts. These frameworks facilitate comparison and benchmarking across industries (Nikolaou & Tsalis, 2013). Standardized reporting enhances accountability and enables stakeholders to make informed decisions based on reliable data. Integrating these practices ensures a holistic approach to sustainability, driving corporate and industry-wide improvements.

## 5.7 Longitudinal Studies

Longitudinal studies tracking the progress of CSR initiatives over time would be valuable in understanding the long-term impact of these efforts on sustainability performance. Such studies could identify trends, best practices, and areas needing improvement, providing a comprehensive view of the evolution of CSR in the fashion industry. By addressing these areas, future research can contribute to the ongoing development and refinement of CSR practices in the fashion industry, ensuring that they effectively promote sustainability and create shared value for all stakeholders.

## 6 Conclusion

The fashion industry faces significant challenges in achieving sustainability, but CSR offers a pathway to address these issues. This study underscores the importance of comprehensive and transparent CSR reporting in fostering a more sustainable fashion sector. By improving quantitative disclosures, addressing holistic environmental impacts, and enhancing social metrics, companies can better align with global sustainability goals. The findings provide valuable insights for developing robust CSR strategies that contribute to a more responsible and sustainable fashion industry. Key findings reveal that fashion companies must integrate comprehensive CSR strategies addressing social, environmental, and economic impacts, focusing on transparency and stakeholder engagement. Consumers drive sustainability through informed purchasing and political consumerism, necessitating enhanced education and feasible sustainable alternatives to bridge the gap between awareness and action. Aligning CSR strategies with global sustainability goals can mitigate negative impacts and enhance corporate reputation. Future research should focus on longitudinal analyses, policy and regulation impacts, and the role of technological advancements in promoting sustainability. Overall, fostering a culture of responsible consumption and production is essential for the fashion industry to achieve sustainability goals and create a more equitable and environmentally conscious society.

**Acknowledgements** VM and RSS thank the Institute of Eminence, University of Delhi, for Faculty Research GrantProjects (Ref. NO./IoE/2023-24/12/FRP, 31.08.2023). AT acknowledges UGC-JRF for Ph.D. programme.

## References

Austgulen, M. H. (2016). Environmentally sustainable textile consumption—what characterizes the political textile consumers? *Journal of Consumer Policy, 39*, Article 441466.

Beckmann, S. C. (2007). Consumers and corporate social responsibility: Matching the unmatchable? *Australasian Marketing Journal, 15*(1), 27–36.

Berg, A., Magnus, K. H., Kappelmark, S., Granskog, A., Lee, L., Sawers, C., Polgampola, P., Lehmann, M., Syrett, H., & Arici, G. (2020). Fashion on climate: How the fashion industry can urgently act to reduce its greenhouse gas emissions. McKinsey & Company.

Bläse, R., Filser, M., Kraus, S., Puumalainen, K., & Moog, P. (2024). Non-sustainable buying behavior: How the fear of missing out drives purchase intentions in the fast fashion industry. *Business Strategy and the Environment, 33*(2), 626–641.

Bostrom, M., & Micheletti, M. (Eds.). (2019). *The oxford handbook of political consumerism.* Oxford University Press.

Brown, S., & Vacca, F. (2022). Cultural sustainability in fashion: Reflections on craft and sustainable development models. *Sustainability: Science, Practice and Policy, 18*(1), 590–600.

Carroll, A. B. (1999). Corporate social responsibility: Evolution of a definitional construct. *Business & Society, 38*(3), 268–295.

Chakraborty, S., & Sadachar, A. (2023). Can a connection with the indigenous cultural values encourage sustainable apparel consumption? *Journal of Fashion Marketing and Management: An International Journal, 27*(1), 80–99.

Cheng, Q. L., Wang, J. J., & Yin, L. G. (2014). A research framework of green consumption behavior based on value-belief-norm theory. *Advanced Materials Research, 962*, 1485–1489.

Connell, K. Y. H. (2019). Utilizing political consumerism to challenge the 21stcentury fast fashion industry. In M. Micheletti & A. Bostrom (Eds.), *The Oxford Handbook of Political Consumerism* (pp. 293–308). Oxford University Press.

Delmas, M. A., & Burbano, V. C. (2011). The drivers of greenwashing. *California Management Review, 54*(1), 64–87.

Desore, A., & Narula, S. A. (2018). An overview on corporate response towards sustainability issues in textile industry. *Environment, Development and Sustainability, 20*, 1439–1459.

Dissanayake, D. (2021a). Sustainability key performance indicators and the global reporting initiative: Usage and challenges in a developing country context. *Meditari Accountancy Research, 29*(3), 543–567.

Dissanayake, D. G. K. (2021b). Adopting the GRI framework for CSR reporting: An analysis of sustainability reports of apparel companies in Sri Lanka. *Journal of Cleaner Production, 32*(4), 188–195.

Domingos, M., Vale, V. T., & Faria, S. (2022). Slow fashion consumer behavior: A literature review. *Sustainability, 14*(5), 2860.

Eccles, R. G., Ioannou, I., & Serafeim, G. (2014). The impact of corporate sustainability on organizational processes and performance. *Management Science, 60*(11), 2835–2857.

Ellen MacArthur Foundation. (2017). *A new textiles economy: Redesigning fashion's future*. https://www.ellenmacarthurfoundation.org/a-newtextiles-economy

Fletcher, K. (2013). *Sustainable fashion and textiles: Design journeys*. Routledge.

Fonseca, L., Carvalho, F., & Santos, G. (2023). Strategic CSR: Framework for sustainability through management systems standards—implementing and disclosing sustainable development goals and results. *Sustainability, 15*(15), 11904.

Freeman, R. E. (2010). *Strategic management: A stakeholder approach*. Cambridge University Press.

Garg, S., Kumar, P., Mishra, V., Guijt, R., Singh, P., Dumée, L. F., & Sharma, R. S. (2020). A review on the sources, occurrence and health risks of per-/poly-fluoroalkyl substances (PFAS) arising from the manufacture and disposal of electric and electronic products. *Journal of Water Process Engineering, 38*, Article 101683.

GRI. (2021). *Consolidated set of GRI standards 2021*. Global Reporting Initiative.

Halder, P., Hansen, E. N., Kangas, J., & Laukkanen, T. (2020). How national culture and ethics matter in consumers' green consumption values. *Journal of Cleaner Production, 265*, Article 121754.

Hur, W. M., & Kim, Y. (2017). How does culture improve consumer engagement in CSR initiatives? The mediating role of motivational attributions. *Corporate Social Responsibility and Environmental Management, 24*(6), 620–633.

Joung, H. M. (2014). Fast fashion consumers' post-purchase behaviors. *International Journal of Retail & Distribution Management, 42*(8), 688–697.

Joung, H. M., & Park-Poaps, H. (2013). Factors motivating and influencing clothing disposal behaviors. *International Journal of Consumer Studies, 37*(1), 105–111.

Kaplan, R. S., & Norton, D. P. (1992). The balanced scorecard: Measures that drive performance. *Harvard Business Review, 70*(1), 71–79.

Kaur, H., Rawat, D., Poria, P., Sharma, U., Gibert, Y., Ethayathulla, A. S., Dumée, L. F., Sharma, R. S., & Mishra, V. (2022). Ecotoxic effects of microplastics and contaminated microplastics—emerging evidence and perspective. *Science of the Total Environment, 841*, Article 156593.

Kozar, J. M., & Hiller Connell, K. Y. (2013). Socially and environmentally responsible apparel consumption: Knowledge, attitudes, and behaviors. *Social Responsibility Journal, 9*(2), 315–324.

Kramer, M. R., & Porter, M. E. (2011). Creating shared value. *Harvard Business Review, 89*(1/2), 62–77.

Latapí Agudelo, M. A., Jóhannsdóttir, L., & Davídsdóttir, B. (2019). A literature review of the history and evolution of corporate social responsibility. *International Journal of Corporate Social Responsibility, 4*(1), 1–23.

Locke, R. M., Amengual, M., & Mangla, A. (2009). Virtue out of necessity? Compliance, commitment, and the improvement of labor conditions in global supply chains. *Politics & Society, 37*(3), 319–351.

Meseguer-Sánchez, V., Gálvez-Sánchez, F. J., López-Martínez, G., & MolinaMoreno, V. (2021). Corporate social responsibility and sustainability: A bibliometric analysis of their interrelations. *Sustainability, 13*(4), 1636.

Mishra, V., Sharma, U., Rawat, D., Benson, D., Singh, M., & Sharma, R. S. (2020). Fast-changing life-styles and ecotoxicity of hair dyes drive the emergence of hidden toxicants threatening environmental sustainability in Asia. *Environmental Research, 184*, Article 109253.

Mishra, V., Mudgal, N., Rawat, D., Poria, P., Mukherjee, P., Sharma, U., & Sharma, R. S. (2023). Integrating microalgae into textile wastewater treatment processes: Advancements and opportunities. *Journal of Water Process Engineering, 55*, Article 104128.

Niinimäki, K. (2010). Eco-clothing, consumer identity, and ideology. *Sustainable Development, 18*(3), 150–162.

Niinimäki, K., & Hassi, L. (2011). Emerging design strategies in sustainable production and consumption of textiles and clothing. *Journal of Cleaner Production, 19*(16), 1876–1883.

Niinimäki, K., Peters, G., Dahlbo, H., Perry, P., Rissanen, T., & Gwilt, A. (2020). The environmental price of fast fashion. *Nature Reviews Earth & Environment, 1*(4), Article 189200.

Nikolaou, I. E., & Tsalis, T. A. (2013). Development of a sustainable balanced scorecard framework. *Ecological Indicators, 34*, 76–86.

Nikolaou, I. E., Tsalis, T. A., & Evangelinos, K. I. (2019). A framework to measure corporate sustainability performance: A strong sustainability-based view of the firm. *Sustainable Production and Consumption, 18*, 1–18.

Orlitzky, M., Siegel, D. S., & Waldman, D. A. (2011). Strategic corporate social responsibility and environmental sustainability. *Business & Society, 50*(1), 6–27.

Rawat, D., Mishra, V., & Sharma, R. S. (2016). Detoxification of azo dyes in the context of environmental processes. *Chemosphere, 155*, 591–605.

Rawat, D., Sharma, R. S., Karmakar, S., Arora, L. S., & Mishra, V. (2018). Ecotoxic potential of a presumably non-toxic azo dye. *Ecotoxicology and Environmental Safety, 148*, 528–537.

Rawat, D., Sharma, U., Poria, P., Finlan, A., Parker, B., Sharma, R. S., & Mishra, V. (2022). Iron-dependent mutualism between *Chlorella sorokiniana* and *Ralstonia pickettii* forms the basis for a sustainable bioremediation system. *ISME Communications, 2*(1), 83.

Rawat, D., Sharma, U., Yadav, A., Poria, P., Farooqi, F., Pani, B., Mukherjee, P., Singh, M., Sharma, R. S., & Mishra, V. (2023). A bottom-up approach to select microbes from textile wastewater for detoxification of aromatic amine and azo dye in a single stage. *Journal of Water Process Engineering, 53*, Article 103668.

Saberi, S., Kouhizadeh, M., Sarkis, J., & Shen, L. (2019). Blockchain technology and its relationships to sustainable supply chain management. *International Journal of Production Research, 57*(7), 2117–2135.

Thorisdottir, T. S., & Johannsdottir, L. (2020). Corporate social responsibility influencing sustainability within the fashion industry: A systematic review. *Sustainability, 12*(21), 9167.

UNGC (2018). *Integrating the SDGs into corporate reporting: A practical guide.*

Vallerand, R. J., Deshaies, P., Cuerrier, J. P., Pelletier, L. G., & Mongeau, C. (1992). Ajzen and Fishbein's theory of reasoned action as applied to moral behavior: A confirmatory analysis. *Journal of Personality and Social Psychology, 62*(1), 98.

The manufacturer's authorised representative in the EU is Springer Nature Customer Service Centre GmbH, Europaplatz 3, 69115 Heidelberg, Germany. If you have any concerns regarding our products, please contact ProductSafety@springernature.com

Printed and bound by CPI Group (UK) Ltd, Croydon, CR0 4YY
01/10/2026
02225921-0012